STERLING
Test Prep

College Physics

Practice Questions

Vol. 1

5th edition

Copyright © 2022 Sterling Test Prep

> **Customer Satisfaction Guarantee**
>
> Your feedback is important because we strive to provide the highest quality prep materials. Email us comments or suggestions.
>
> info@sterling–prep.com
>
> We reply to emails – check your spam folder

All rights reserved. This publication's content, including the text and graphic images or part thereof, may not be reproduced, downloaded, disseminated, published, converted to electronic media, or distributed by any means whatsoever without prior written consent from the publisher. Copyright infringement violates federal law and is subject to criminal and civil penalties.

5 4 3 2 1

ISBN-13: 978-1-9547257-6-8

Sterling Test Prep materials are available at quantity discounts.
Contact info@sterling–prep.com

Sterling Test Prep
6 Liberty Square #11
Boston, MA 02109

© 2022 Sterling Test Prep

Published by Sterling Test Prep

 Printed in the U.S.A.

Thousands of students use our study aids to achieve higher grades!

To achieve a high grade in college physics, you need to do well on your tests and final exam. This book helps you develop and apply knowledge to quickly choose the correct answers to questions typically tested in college physics courses. Solving targeted practice questions builds your understanding of fundamental concepts and is a more effective strategy than merely memorizing terms.

This book has over 730 high-yield practice questions covering college physics topics. Physics instructors with years of teaching and applied physics experience prepared this practice material to build your knowledge and skills crucial for success in a college physics course. Our editorial team reviewed and systematized the content for targeted preparation.

The detailed explanations describe why an answer is correct and – more important for your learning – why another attractive choice is wrong. They provide step-by-step solutions for quantitative questions and teach the scientific foundations and details of essential physics topics needed to answer conceptual questions. Read the explanations carefully to understand how they apply to the question and learn important principles and the relationships between them. With the practice material contained in this book, you will significantly improve your understanding, test scores, and your grade.

We wish you great success in your academic achievements and look forward to being an important part of your preparation!

College study aids by Sterling Test Prep

Cell and Molecular Biology Review

Organismal Biology Review

Cell and Molecular Biology Practice Questions

Organismal Biology Practice Questions

Physics Review (Paert 1 and 2)

Physics Practice Questions (Vol. 1 and 2)

Organic Chemistry Practice Questions

United States History 101

American Government and Politics 101

Environmental Science 101

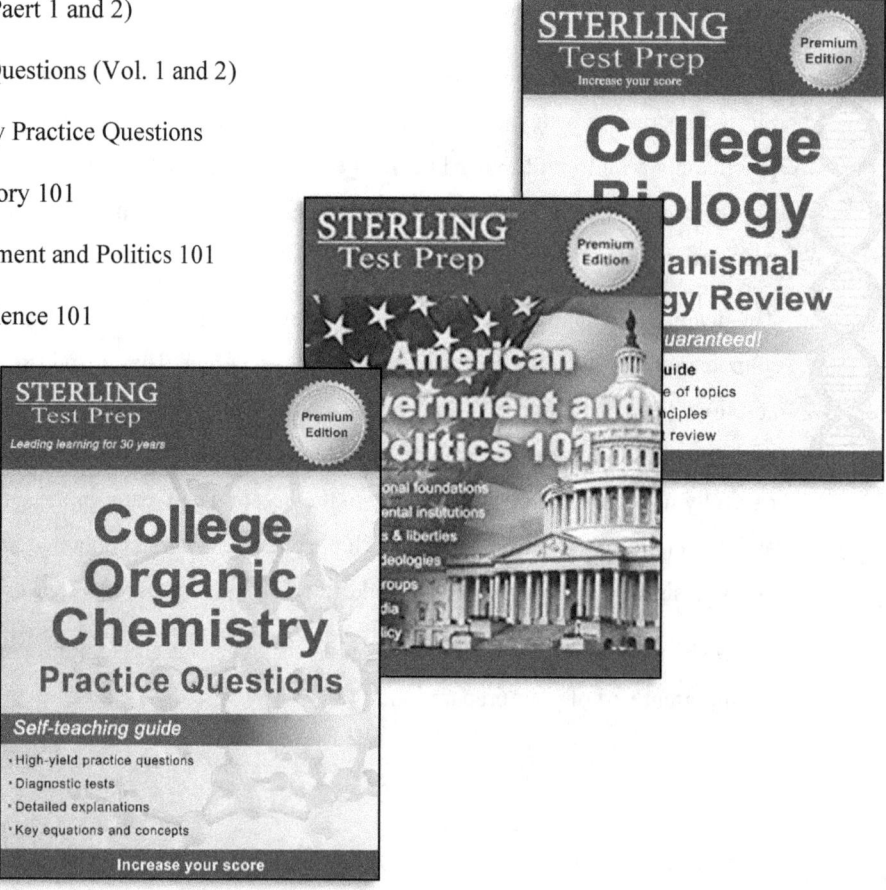

Visit our Amazon store

Table of Contents

Common Physics Equations and Conversions **7**

Topical Practice Questions **15**
    Kinematics and Dynamics 17
    Force, Motion, Gravitation 29
    Equilibrium and Momentum 41
    Rotational Motion 53
    Work and Energy 63
    Waves and Periodic Motion 75
    Sound 87
    Light and Optics 99

Diagnostic Tests **111**
    Diagnostic Test 1 113
    Diagnostic Test 2 135
    Diagnostic Test 3 157

Topical Practice Questions: Answer Keys & Detailed Explanations **179**
    Answer Keys 181
    Kinematics and Dynamics 185
    Force, Motion, Gravitation 201
    Equilibrium and Momentum 217
    Rotational Motion 237
    Work and Energy 263
    Waves and Periodic Motion 281
    Sound 295
    Light and Optics 311

Diagnostic Tests: Answer Keys & Detailed Explanations **327**
    Diagnostic Test 1 329
    Diagnostic Test 2 361
    Diagnostic Test 3 395

Physics Glossary of Terms **433**

College Level Examination Program (CLEP)

Biology Review

Biology Practice Questions

Chemistry Review

Chemistry Practice Questions

Introductory Business Law Review

College Algebra Practice Questions

College Mathematics Practice Questions

History of the United States I Review

History of the United States II Review

Western Civilization I Review

Western Civilization II Review

Social Sciences and History Review

American Government Review

Introductory Psychology Review

Visit our Amazon store

Common Physics Equations and Conversions

Constants and Conversion Factors

1 unified atomic mass unit	$1\text{ u} = 1.66 \times 10^{-27}\text{ kg}$
	$1\text{ u} = 931\text{ MeV}/c^2$
Proton mass	$m_p = 1.67 \times 10^{-27}\text{ kg}$
Neutron mass	$m_n = 1.67 \times 10^{-27}\text{ kg}$
Electron mass	$m_e = 9.11 \times 10^{-31}\text{ kg}$
Electron charge magnitude	$e = 1.60 \times 10^{-19}\text{ C}$
Avogadro's number	$N_0 = 6.02 \times 10^{23}\text{ mol}^{-1}$
Universal gas constant	$R = 8.31\text{ J/(mol·K)}$
Boltzmann's constant	$k_B = 1.38 \times 10^{-23}\text{ J/K}$
Speed of light	$c = 3.00 \times 10^8\text{ m/s}$
Planck's constant	$h = 6.63 \times 10^{-34}\text{ J·s}$
	$h = 4.14 \times 10^{-15}\text{ eV·s}$
	$hc = 1.99 \times 10^{-25}\text{ J·m}$
	$hc = 1.24 \times 10^3\text{ eV·nm}$
Vacuum permittivity	$\varepsilon_0 = 8.85 \times 10^{-12}\text{ C}^2/\text{N·m}^2$
Coulomb's law constant	$k = 1/4\pi\varepsilon_0 = 9.0 \times 10^9\text{ N·m}^2/\text{C}^2$
Vacuum permeability	$\mu_0 = 4\pi \times 10^{-7}\text{ (T·m)/A}$
Magnetic constant	$k' = \mu_0/4\pi = 10^{-7}\text{ (T·m)/A}$
Universal gravitational constant	$G = 6.67 \times 10^{-11}\text{ m}^3/\text{kg·s}^2$
Acceleration due to gravity at Earth's surface	$g = 9.8\text{ m/s}^2$
1 atmosphere pressure	$1\text{ atm} = 1.0 \times 10^5\text{ N/m}^2$
	$1\text{ atm} = 1.0 \times 10^5\text{ Pa}$
1 electron volt	$1\text{ eV} = 1.60 \times 10^{-19}\text{ J}$
Balmer constant	$B = 3.645 \times 10^{-7}\text{ m}$
Rydberg constant	$R = 1.097 \times 10^7\text{ m}^{-1}$
Stefan constant	$\sigma = 5.67 \times 10^{-8}\text{ W/m}^2\text{K}^4$

Units			Prefixes	
Name	**Symbol**	**Factor**	**Prefix**	**Symbol**
meter	m	10^{12}	tera	T
kilogram	kg	10^{9}	giga	G
second	s	10^{6}	mega	M
ampere	A	10^{3}	kilo	k
kelvin	K	10^{-2}	centi	c
mole	mol	10^{-3}	mili	m
hertz	Hz	10^{-6}	micro	μ
newton	N	10^{-9}	nano	n
pascal	Pa	10^{-12}	pico	p
joule	J			
watt	W			
coulomb	C			
volt	V			
ohm	Ω			
henry	H			
farad	F			
tesla	T			
degree Celsius	°C			
electronvolt	eV			

Newtonian Mechanics

Translational Motion	$v = v_0 + a\Delta t$ $x = x_0 + v_0 \Delta t + \frac{1}{2} a\Delta t^2$ $v^2 = v_0^2 + 2a\Delta x$ $\vec{a} = \frac{\sum \vec{F}}{m} = \frac{\vec{F}_{net}}{m}$	a = acceleration A = amplitude E = energy F = force f = frequency h = height
Rotational Motion	$\omega = \omega_0 + \alpha t$ $\theta = \theta_0 + \omega_0 t + \frac{1}{2} \alpha t^2$ $\omega^2 = \omega_0^2 + 2\alpha \Delta \theta$ $\vec{\alpha} = \frac{\sum \vec{\tau}}{I} = \frac{\vec{\tau}_{net}}{I}$	I = rotational inertia J = impulse K = kinetic energy k = spring constant ℓ = length
Force of Friction	$|\vec{F}_f| \leq \mu |\vec{F}_n|$	m = mass
Centripetal Acceleration	$a_c = \frac{v^2}{r}$	N = normal force P = power
Torque	$\tau = r_\perp F = rF \sin\theta$	p = momentum
Momentum	$\vec{p} = m\vec{v}$	L = angular momentum r = radius of distance
Impulse	$\vec{J} = \Delta \vec{p} = \vec{F} \Delta t$	T = period
Kinetic Energy	$K = \frac{1}{2} mv^2$	t = time U = potential energy
Potential Energy	$\Delta U_g = mg\Delta y$	v = velocity or speed
Work	$\Delta E = W = F_\parallel d = Fd \cos\theta$	W = work done on system x = position
Power	$P = \frac{\Delta E}{\Delta t} = \frac{\Delta W}{\Delta t}$	y = height α = angular acceleration
Simple Harmonic Motion	$x = A \cos(\omega t)$ $x = A \cos(2\pi f t)$	

Center of Mass	$x_{cm} = \dfrac{\sum m_i x_i}{\sum m_i}$	μ = coefficient of friction
		θ = angle
Angular Momentum	$L = I\omega$	τ = torque
		ω = angular speed
Angular Impulse	$\Delta L = \tau \Delta t$	
Angular Kinetic Energy	$K = \dfrac{1}{2} I \omega^2$	
Work	$W = F \Delta r \cos\theta$	
Power	$P = Fv \cos\theta$	
Spring Force	$|\vec{F}_s| = k|\vec{x}|$	
Spring Potential Energy	$U_s = \dfrac{1}{2} k x^2$	
Period of Spring Oscillator	$T_s = 2\pi \sqrt{m/k}$	
Period of Simple Pendulum	$T_p = 2\pi \sqrt{\ell/g}$	
Period	$T = \dfrac{2\pi}{\omega} = \dfrac{1}{f}$	
Gravitational Body Force	$|\vec{F}_g| = G \dfrac{m_1 m_2}{r^2}$	
Gravitational Potential Energy of Two Masses	$U_G = -\dfrac{G m_1 m_2}{r}$	

Optics

Wavelength to Frequency	$\lambda = \dfrac{v}{f}$	d = separation
		f = frequency or focal length
Index of Refraction	$n = \dfrac{c}{v}$	h = height
		L = distance
Snell's Law	$n_1 \sin\theta_1 = n_2 \sin\theta_2$	M = magnification
Thin Lens Equation	$\dfrac{1}{s_i} + \dfrac{1}{s_o} = \dfrac{1}{f}$	m = an integer
		n = index of refraction
Magnification Equation	$\lvert M \rvert = \left\lvert \dfrac{h_i}{h_o} \right\rvert = \left\lvert \dfrac{s_i}{s_o} \right\rvert$	R = radius of curvature
		s = distance
Double Slit Diffraction	$d \sin\theta = m\lambda$	v = speed
	$\Delta L = m\lambda$	x = position
		λ = wavelength
Critical Angle	$\sin\theta_c = \dfrac{n_2}{n_1}$	θ = angle
Focal Length of Spherical Mirror	$f = \dfrac{R}{2}$	

Sound

Standing Wave/ Open Pipe Harmonics	$\lambda = \dfrac{2L}{n}$	f = frequency
		L = length
Closed Pipe Harmonics	$\lambda = \dfrac{4L}{n}$	m = mass
		M = molecular mass
Harmonic Frequencies	$f_n = nf_1$	n = harmonic number
		R = gas constant
Speed of Sound in Ideal Gas	$v_{sound} = \sqrt{\dfrac{yRT}{M}}$	T = tension
		v = velocity
Speed of Wave Through Wire	$v = \sqrt{\dfrac{T}{m/L}}$	y = adiabatic constant
		λ = wavelength
Doppler Effect (approaching stationary observer)	$f_{observed} = \left(\dfrac{v}{v - v_{source}}\right) f_{source}$	
Doppler Effect (receding stationary observer)	$f_{observed} = \left(\dfrac{v}{v + v_{source}}\right) f_{source}$	
Doppler Effect (observer moving towards source)	$f_{observed} = \left(1 + \dfrac{v_{observer}}{v}\right) f_{source}$	
Doppler Effect (observer moving away from source)	$f_{observed} = \left(1 - \dfrac{v_{observer}}{v}\right) f_{source}$	

Geometry and Trigonometry

Rectangle	$A = bh$	A = area
		C = circumference
Triangle	$A = \frac{1}{2}bh$	V = volume
		S = surface area
Circle	$A = \pi r^2$	b = base
	$C = 2\pi r$	h = height
Rectangular Solid	$V = lwh$	l = length
		w = width
Cylinder	$V = \pi r^2 l$	r = radius
	$S = 2\pi r l + 2\pi r^2$	θ = angle
Sphere	$V = \frac{4}{3}\pi r^3$	
	$S = 4\pi r^2$	
Right Triangle	$a^2 + b^2 = c^2$	
	$\sin\theta = \frac{a}{c}$	
	$\cos\theta = \frac{b}{c}$	
	$\tan\theta = \frac{a}{b}$	

Trigonometric Functions for Common Angles

θ	$\sin\theta$	$\cos\theta$	$\tan\theta$
0°	0	1	0
30°	1/2	$\sqrt{3}/2$	$\sqrt{3}/3$
37°	3/5	4/5	3/4
45°	$\sqrt{2}/2$	$\sqrt{2}/2$	1
53°	4/5	3/5	4/3
60°	$\sqrt{3}/2$	1/2	$\sqrt{3}$
90°	1	0	∞

Notes for active learning

Topical Practice Questions

Kinematics and Dynamics

1. Starting from rest, how long does it take for a sports car to reach 60 mi/h if it has an average acceleration of 13.1 mi/h·s?

    A. 6.6 s
    B. 3.1 s
    C. 4.5 s
    D. 4.6 s
    E. 13.1 s

2. A cannonball is fired with an initial 20 m/s at a 30° angle with the horizontal. Ignoring air resistance, how long does it take the cannonball to reach the top of its trajectory? (Use acceleration due to gravity $g = 10$ m/s^2)

    A. 0.5 s
    B. 1 s
    C. 1.5 s
    D. 2 s
    E. 2.5 s

3. Darlene starts her car from rest and accelerates at a constant 2.5 m/s^2 for 9 s to get to her cruising speed. She then drives for 15 minutes at a constant speed. Darlene arrived at her destination, a straight-line distance of 31.5 km, exactly 1.25 hours later. What is her average velocity during 1.25 hours?

    A. 3 m/s
    B. 7 m/s
    C. 18 m/s
    D. 22.5 m/s
    E. 2.5 m/s

4. Which of the following cannot be negative?

    A. Instantaneous speed
    B. Instantaneous acceleration
    C. Acceleration of gravity
    D. Displacement
    E. Position

5. How far does a car travel while accelerating from 5 m/s to 21 m/s at a rate of 3 m/s^2?

    A. 15 m
    B. 21 m
    C. 69 m
    D. 105 m
    E. 210 m

6. Acceleration is sometimes expressed in multiples of g, where g is the acceleration due to gravity. How many g are experienced, on average, by the driver in a car crash if the car's velocity changes from 30 m/s to 0 m/s in 0.15 s? (Use acceleration due to gravity $g = 9.8$ m/s^2)

    A. 22 g
    B. 28 g
    C. 16 g
    D. 14 g
    E. 20 g

7. Ignoring air resistance, how many forces act on a bullet fired horizontally after leaving the rifle?

   A. Two (one from the gunpowder explosion and one from gravity)
   B. One (from the motion of the bullet)
   C. One (from the gunpowder explosion)
   D. One (from the pull of gravity)
   E. None; it is in freefall and unaffected by any forces

8. Suppose that a car traveling to the East begins to slow down as it approaches a traffic light. Which of the following statements about its acceleration is correct?

   A. The acceleration is towards the East
   B. The acceleration is towards the West
   C. Since the car is slowing, its acceleration is positive
   D. The acceleration is zero
   E. Since the car is slowing, its acceleration cannot be determined

9. On a planet where the acceleration due to gravity is 20 m/s², a freely falling object increases its speed each second by about:

   A. 20 m/s
   B. 10 m/s
   C. 30 m/s
   D. 40 m/s
   E. depends on its initial speed

10. What is a car's acceleration if it accelerates uniformly in one direction from 15 m/s to 40 m/s in 10 s?

    A. 1.75 m/s²
    B. 2.5 m/s²
    C. 3.5 m/s²
    D. 7.6 m/s²
    E. 4.75 m/s²

11. If the fastest a person can drive is 65 mi/h, what is the longest time she can stop for lunch if she wants to travel 540 mi in 9.8 h?

    A. 1 h
    B. 2.4 h
    C. 1.5 h
    D. 2 h
    E. 0.5 h

12. What is a racecar's average velocity if it completes one lap around a 500 m track in 10 s?

    A. 10 m/s
    B. 0 m/s
    C. 5 m/s
    D. 20 m/s
    E. 15 m/s

13. What is a ball's net displacement after 5 s if it initially rolls up a slight incline at 0.2 m/s and decelerates uniformly at 0.05 m/s²?

   A. 0.38 m
   B. 0.6 m
   C. 0.9 m
   D. 1.2 m
   E. 2.4 m

14. What does the slope of a line connecting two points on a velocity *vs.* time graph represent?

   A. Change in acceleration
   B. Instantaneous acceleration
   C. Average acceleration
   D. Instantaneous velocity
   E. Displacement

15. An airplane needs to reach a speed of 210.0 km/h to take off. On a 1,800.0 m runway, what minimum acceleration is necessary for the plane to reach this speed, assuming acceleration is constant?

   A. 0.78 m/s²
   B. 0.95 m/s²
   C. 1.47 m/s²
   D. 1.1 m/s²
   E. 2.5 m/s²

16. A test rocket is fired straight up from rest with a net acceleration of 22 m/s². What maximum elevation does the rocket reach if the motor turns off after 4 s, but the rocket continues to coast upward? (Use the acceleration due to gravity $g = 10$ m/s²)

   A. 408 m
   B. 320 m
   C. 357 m
   D. 563 m
   E. 260 m

17. Without a reference to direction, how fast an object moves refers to its:

   A. acceleration
   B. impulse
   C. momentum
   D. velocity
   E. speed

18. Ignoring air resistance, a 10 kg rock and a 20 kg rock are dropped simultaneously. If the 10 kg rock falls with acceleration *a*, what is the acceleration of the 20 kg rock?

   A. $a/2$
   B. a
   C. $2a$
   D. $4a$
   E. $a/4$

19. As an object falls freely, its magnitude of:

   I. velocity increases   II. acceleration increases   III. displacement increases

- **A.** I only
- **B.** I and II only
- **C.** II and III only
- **D.** I and III only
- **E.** I, II and III

20. A man stands in an elevator ascending at a constant velocity. What forces are exerted on the man, and in which direction does the net force point?

- **A.** Gravity pointing downward, normal force from the floor pointing upward, and tension force from the elevator cable pointing upward; net force points upward
- **B.** Gravity pointing downward and the normal force from the floor pointing upward; net force points upward
- **C.** Gravity pointing downward and normal force from the floor pointing upward; net force is zero
- **D.** Gravity pointing downward; net force is zero
- **E.** Gravity pointing downward, normal force from the floor pointing upward, and tension force from the elevator cable pointing upward; net force is zero

21. A football kicker is attempting a field goal from 44 m, and the ball just clears the lower bar with a time of flight of 2.9 s. What was the ball's initial speed if the kick's angle was 45° with the horizontal?

- **A.** 37 m/s
- **B.** 2.5 m/s
- **C.** 18.3 m/s
- **D.** 7.2 m/s
- **E.** 21.4 m/s

22. Ignoring air resistance, if a rock, starting at rest, is dropped from a cliff and strikes the ground with an impact velocity of 14 m/s, from what height was it dropped? (Use the acceleration due to gravity $g = 10$ m/s^2)

- **A.** 10 m
- **B.** 30 m
- **C.** 45 m
- **D.** 70 m
- **E.** 90 m

23. An SUV is traveling at 20 m/s. Then Joseph steps on the accelerator pedal, accelerating at a constant 1.4 m/s^2 for 7 s. How far does he travel during these 7 s?

- **A.** 205 m
- **B.** 174 m
- **C.** 143 m
- **D.** 158 m
- **E.** 115 m

24. Which of the following is NOT a scalar?

A. temperature
B. distance
C. mass
D. force
E. time

25. Two identical balls (A and B) fall from rest from different heights to the ground. Ignoring air resistance, what is the ratio of the heights from which A and B fall if ball B takes twice as long as ball A to reach the ground?

A. 1 : √2
B. 1 : 4
C. 1 : 2
D. 1 : 8
E. 1 : 9

26. How far does a car travel in 10 s when it accelerates uniformly in one direction from 5 m/s to 30 m/s?

A. 65 m
B. 25 m
C. 250 m
D. 650 m
E. 175 m

27. Which graph represents an acceleration of zero?

I.
II.
III. V
  0 ⎯⎯⎯ t

A. I only
B. II only
C. I and II only
D. II and III only
E. I and III only

28. Doubling the distance between an orbiting satellite and the Earth results in what change in the gravitational attraction between the two?

A. Twice as much
B. Four times as much
C. One half as much
D. One fourth as much
E. Remains the same

29. An object is moving in a straight line. Consider its motion during some interval of time: under what conditions is it possible for the instantaneous velocity of the object at some point during the interval to be equal to the average velocity over the interval?

    I. When velocity is constant during the interval
    II. When velocity is increasing at a constant rate during the interval
    III. When velocity is increasing at an irregular rate during the interval

A. II only
B. I and III only
C. II and III only
D. I, II and III
E. I and II only

30. A freely falling object on Earth, 10 s after starting from rest, has a speed of about: (Use the acceleration due to gravity $g = 10$ m/s^2)

A. 10 m/s
B. 20 m/s
C. 80 m/s
D. 150 m/s
E. 100 m/s

31. A truck travels a certain distance at a constant velocity v for time t. If the truck travels three times as fast, covering the same distance, then by what factor does the time of travel in relation to t change?

A. Increases by 3
B. Decreases by 3
C. Decreases by $\sqrt{3}$
D. Increases by 9
E. Decreases by 1/9

32. Assuming equal rates of acceleration, how much farther would Steve travel if he braked from 59 mi/h to rest than from 29 mi/h to rest?

A. 2 times farther
B. 16 times farther
C. 4 times farther
D. 3.2 times farther
E. 1.5 times farther

33. What is the average speed if the horse does one lap around a 400 m track in 20 s?

A. 0 m/s
B. 7.5 m/s
C. 15 m/s
D. 20 m/s
E. 25 m/s

34. What was a car's initial velocity if the car traveled up a slight slope while decelerating at 0.1 m/s^2 and came to a stop after 5 s?

A. 0.02 m/s
B. 0.25 m/s
C. 2 m/s
D. 1.5 m/s
E. 0.5 m/s

35. Average velocity equals the average of an object's initial and final velocity when acceleration is:

   A. constantly decreasing
   B. constantly increasing
   C. constant
   D. equal to zero
   E. equal to the reciprocal of the initial velocity

36. Ignoring air resistance, compared to a rock dropped from the same point, how much earlier does a thrown rock strike the ground if thrown downward with an initial velocity of 10 m/s from the top of a 300 m building? (Use acceleration due to gravity $g = 9.8$ m/s^2)

   A. 0.75 s
   B. 0.33 s
   C. 0.66 s
   D. 0 s
   E. 0.95 s

37. With other factors equal, what happens to the acceleration if the unbalanced force on an object of a given mass is doubled?

   A. Increased by one-fourth
   B. Increased by one-half
   C. Increased fourfold
   D. Doubled
   E. Remains the same

38. How fast an object is changing speed or direction of travel is a property of motion known as:

   A. velocity
   B. acceleration
   C. speed
   D. flow
   E. momentum

39. Which statement concerning a car's acceleration must be correct if a car traveling to the North (+y direction) begins to slow down as it approaches a stop sign?

   A. Acceleration is positive
   B. Acceleration is zero
   C. Cannot be determined from the data provided
   D. Acceleration decreases in magnitude as the car slows
   E. Acceleration is negative

40. For the velocity *vs.* time graph of a basketball player traveling up and down the court in a straight-line path, what is the total distance run by the player in the 10 s?

   A. 24 m
   B. 22 m
   C. 14 m
   D. 18 m
   E. 20 m

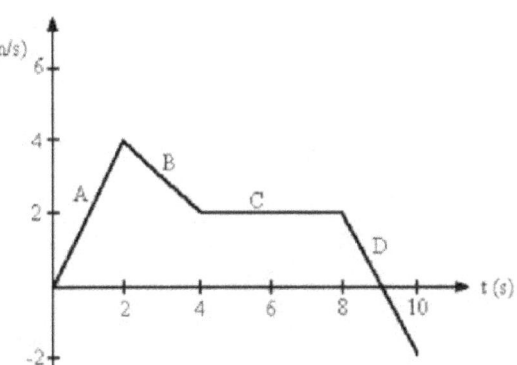

41. At the same time that a bullet is dropped into a river from a high bridge, another bullet is fired from a gun, straight down towards the water. Ignoring air resistance, the acceleration just before striking the water:

A. is higher for the dropped bullet
B. is higher for the fired bullet
C. is the same for each bullet
D. depends on how high the bullets started
E. depends on the mass of the bullets

42. Sarah starts her car from rest and accelerates at a constant 2.5 m/s² for 9 s to get to her cruising speed. What was her final velocity?

A. 22.5 m/s
B. 12.3 m/s
C. 4.6 m/s
D. 8.5 m/s
E. 1.25 m/s

43. A bat hits a baseball, and the baseball's direction is completely reversed, and its speed is doubled. If the actual time of contact with the bat is 0.45 s, what is the ratio of the acceleration to the original velocity?

A. -2.5 s^{-1} : 1
B. -0.15 s^{-1} : 1
C. -9.8 s^{-1} : 1
D. -4.1 s^{-1} : 1
E. -6.7 s^{-1} : 1

44. A 2 kg weight is thrown vertically upward from the surface of the Moon at a speed of 3.2 m/s, and it returns to its starting point in 4 s. What is the magnitude of the acceleration due to gravity on the Moon?

A. 0.8 m/s²
B. 1.6 m/s²
C. 3.7 m/s²
D. 8.4 m/s²
E. 12.8 m/s²

45. What is the change in velocity for a bird cruising at 1.5 m/s and then accelerating at a constant 0.3 m/s² for 3 s?

A. 0.9 m/s
B. 0.6 m/s
C. 1.6 m/s
D. 0.3 m/s
E. 1.9 m/s

46. All of the following are vectors, except:

A. velocity
B. displacement
C. acceleration
D. mass
E. force

Questions **47-49** are based on the following:

A toy rocket is launched vertically from ground level where $y = 0$ m, at time $t = 0$ s. The rocket engine provides constant upward acceleration during the burn phase. At the instant of engine burnout, the rocket has risen to 64 m and acquired a velocity of 60 m/s. The rocket rises in unpowered flight reaches the maximum height and falls back to the ground. (Use the acceleration due to gravity $g = 9.8$ m/s²)

47. What is the maximum height reached by the rocket?

    A. 274 m
    B. 205 m
    C. 223 m
    D. 120 m
    E. 248 m

48. What is the upward acceleration of the rocket during the burn phase?

    A. 9.9 m/s²
    B. 4.8 m/s²
    C. 28 m/s²
    D. 11.8 m/s²
    E. 8.6 m/s²

49. What is the time interval during which the rocket engine provides upward acceleration?

    A. 1.5 s
    B. 1.9 s
    C. 2.3 s
    D. 2.1 s
    E. 2.6 s

50. A car accelerates uniformly from rest along a straight track that has markers spaced at equal distances along it. As it passes Marker 2, the car reaches a speed of 140 km/h. Where on the track is the car when it is traveling at 70 km/h?

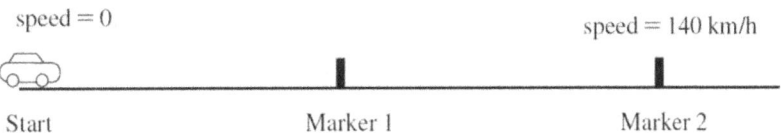

    A. Close to Marker 2
    B. Between Marker 1 and Marker 2
    C. At Marker 1
    D. Close to the starting point
    E. Before Marker 1

51. What are the two measurements necessary for calculating average speed?

    A. Distance and time
    B. Distance and acceleration
    C. Velocity and time
    D. Velocity and acceleration
    E. Acceleration and time

52. A pedestrian traveling at speed *v* covers a distance *x* during a time interval *t*. If a bicycle travels at speed 3*v*, how much time does it take to travel the same distance?

A. $t/3$
B. $t-3$
C. $t+3^2$
D. $3t$
E. $t+3^3$

53. Ignoring air resistance, how much time passes before a ball strikes the ground if it is thrown straight upward with a velocity of 39 m/s? (Use the acceleration due to gravity $g = 9.8$ m/s^2)

A. 2.2 s
B. 1.4 s
C. 12 s
D. 4 s
E. 8 s

54. A particle travels to the right along a horizontal axis with a constantly decreasing speed. Which one of the following describes the direction of the particle's acceleration?

A. ↑
B. ↓
C. →
D. ←
E. None of the above

55. Larry is carrying a 25 kg package at a constant velocity of 1.8 m/s across a room for 12 s. What is the work done by Larry on the package during the 12 s? (Use the acceleration due to gravity $g = 10$ m/s^2)

A. 0 J
B. 280 J
C. 860 J
D. 2,200 J
E. 1,125 J

Notes for active learning

Notes for active learning

Force, Motion, Gravitation

1. A boy attaches a weight to a string, which he swings counterclockwise in a horizontal circle. Which path does the weight follow when the string breaks at point P?

   A. path A
   B. path B
   C. path C
   D. path D
   E. path E

   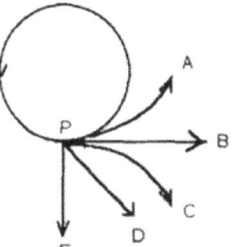

2. A garment bag hangs from a clothesline. The tension in the clothesline is 10 N on the right side of the garment bag and 10 N on the left side of the garment bag. The clothesline makes an angle of 60° from vertical. What is the mass of the garment bag? (Use the acceleration due to gravity g = 10.0 m/s²)

   A. 0.5 kg
   B. 8 kg
   C. 4 kg
   D. 10 kg
   E. 1 kg

3. A sheet of paper can be withdrawn from under a milk carton without toppling the carton if the paper is jerked away quickly. This demonstrates:

   A. the inertia of the milk carton
   B. that gravity tends to hold the milk carton secure
   C. there is an action-reaction pair of forces
   D. that the milk carton has no acceleration
   E. none of the above

4. A car of mass m goes up a shallow slope with an angle θ to the horizontal when the driver suddenly applies the brakes. The car skids as it comes to a stop. The coefficient of static friction between the tires and the road is μ_s, and the coefficient of kinetic friction is μ_k. Which expression represents the normal force on the car?

   A. $mg \tan \theta$
   B. $mg \sin \theta$
   C. $mg \cos \theta$
   D. mg
   E. $mg \sec \theta$

5. A 27 kg object is accelerated at a rate of 1.7 m/s². How much force does the object experience?

   A. 62 N
   B. 74 N
   C. 7 N
   D. 18 N
   E. 46 N

6. How are two identical masses moving if they are attached by a light string that passes over a small pulley? Assume that the table and the pulley are frictionless.

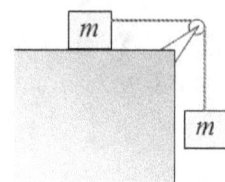

A. With an acceleration equals g
B. With an acceleration greater than g
C. At a constant speed
D. With an acceleration less than g
E. Not moving because the masses are equal

7. An object is moving to the right in a straight line. The net force acting on the object is directed to the right, but the magnitude of the force decreases with time. What happens to the object?

A. Continues to move to the right with its speed increasing with time
B. Continues to move to the right with a constant speed
C. Continues to move to the right with its speed decreasing with time
D. Continues to move to the right, slowing quickly to a stop
E. Stops and then begins moving to the left with its speed decreasing with time

8. A crate is sliding down an inclined ramp at a constant speed of 0.55 m/s. Where does the vector sum of the forces acting on this crate point?

A. Perpendicular to the ramp
B. Vertically downward
C. Vertically upward
D. Across the ramp
E. None of the above

9. Consider an inclined plane that makes an angle θ with the horizontal. What is the relationship between the length of the ramp L and the vertical height of the ramp h?

A. $h = L \sin \theta$
B. $h = L \tan \theta$
C. $L = h \sin \theta$
D. $h = L \cos \theta$
E. $L = h \cos \theta$

10. Why is it just as difficult to accelerate a car on the Moon to accelerate the same car on Earth?

    I. Moon and Earth have the same gravity
    II. weight of the car is independent of gravity
    III. mass of the car is independent of gravity

A. I only
B. II only
C. III only
D. I and II only
E. I and III only

11. Sean is pulling his son in a toy wagon. His son and the wagon are 60 kg. For 3 s Sean exerts a force that uniformly accelerates the wagon from 1.5 m/s to 3.5 m/s. What is the acceleration of the wagon with his son?

   A. 0.67 m/s^2
   B. 0.84 m/s^2
   C. 1.66 m/s^2
   D. 15.32 m/s^2
   E. 20.84 m/s^2

12. When an object moves in a uniform circular motion, the direction of its acceleration is:

   A. directed away from the center of its circular path
   B. dependent on its speed
   C. in the opposite direction of its velocity vector
   D. in the same direction as its velocity vector
   E. directed toward the center of its circular path

13. What happens to a moving object in the absence of an external force?

   A. Gradually accelerates until it reaches its terminal velocity, then continues at a constant velocity
   B. Moves with constant velocity
   C. Stops immediately
   D. Slows and eventually stops
   E. Moves with a constant speed in a circular orbit

14. A force of 1 N causes a 1 kg mass to accelerate 1 m/s^2. A force of 9 N applied to a 9 kg mass would have what magnitude of acceleration?

   A. 18 m/s^2
   B. 9 m/s^2
   C. 1 m/s^2
   D. 3 m/s^2
   E. 27 m/s^2

15. Which of the following statements is true about an object in two-dimensional projectile motion with no air resistance?

   A. The acceleration of the object is zero at its highest point
   B. The horizontal acceleration is always positive, regardless of the vertical acceleration
   C. The velocity is always in the same direction as the acceleration
   D. The acceleration of the object is +g when the object is rising and –g when it is falling
   E. The horizontal acceleration is always zero, and vertical acceleration is always a nonzero constant downward

16. A can of paint with a mass of 10 kg hangs from a rope. If the can is to be pulled up to a rooftop with a constant velocity of 0.5 m/s, what must the tension on the rope be? (Use acceleration due to gravity $g = 10$ m/s^2)

A. 100 N
B. 40 N
C. 0 N
D. 120 N
E. 160 N

17. What is the magnitude of the force exerted on a 1,000 kg object that accelerates at 2 m/s^2?

A. 500 N
B. 1,000 N
C. 1,200 N
D. 2,000 N
E. 2,200 N

18. A 1,300 kg car is driven at a constant speed of 4 m/s and turns to the right on a curve on the road, an effective radius of 4 m. What is the acceleration of the car?

A. 0 m/s^2
B. 3 m/s^2
C. 4 m/s^2
D. 9.8 m/s^2
E. 8 m/s^2

19. A block of mass *m* is resting on a 20° slope. The block has coefficients of friction $\mu_s = 0.55$ and $\mu_k = 0.45$ with the surface.

Block *m* is connected via a massless string over a massless, frictionless pulley to a hanging 2 kg block. What is the minimum mass of block *m* so that it does not slip? (Use the acceleration due to gravity $g = 9.8$ m/s^2)

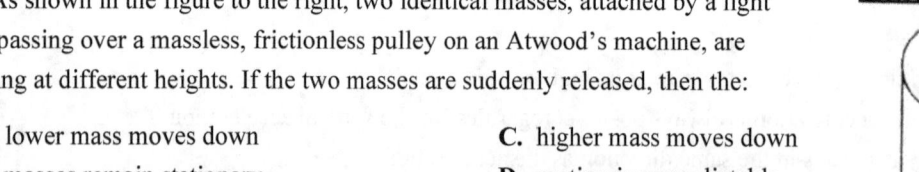

A. 0.8 kg
B. 1.3 kg
C. 3.7 kg
D. 4.1 kg
E. 2.3 kg

20. As shown in the figure to the right, two identical masses, attached by a light cord passing over a massless, frictionless pulley on an Atwood's machine, are hanging at different heights. If the two masses are suddenly released, then the:

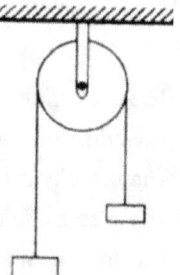

A. lower mass moves down
B. masses remain stationary
C. higher mass moves down
D. motion is unpredictable
E. masses oscillate uniformly

21. When Victoria jumps up in the air, which of the following statements is the most accurate?

  A. The ground cannot exert the upward force necessary to lift her into the air because the ground is stationary. Instead, Victoria is propelled into the air by the internal force of her muscles acting
  B. When Victoria pushes down on the Earth with force greater than her weight, the Earth pushes back with the same magnitude force and propels her into the air
  C. Victoria is propelled up by the upward force exerted by the ground, but this force cannot be greater than her weight
  D. The Earth exerts an upward force on Victoria stronger than the downward force she exerts on the Earth: therefore, Victoria is able to spring up
  E. Because gravity is what keeps her on the ground, the internal force of her muscles acting on her body, needs to be greater than the force of gravity to propel her into the air

22. If a feather is pounded with a hammer, which experiences a greater force?

  A. The magnitude of the force is always the same on both
  B. If the feather moves, then it feels the greater force
  C. Depends on the force with which the hammer strikes the feather
  D. Always the hammer
  E. Always the feather

23. A block is moving down a slope of a frictionless inclined plane. Compared to the weight of the block, what is the force parallel to the surface of the plane experienced by the block?

  A. Greater
  B. Unrelated
  C. Less than
  D. Equal
  E. Requires more information

24. A package falls off a truck moving at 30 m/s. Ignoring air resistance, the horizontal speed of the package just before it hits the ground is:

  A. 0 m/s
  B. 15 m/s
  C. √60 m/s
  D. √30 m/s
  E. 30 m/s

25. A carousel with the radius r is turning counterclockwise at a frequency f. How does the velocity of a seat on the carousel change when f is doubled?

  A. Increases by a factor of $2r$
  B. Increases by a factor of r
  C. Remains unchanged
  D. Doubles
  E. It depends on the mass of the chair

26. What is the mass of a car if it takes 4,500 N to accelerate it at a rate of 5 m/s²?

A. 900 kg
B. 1,320 kg
C. 620 kg
D. 460 kg
E. 1,140 kg

27. Steve is standing facing forward on a moving bus. What force causes Steve to suddenly move forward when the bus comes to an abrupt stop?

A. Force due to the air pressure inside the previously moving bus
B. Force due to kinetic friction between Steve and the floor of the bus
C. Force due to stored kinetic energy
D. Force of gravity
E. No forces were responsible for Steve's movement

28. A plastic ball in a liquid is acted upon by its weight and a buoyant force. The weight of the ball is 4.4 N. The buoyant force of 8.4 N acts vertically upward. An external force acting on the ball maintains it in a state of rest. What is the magnitude and direction of the external force?

A. 4 N, upward
B. 8.4 N, downward
C. 4.4 N, upward
D. 4 N, downward
E. 2 N, downward

29. A passenger on a train traveling in the forward direction notices that a piece of luggage starts to slide directly toward the front of the train. From this, it can be concluded that the train is:

A. slowing down
B. speeding up
C. moving at a constant velocity forward
D. changing direction
E. moving at a constant velocity in the reverse direction

30. An object has a mass of 36 kg and weighs 360 N at the surface of the Earth. If this object is transported to an altitude twice the Earth's radius, what is the object's mass and weight, respectively?

A. 9 kg and 90 N
B. 36 kg and 90 N
C. 4 kg and 90 N
D. 36 kg and 40 N
E. 9 kg and 40 N

31. A truck is moving at a constant velocity. Inside the storage compartment, a rock is dropped from the midpoint of the ceiling and strikes the floor below. The rock hits the floor:

A. just behind the midpoint of the ceiling
B. exactly halfway between the midpoint and the front of the truck
C. exactly below the midpoint of the ceiling
D. just ahead of the midpoint of the ceiling
E. exactly halfway between the midpoint and the rear of the truck

32. Jason takes off across level water on his jet-powered skis. The combined mass of Jason and his skis is 75 kg (the mass of the fuel is negligible). The skis have a thrust of 200 N and a coefficient of kinetic friction on the water of 0.1. If the skis run out of fuel after only 67 s, how far has Jason traveled before he stops?

A. 5,428 m
B. 3,793 m
C. 8,224 m
D. 7,642 m
E. 10,331 m

33. A 200 g hockey puck is launched up a metal ramp inclined at a 30° angle. The puck's initial speed is 63 m/s. What vertical height does the puck reach above its starting point? (Use the acceleration due to gravity $g = 9.8$ m/s^2, the coefficient of static friction $\mu_s = 0.40$ and the kinetic friction $\mu_k = 0.30$ between the hockey puck and the metal ramp)

A. 66 m
B. 200 m
C. 170 m
D. 130 m
E. 48 m

34. When a 4 kg mass and a 10 kg mass are pushed from rest with equal force:

A. 4 kg mass accelerates 2.5 times faster than the 10 kg mass
B. 10 kg mass accelerates 10 times faster than the 4 kg mass
C. 4 kg mass accelerates at the same rate as the 10 kg mass
D. 10 kg mass accelerates 2.5 times faster than the 4 kg mass
E. 4 kg mass accelerates 10 times faster than the 10 kg mass

35. On a different planet, a person's:

A. weight and mass decrease
B. weight and mass remain the same
C. weight remains the same, but the mass changes
D. weight changes, but the mass remains the same
E. weight and mass increase

36. Which of the following statements must be true when a 20-ton truck collides with a 1,500 lb car?

   A. During the collision, the force on the truck equals the force on the car
   B. The truck did not slow down during the collision, but the car did
   C. During the collision, the force on the truck is greater than the force on the car
   D. During the collision, the force on the truck is smaller than the force on the car
   E. The car did not slow down during the collision, but the truck did

37. A block is on a frictionless table on Earth. The block accelerates at 3 m/s² when a 20 N horizontal force is applied to it. The block and table are then transported to the Moon. What is the weight of the block on the Moon? (Use the acceleration due to gravity at the surface of the Moon = 1.62 m/s²)

   A. 5.8 N
   B. 14.2 N
   C. 8.5 N
   D. 11 N
   E. 17.5 N

38. What is the weight of a 0.4 kg bottle of wine? (Use acceleration due to gravity $g = 9.8$ m/s²)

   A. 0.4 N
   B. 4 N
   C. 40 N
   D. 20 N
   E. 2 N

39. Car A starts from rest and accelerates uniformly for time t to travel a distance of d. Car B, which has four times the mass of car A, starts from rest and accelerates uniformly. If the magnitudes of the forces accelerating car A and car B are the same, how long does it take car B to travel the same distance d?

   A. t
   B. $2t$
   C. $t/2$
   D. $16t$
   E. $4t$

40. A 1,100 kg vehicle is traveling at 27 m/s when it starts to decelerate. What is the average braking force acting on the vehicle, if after 578 m, it comes to a complete stop?

   A. –440 N
   B. –740 N
   C. –690 N
   D. –540 N
   E. –880 N

41. A string suspends an ornament of mass M from the ceiling inside an elevator. What is the tension in the string holding the ornament when the elevator travels upward at a constant speed?

   A. Equal to Mg
   B. Less than Mg
   C. Greater than Mg
   D. Equal to M/g
   E. Less than M/g

42. An object that weighs 75 N is pulled on a horizontal surface by a force of 50 N to the right. The friction force on this object is 30 N to the left. What is the acceleration of the object? (Use the acceleration due to gravity $g = 9.8$ m/s^2)

A. 0.46 m/s^2
B. 1.7 m/s^2
C. 2.6 m/s^2
D. 10.3 m/s^2
E. 12.1 m/s^2

43. While flying horizontally in an airplane, a string attached from the overhead luggage compartment hangs at rest 15° from the vertical toward the front of the plane. From this observation, it can be concluded that the airplane is:

A. accelerating forward
B. accelerating backward
C. accelerating upward at 15° from horizontal
D. moving backward
E. not moving

44. An object slides down an inclined ramp at a constant speed. If the ramp's incline angle is θ, what is the coefficient of kinetic friction (μ_k) between the object and the ramp?

A. $\mu_k = 1$
B. $\mu_k = \cos\theta / \sin\theta$
C. $\mu_k = \sin\theta / \cos\theta$
D. $\mu_k = \sin\theta$
E. $\mu_k = \cos\theta$

45. What is the magnitude of the net force on a 1 N apple when it is in free fall?

A. 1 N
B. 0.1 N
C. 0.01 N
D. 10 N
E. 100 N

46. What is the acceleration of a 105 kg tiger that accelerates uniformly from rest to 20 m/s in 10 s?

A. 4.7 m/s^2
B. 1.5 m/s^2
C. 2 m/s^2
D. 3.4 m/s^2
E. 16.7 m/s^2

47. Yana tries to pull an object by tugging on a rope attached to the object with a force of F. If the object does not move, what does this imply?

A. The object has reached its natural state of rest and can no longer be set into motion
B. The rope is not transmitting the force to the object
C. No other forces are acting on the object
D. The inertia of the object prevents it from accelerating
E. There are one or more other forces that act on the object with a sum of $-F$

48. If a force F is exerted on an object, the force which the object exerts back:

A. depends on the mass of the object
B. depends on the density of the object
C. depends on if the object is moving
D. depends on if the object is stationary
E. equals –F

49. What is the mass of an object that experiences a gravitational force of 685 N near Earth's surface? (Use the acceleration due to gravity $g = 9.8$ m/s^2)

A. 76 kg
B. 62 kg
C. 70 kg
D. 81 kg
E. 54 kg

50. Sarah and her father Bob (who weighs four times as much) are standing on identical skateboards (with frictionless ball bearings), both initially at rest. For a short time, Bob pushes Sarah on the skateboard. When Bob stops pushing:

A. Sarah and Bob move away from each other, and Sarah's speed is four times that of Bob's
B. Sarah and Bob move away from each other, and Sarah's speed is one-fourth of Bob's
C. Sarah and Bob move away from each other with equal speeds
D. Sarah moves away from Bob, and Bob is stationary
E. Sarah and Bob move away from each other, and Bob's speed is less than one fourth that of Sarah's

51. Considering the effects of friction, which statement best describes the motion of an object along a surface?

A. Less force is required to start than to keep the object in motion at a constant velocity
B. The same force is required to start as to keep the object in motion at a constant velocity
C. More force is required to start than to keep the object in motion at a constant velocity
D. Once the object is set in motion, no force is required to keep it in motion at constant velocity
E. More information is needed about the surface before the amount of force can be determined

52. On the surface of Jupiter, the acceleration due to gravity is about three times that of Earth. What is the weight of a 100 kg rock when it is taken from Earth to Jupiter? (Use the acceleration due to gravity $g = 10$ m/s^2)

A. 1,800 N
B. 3,000 N
C. 3,300 N
D. 4,000 N
E. 9,000 N

53. Joe and Bill are playing tug-of-war. Joe is pulling with a force of 200 N, while Bill is merely holding onto the rope. What is the tension of the rope if neither person is moving?

A. 75 N
B. 0 N
C. 100 N
D. 200 N
E. 50 N

54. A 4 kg wooden block A slides on a frictionless table pulled by a hanging 5 kg block B via a massless string and pulley system as shown. What is the acceleration of block A as it slides? (Use the acceleration due to gravity $g = 9.8$ m/s^2)

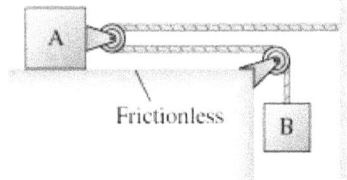

A. 2.8 m/s^2
B. 1.6 m/s^2
C. 3.4 m/s^2
D. 4.9 m/s^2
E. 4.1 m/s^2

55. Which of the following best describes the direction in which the force of kinetic friction acts relative to the interface between the interacting bodies?

A. Parallel to the interface and in the same direction as the relative velocity
B. Parallel to the interface and in the opposite direction of the relative velocity
C. Perpendicular to the interface and in the same direction as the relative velocity
D. Perpendicular to the interface and in the opposite direction of the relative velocity
E. Because kinetic friction depends on movement, cannot estimate it unless described a body's velocity

Notes for active learning

Equilibrium and Momentum

1. When is the angular momentum of a system constant?

   A. When no net external torque acts on the system
   B. When the linear momentum and the energy are constant
   C. When no net external force acts on the system
   D. When the total kinetic energy is positive
   E. When the moment of inertia is positive

2. When a rock rolls down a mountainside at 7 m/s, the horizontal component of its velocity vector is 1.8 m/s. What was the angle of the mountain surface above the horizontal?

   A. 15°
   B. 63°
   C. 40°
   D. 75°
   E. 9.5°

3. A 200 N sled slides down a frictionless hill at an angle of 37° to the horizontal. What is the magnitude of the force that the hill exerts on the sled parallel to the surface of the hill?

   A. 170 N
   B. 200 N
   C. 74 N
   D. 37 N
   E. 0 N

4. Water causes a water wheel to turn as it passes by. The force of the water is 300 N, and the radius of the wheel is 10 m. What is the torque around the center of the wheel?

   A. 0 N·m
   B. 300 N·m
   C. 3,000 N·m
   D. 3 N·m
   E. 30 N·m

5. Through what angle, in degrees, does a 33-rpm record turn in 0.32 s?

   A. 44°
   B. 94°
   C. 113°
   D. 32°
   E. 63°

6. A freight train rolls along a track with considerable momentum. What is its momentum if it rolls at the same speed but has twice the mass?

   A. Zero
   B. Doubled
   C. Quadrupled
   D. Unchanged
   E. Cannot be estimated

Questions **7-9** are based on the following:

Three carts run along a level, frictionless one-dimensional track. Furthest to the left is a 1 kg cart I, moving at 0.5 m/s to the right. In the middle is a 1.5 kg cart II, moving at 0.3 m/s to the left. Furthest to the right is a 3.5 kg cart III moving at 0.5 m/s to the left. The carts collide in sequence, sticking. (Assume the direction to the right is the positive direction)

7. What is the total momentum of the system before the collision?

    A. −2.6 kg·m/s   **C.** 0.6 kg·m/s
    B. 1.4 kg·m/s   **D.** −1.7 kg·m/s
                                                                       E. 1.1 kg·m/s

8. Assuming cart I and cart II collide first, and cart III is still independent, what is the total momentum of the system just after cart I and cart II collide?

    A. −1.7 kg·m/s   **C.** 0.9 kg·m/s
    B. 0.1 kg·m/s   **D.** −0.9 kg·m/s
                                                             E. −0.11 kg·m/s

9. What is the final velocity of the three carts?

    A. −0.35 m/s   **C.** −0.87 m/s
    B. −0.28 m/s   **D.** 0.35 m/s
                                                         E. 0.15 m/s

10. A 480 kg car moves at 14.4 m/s when it collides with another car moving at 13.3 m/s in the same direction. If the second car has a mass of 570 kg and a new velocity of 17.9 m/s after the collision, what is the velocity of the first car after the collision?

    A. 19 m/s   **C.** 9 m/s
    B. −9 m/s   **D.** 14 m/s
                                               E. −14 m/s

11. An 8 g bullet is shot into a 4 kg block at rest on a frictionless horizontal surface. The bullet remains lodged in the block. The block moves into a spring and compresses it by 8.9 cm. After the block comes to a stop, the spring fully decompresses and sends the block in the opposite direction. What is the magnitude of the impulse of the block (including the bullet), due to the spring, during the entire time interval in which the block and spring are in contact? (Use the value of the spring constant = 1,400 N/m)

    A. 11 N·s   **C.** 6.4 N·s
    B. 8.3 N·s   **D.** 12 N·s
                                                E. 13 N·s

12. An ice skater performs a fast spin by pulling her outstretched arms close to her body. What happens to her rotational kinetic energy about the axis of rotation?

- **A.** Decreases
- **B.** Remains the same
- **C.** Increases
- **D.** It changes, but it depends on her body mass
- **E.** It decreases in proportion to √(length of her arms)

13. A toy car is traveling in a circular path. The force required to maintain this motion is F. If the velocity of the object is doubled, what is the force required to maintain its motion?

- **A.** $2F$
- **B.** F
- **C.** $½F$
- **D.** $4F$
- **E.** $\sqrt{2}F$

14. Which of the following are units of momentum?

- **A.** kg·m/s²
- **B.** J·s/m
- **C.** N·m
- **D.** kg·s
- **E.** kg·m²/s²

15. The impulse on an apple hitting the ground depends on:

   I. the speed of the apple just before it hits
   II. whether the apple bounces
   III. the time of impact with the ground

- **A.** I only
- **B.** II only
- **C.** III only
- **D.** I and III only
- **E.** I, II and III

16. A 55 kg girl throws a 0.8 kg ball against a wall. The ball strikes the wall horizontally with a speed of 25 m/s and bounces back at the same speed. The ball is in contact with the wall for 0.05 s. What is the average force exerted on the wall by the ball?

- **A.** 27,500 N
- **B.** 55,000 N
- **C.** 400 N
- **D.** 800 N
- **E.** 13,750 N

17. Three objects are moving along a straight line, as shown. If the positive direction is to the right, what is the total momentum of this system?

- **A.** −70 kg·m/s
- **B.** +70 kg·m/s
- **C.** +86 kg·m/s
- **D.** −86 kg·m/s
- **E.** 0 kg·m/s

6 m/s → 7 kg ; 3 m/s → 12 kg ; 2 m/s ← 4 kg

Questions **18-19** are based on the following:

Two ice skaters, Vladimir (60 kg) and Olga (40 kg), collide in midair. Before the collision, Vladimir went North at 0.5 m/s, and Olga was going West at 1 m/s. Right after the collision and well before they land on the ground, they stick. Assume they have no vertical velocity.

18. What is the magnitude of their velocity just after the collision?

   A. 0.1 m/s
   B. 1.8 m/s
   C. 0.9 m/s
   D. 1.5 m/s
   E. 0.5 m/s

19. What is the magnitude of the total momentum just after the collision?

   A. 25 kg·m/s
   B. 50 kg·m/s
   C. 65 kg·m/s
   D. 80 kg·m/s
   E. 40 kg·m/s

20. A horse is running in a straight line. If the mass and the horse's speed are doubled, by what factor does its momentum increase?

   A. $\sqrt{2}$
   B. 2
   C. 4
   D. 8
   E. 16

21. The mass of box P is greater than the mass of box Q. Both boxes are on a frictionless horizontal surface and connected by a light cord. A horizontal force F is applied to box Q, accelerating the boxes to the right. What is the magnitude of the force exerted by the connecting cord on box P?

   A. equal to F
   B. equal to $2F$
   C. zero
   D. less than F but > 0
   E. equal to $3F$

22. Which of the following is true when Melissa and Samantha ride on a merry-go-round, as viewed above?

   A. They have the same speed, but different angular velocity
   B. They have different speeds, but the same angular velocity
   C. They have the same speed and the same angular velocity
   D. They have different speeds and different angular velocities
   E. Requires the radius of the merry-go-round

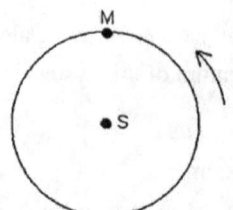

23. The relationship between impulse and impact force involves the:

A. time the force acts
B. distance the force acts
C. difference between acceleration and velocity
D. mass and its effect on resisting a change in velocity
E. difference between acceleration and speed

24. Angular momentum cannot be conserved if the:

A. moment of inertia changes
B. system is experiencing a net force
C. angular velocity changes
D. angular displacement changes
E. system has a net torque

25. A 6.8-kg block moves on a frictionless surface with a speed of $v_i = 5.4$ m/s and makes a perfectly elastic collision with a 4.8-kg stationary block. After the collision, the 6.8-kg block recoils with a speed of $v_f = 3.2$ m/s. What is the magnitude of the average force on the 6.8-kg block while the two blocks contact 2 s?

A. 4.4 N
B. 46.1 N
C. 32.6 N
D. 29.2 N
E. 18.4 N

Questions **26-27** are based on the following:

A 4 kg rifle imparts a high velocity to a small 10 g bullet by exploding a charge that causes the bullet to leave the barrel at 300 m/s. Take the system as the combination of the rifle and bullet. Typically, the rifle is fired with the butt of the gun pressed against the shooter's shoulder. Ignore the force of the shoulder on the rifle.

26. What is the momentum of the system just after the bullet leaves the barrel?

A. 0 kg·m/s
B. 3 kg·m/s
C. 9 kg·m/s
D. 30 kg·m/s
E. 120 kg·m/s

27. What is the recoil velocity of the rifle (i.e., the velocity of the rifle just after firing)?

A. 23 m/s
B. 1.5 m/s
C. 5.6 m/s
D. 12.4 m/s
E. 0.75 m/s

28. A ball thrown horizontally from a point 24 m above the ground strikes the ground after traveling a distance of 18 m horizontally. With what speed was it thrown, assuming negligible air resistance? (Use the acceleration due to gravity $g = 9.8$ m/s^2)

A. 6.8 m/s
B. 7.5 m/s
C. 8.1 m/s
D. 8.6 m/s
E. 9.7 m/s

29. An object is moving in a circle at a constant speed. Its acceleration vector is directed:

A. toward the center of the circle
B. away from the center of the circle
C. tangent to the circle and in the direction of the motion
D. behind the normal and toward the center of the circle
E. ahead of the normal and toward the center of the circle

30. Impulse is equal to the:

   I. force multiplied by the distance over which the force acts
   II. change in momentum
   III. momentum

A. I only
B. II only
C. III only
D. I and II only
E. I and III only

31. A 4 kg object is at the height of 10 m above the Earth's surface. Ignoring air resistance, what is its kinetic energy immediately before impacting the ground if it is thrown straight downward with an initial speed of 20 m/s? (Use acceleration due to gravity $g = 10$ m/s^2)

A. 150 J
B. 300 J
C. 1,200 J
D. 900 J
E. 600 J

32. A car traveling along the highway needs a certain amount of force to stop. More stopping force may be required when the car has:

   I. less stopping distance   II. more momentum   III. more mass

A. I only
B. II only
C. III only
D. I and III only
E. I, II and III

33. A table tennis ball moving East at a speed of 4 m/s collides with a stationary bowling ball. The table tennis ball bounces back to the West, and the bowling ball moves very slowly to the East. Which ball experiences the greater magnitude of impulse during the collision?

- A. Bowling ball
- B. Table tennis ball
- C. Neither because both experience the same magnitude of the impulse
- D. It is not possible to determine since the velocities after the collision are unknown
- E. It is not possible to determine since the masses of the objects are unknown

34. Assume that a massless bar of 5 m is suspended from a rope attached to the bar at a distance of x from the bar's left end. If a 30 kg mass hangs from the right side of the bar and a 6 kg mass hangs from the left side of the bar, what value of x results in equilibrium? (Use the acceleration due to gravity $g = 9.8$ m/s^2)

- A. 2.8 m
- B. 4.2 m
- C. 3.2 m
- D. 1.6 m
- E. 4.5 m

35. A block of mass m sits at rest on a rough inclined ramp that makes an angle θ with the horizontal. What must be true about the force of static friction (f) on the block?

- A. $f > mg \sin \theta$
- B. $f = mg \cos \theta$
- C. $f = mg$
- D. $f < mg \cos \theta$
- E. $f = mg \sin \theta$

36. A 30 kg block is pushed in a straight line across a horizontal surface. What is the coefficient of kinetic friction μ_k between the block and the surface if a constant force of 45 N must be applied to the block to maintain a constant velocity of 3 m/s? (Use the acceleration due to gravity $g = 10$ m/s^2)

- A. 0.1
- B. 0.33
- C. 0.15
- D. 0.5
- E. 0.66

37. The impulse-momentum relationship is a direct result of:

    I. Newton's First Law   II. Newton's Second Law   III. Newton's Third Law

- A. I only
- B. II only
- C. III only
- D. I and II only
- E. I and III only

Questions **38-40** are based on the following:

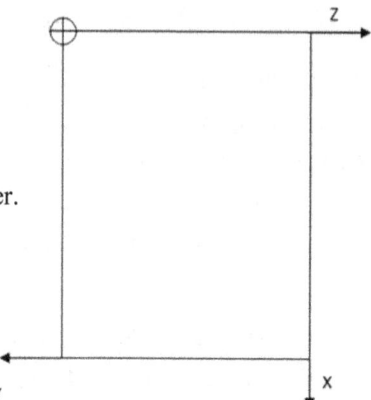

A 0.5 m by 0.6 m rectangular piece of metal is hinged (⊗) (as shown) in the upper left corner, hanging so that the long edge is vertical. A 25 N force (Y) acts to the left at the lower-left corner. A 15 N force (X) acts down at the lower right corner. A 30 N force (Z) acts to the right at the upper right corner. Each force vector is in the plane of the metal. Use counterclockwise as the positive direction.

38. What is the torque of force X about the pivot?

A. 5 N·m
B. 3 N·m
C. −7.5 N·m
D. 0 N·m
E. −5 N·m

39. What is the torque of force Z about the pivot?

A. −10 N·m
B. −4.5 N·m
C. 4.5 N·m
D. 10 N·m
E. 0 N·m

40. What is the torque of force Y about the pivot?

A. −15 N·m
B. −3 N·m
C. 0 N·m
D. 3 N·m
E. 7.5 N·m

41. A 50 g weight is tied to the end of a string and whirled at 20 m/s in a horizontal circle with a radius of 2 m. Ignoring the force of gravity, what is the tension in the string?

A. 5 N
B. 10 N
C. 50 N
D. 150 N
E. 20 N

42. A small car collides with a large truck in a head-on collision. Which of the following statements concerning the magnitude of the average force during the collision is correct?

A. The small car and the truck experience the same average force
B. The force experienced by each one is inversely proportional to its velocity
C. The truck experiences a greater average force
D. The small car experiences the greater average force
E. The force experienced by each one is directly proportional to its velocity

43. A 10 kg bar 2 m long extends perpendicularly from a vertical wall. The free end of the bar is attached to a point on the wall by a light cable, which makes an angle of 30° with the bar. What is the tension in the cable? (Use the acceleration due to gravity $g = 10$ m/s^2)

    A. 75 N   **C.** 100 N
    B. 150 N   **D.** 125 N
                                                          E. 50 N

44. Object A has the same size and shape as object B but is twice as heavy. When objects A and B are dropped simultaneously from a tower, they reach the ground simultaneously. Object A has greater:

        I. speed   II. momentum   III. acceleration

    A. I only   **C.** III only
    B. II only   **D.** I and II only
                                                     E. I and III only

45. Two vehicles approach a right-angle intersection and then collide. After the collision, they become entangled. If their mass ratio was 1 : 4 and their respective speeds as they approached were 12 m/s, what is the magnitude of the velocity immediately following the collision?

    A. 16.4 m/s   **C.** 13.4 m/s
    B. 11.9 m/s   **D.** 9.9 m/s
                                                         E. 8.5 m/s

46. A skater stands stationary on frictionless ice, and she throws a heavy ball to the right at an angle of 5° above the horizontal. With respect to the ice, if the ball weighs one-third as much as the skater and she is measured to be moving with a speed of 2.9 m/s to the left after the throw, how fast did she throw the ball?

    A. 10.2 m/s   **C.** 8.73 m/s
    B. 7.2 m/s   **D.** 9.8 m/s
                                                       E. 8.1 m/s

47. Ignoring the forces of friction, what horizontal force must be applied to an object with a weight of 98 N to give it a horizontal acceleration of 10 m/s^2? (Use the acceleration due to gravity $g = 9.8$ m/s^2)

    A. 9.8 N   **C.** 79 N
    B. 100 N   **D.** 125 N
                                                       E. 4.9 N

48. Consider a winch that pulls a cart at constant speed up an incline. Point A is at the bottom of the incline, and point B is at the top. Which of the following statements is/are true from point A to B?

    I. The KE of the cart is constant
    II. The PE of the cart is constant
    III. The sum of the KE and PE of the cart is constant

- **A.** I only
- **B.** II only
- **C.** III only
- **D.** I and II only
- **E.** I, II and III

49. A high-speed dart is shot from ground level with a speed of 140 m/s at an angle of 35° above the horizontal. What is the vertical component of its velocity after 4 s if air resistance is ignored? (Use the acceleration due to gravity $g = 9.8$ m/s^2)

- **A.** 59 m/s
- **B.** 75 m/s
- **C.** 34 m/s
- **D.** 41 m/s
- **E.** 38 m/s

50. What does the area under the force *vs.* time graph curve represent a diver as she leaves the platform during her approach to the water below?

- **A.** Work
- **B.** Momentum
- **C.** Impulse
- **D.** Displacement
- **E.** Force

51. Strings suspend a rifle of mass 2 kg. The rifle fires a bullet of mass 0.01 kg at a speed of 220 m/s. What is the recoil velocity of the rifle?

- **A.** 0.001 m/s
- **B.** 0.01 m/s
- **C.** 0.1 m/s
- **D.** 1.1 m/s
- **E.** 10.1 m/s

52. How do automobile airbags reduce injury during a collision?

- **A.** They reduce the kinetic energy transferred to the passenger
- **B.** They reduce the momentum transferred to the passenger
- **C.** They reduce the acceleration of the automobile
- **D.** They reduce the forces exerted upon the passenger
- **E.** All of the above

Questions **53-55** are based on the following:

Tim nails a meter stick to a board at the meter stick's 0 m mark. Force I acts at the 0.5 m mark perpendicular to the meter stick with a force of 10 N, as shown in the figure. Force II acts at the end of the meter stick with a force of 5 N, making a 35° angle. Force III acts at the same point with a force of 20 N, providing tension but no shear stress. Use counterclockwise as the positive direction.

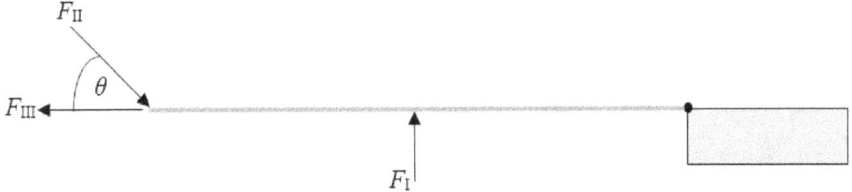

53. What is the torque of Force I about the fixed point?

A. −5 N·m
B. 0 N·m
C. 5 N·m
D. 10 N·m
E. −10 N·m

54. What is the torque of Force II about the fixed point?

A. −4.8 N·m
B. −2.9 N·m
C. 4.8 N·m
D. 6.6 N·m
E. 2.9 N·m

55. What is the torque of Force III about the fixed point?

A. −20 N·m
B. 0 N·m
C. 10 N·m
D. 20 N·m
E. −10 N·m

Notes for active learning

Rotational Motion

1. Suppose a uniform solid sphere of mass M and radius R rolls without slipping down an inclined plane starting from rest. The linear velocity of the sphere at the bottom of the incline depends on:

- **A.** the radius of the sphere
- **B.** the mass of the sphere
- **C.** the mass and the radius of the sphere
- **D.** neither the mass nor the radius of the sphere
- **E.** the mass and the square root of the radius of the sphere

2. A solid, uniform sphere of mass 2.0 kg and radius 1.7 m rolls from rest without slipping down an inclined plane of height 5.3 m. What is the angular velocity of the sphere at the bottom of the inclined plane?

- **A.** 3.7 rad/s
- **B.** 5.1 rad/s
- **C.** 6.7 rad/s
- **D.** 8.3 rad/s
- **E.** 11.9 rad/s

3. A solid uniform ball with a mass of 125.0 g is rolling without slipping along the horizontal surface of a table with a speed of 4.5 m/s when it rolls off the edge and falls towards the floor, 1.1 m below. What is the rotational kinetic energy of the ball just before it hits the floor?

- **A.** 0.51 J
- **B.** 0.87 J
- **C.** 1.03 J
- **D.** 2.26 J
- **E.** Requires the radius of the ball

4. David swings a 0.38 kg ball in a circle on a string 1.3 m long. What is the magnitude of the ball's angular momentum if the ball makes 1.2 rev/s?

- **A.** 0.6 kg·m^2/s
- **B.** 2.2 kg·m^2/s
- **C.** 3.6 kg·m^2/s
- **D.** 4.8 kg·m^2/s
- **E.** 6.2 kg·m^2/s

5. An ice skater has a moment of inertia of 5.0 kg·m^2 when her arms are outstretched, and at this time, she is spinning at 3.0 rev/s. If she pulls in her arms and decreases her moment of inertia to 2.0 kg·m^2, how fast will she be spinning?

- **A.** 1.8 rev/s
- **B.** 4.5 rev/s
- **C.** 7.5 rev/s
- **D.** 10.5 rev/s
- **E.** 12.5 rev/s

6. The angular momentum of a system remains constant when:

A. its total kinetic energy is constant

B. the moment of inertia is constant

C. no net external torque acts on the system

D. no net external force acts on the system

E. the linear momentum and the energy are constant

7. A bicycle has wheels 60.0 cm in diameter. What is the angular speed of these wheels when it is moving at 4.0 m/s?

A. 0.28 rad/s

B. 1.6 rad/s

C. 3.4 rad/s

D. 6.7 rad/s

E. 13.3 rad/s

8. What is the kinetic energy of a thin uniform rod of length 120.0 cm with a mass of 450.0 g rotating about its center along the short axis at 3.60 rad/s? (The short axis is perpendicular to the axis of the rod. Imagine spinning the rod like an airplane propeller.)

A. 0.350 J

B. 1.30 J

C. 2.70 J

D. 4.96 J

E. 6.10 J

9. A rope is wrapped around a wheel of radius R = 2.0 meters. The wheel is mounted with frictionless bearings on an axle through its center. A block of mass 14.0 kg is suspended from the end of the rope. When the system is released from rest, it is observed that the block descends 10.0 meters in 2.0 seconds. What is the moment of inertia of the wheel?

A. 300.0 kg·m²

B. 185.0 kg·m²

C. 53.8 kg·m²

D. 521.0 kg·m²

E. 88.5 kg·m²

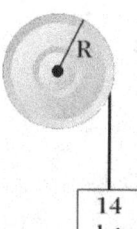

10. A string is wrapped tightly around a fixed frictionless pulley with a moment of inertia of 0.0352 kg·m² and a radius of 12.5 cm. The string is pulled from the pulley with a constant force of 5.00 N, causing the pulley to rotate. If the string does not slip on the pulley, what is the speed of the string after it has unwound 1.25 m? Consider the string to be massless.

A. 0.69 m/s

B. 2.36 m/s

C. 3.62 m/s

D. 4.90 m/s

E. 6.12 m/s

11. When a rigid object rotates about a fixed axis, what is true about the points in the object?

    I. They have the same angular acceleration
    II. They have the same tangential acceleration
    III. They have the same radial acceleration

   A. I only
   B. II only
   C. III only
   D. I and II only
   E. I, II and III

12. A small mass is placed on a record turntable rotating at 33.33 rpm. The linear velocity of the mass is:

   A. zero
   B. directed parallel to the line joining the mass and the center of rotation
   C. independent (in magnitude) of the position of the mass on the turntable
   D. greater the closer the mass is to the center
   E. greater the farther the mass is from the center

13. To drive a midsize car at 40.0 mph on a level road for one hour requires about 3.2×10^7 J of energy. Suppose this much energy was attempted to be stored in a spinning, solid, uniform, cylindrical flywheel. If a flywheel with a diameter of 1.2 m and mass of 400.0 kg were used, what angular speed would be required to store 3.2×10^7 J?

   A. 380 rad/s
   B. 620 rad/s
   C. 940 rad/s
   D. 1,450 rad/s
   E. 2,860 rad/s

14. A wheel having a moment of inertia of 5.0 kg·m² starts from rest and accelerates for 8.0 s under a constant torque of 3.0 N·m. What is the wheel's rotational kinetic energy at the end of 8.0 s?

   A. 29 J
   B. 58 J
   C. 83 J
   D. 112 J
   E. 146 J

15. When a rigid object rotates about a fixed axis, what is true about the points in the object?

    I. They have the same angular speed
    II. They have the same tangential speed
    III. They have the same angular acceleration

   A. I only
   B. II only
   C. III only
   D. I and III only
   E. I, II and III

16. A uniform, solid cylindrical flywheel of radius 1.4 m and mass 15.0 kg rotates at 2.4 rad/s. What is the magnitude of the flywheel's angular momentum?

A. 11 kg·m²/s
B. 18 kg·m²/s
C. 25 kg·m²/s
D. 35 kg·m²/s
E. 64 kg·m²/s

17. A uniform solid disk is released from rest and rolls without slipping down an inclined plane that makes an angle of 25° with the horizontal. What is the forward speed of the disk after it has rolled 3.0 m, measured along the plane?

A. 0.8 m/s
B. 1.8 m/s
C. 2.9 m/s
D. 3.5 m/s
E. 4.1 m/s

18. A tire is rolling along a road, without slipping, with a center-of-mass velocity v. A piece of tape is attached to the tire. When the tape is opposite the road (at the top of the tire), what is its velocity with respect to the road?

A. $2v$
B. v
C. $1.5v$
D. $\sqrt{v}$
E. The velocity depends on the radius of the tire

19. A string is wound tightly around a fixed pulley having a radius of 5.0 cm. As the string is pulled, the pulley rotates without slipping of the string. What is the angular speed of the pulley when the string moves at 5.0 m/s?

A. 10.0 rad/s
B. 25.0 rad/s
C. 75.0 rad/s
D. 100.0 rad/s
E. 150.0 rad/s

20. A 1.4 kg object at $x = 2.00$ m, $y = 3.10$ m moves at 4.62 m/s at an angle 45° north of east. What is the magnitude of the object's angular momentum about the origin?

A. 1.2 kg·m²/s
B. 2.6 kg·m²/s
C. 3.8 kg·m²/s
D. 5.0 kg·m²/s
E. 8.4 kg·m²/s

21. When a fan is turned off, its angular speed decreases from 10.0 rad/s to 6.3 rad/s in 5.0 s. What is the magnitude of the average angular acceleration of the fan?

A. 0.46 rad/s²
B. 0.74 rad/s²
C. 1.86 rad/s²
D. 2.80 rad/s²
E. 3.62 rad/s²

22. At time $t = 0$ s, a wheel has an angular displacement of 0 radians and an angular velocity of +26.0 rad/s. The wheel has a constant acceleration of -0.43 rad/s^2. In this situation, what is the time t (after $t = 0$ s), at which the kinetic energy of the wheel is twice the initial value?

A. 48 s
B. 86 s
C. 115 s
D. 146 s
E. 185 s

23. A solid uniform disk of diameter 3.20 m and mass 42.0 kg rolls without slipping to the bottom of a hill, starting from rest. If the angular speed of the disk is 4.27 rad/s at the bottom, how high vertically did it start on the hill above the bottom?

A. 2.46 m
B. 3.57 m
C. 4.85 m
D. 6.24 m
E. 8.44 m

24. When Steve rides a bicycle, in what direction is the angular velocity of the wheels?

A. to his left
B. to his right
C. forwards
D. backward
E. up

25. A rolling wheel of a diameter of 68.0 cm slows down uniformly from 8.4 m/s to rest over 115.0 m. What is the magnitude of its angular acceleration if there was no slipping?

A. 0.90 rad/s^2
B. 1.6 rad/s^2
C. 4.2 rad/s^2
D. 7.8 rad/s^2
E. 11.4 rad/s^2

26. A uniform solid cylinder with a radius of 10.0 cm and a mass of 3.0 kg is rotating about its center axis with an angular speed of 33.4 rpm. What is the kinetic energy of the uniform solid cylinder?

A. 0.091 J
B. 0.19 J
C. 0.66 J
D. 1.14 J
E. 2.46 J

27. A uniform 135.0-g meter stick rotates about an axis perpendicular to the stick, passing through its center with an angular speed of 3.50 rad/s. What is the magnitude of the angular momentum of the stick?

A. 0.0394 kg·m^2/s
B. 0.0848 kg·m^2/s
C. 0.286 kg·m^2/s
D. 0.458 kg·m^2/s
E. 0.826 kg·m^2/s

28. A 23.0 kg mass is connected to a nail on a frictionless table by a massless string of length 1.3 m. If the tension in the string is 51.0 N while the mass moves in a uniform circle on the table, how long does it take for the mass to make one complete revolution?

A. 2.8 s
B. 3.6 s
C. 4.8 s
D. 5.4 s
E. 6.2 s

29. A machinist turns on the power to a grinding wheel at time $t = 0$ s. The wheel accelerates uniformly from rest for 10.0 s and reaches the operating angular speed of 38.0 rad/s. The wheel is run at that angular speed for 30.0 s, and then the power is shut off. The wheel slows down uniformly at 2.1 rad/s² until the wheel stops. What is the angular acceleration of the wheel between $t = 0$ s and $t = 10.0$ s?

A. 1.21 rad/s²
B. 2.63 rad/s²
C. 3.80 rad/s²
D. 5.40 rad/s²
E. 6.81 rad/s²

30. A force of 17.0 N is applied to the end of a 0.63 m long torque wrench at an angle 45° from a line joining the pivot point to the handle. What is the magnitude of the torque generated about the pivot point?

A. 4.3 N·m
B. 8.2 N·m
C. 7.6 N·m
D. 11.8 N·m
E. 14.0 N·m

31. A solid disk of radius 1.60 m and mass 2.30 kg rolls from rest without slipping to the bottom of an inclined plane. If the angular velocity of the disk is 4.27 rad/s at the bottom, what is the height of the inclined plane?

A. 0.57 m
B. 1.08 m
C. 2.84 m
D. 3.57 m
E. 5.66 m

32. A merry-go-round spins freely when Paul moves quickly to the center along a radius of the merry-go-round. As he does this, the moment of inertia of the system:

A. increases and the angular speed increases
B. decreases and the angular speed remains the same
C. decreases and the angular speed decreases
D. decreases and the angular speed increases
E. increases and the angular speed decreases

33. What is the angular speed of a compact disc that, at a specific instant, is rotating at 210.0 rpm?
- A. 8.5 rad/s
- B. 22.0 rad/s
- C. 36.4 rad/s
- D. 52.6 rad/s
- E. 68.2 rad/s

34. A solid uniform sphere is rolling without slipping along a horizontal surface with a speed of 5.5 m/s when it starts up a ramp that makes an angle of 25° with the horizontal. What is the speed of the sphere after it has rolled 3.0 m up as measured along the surface of the ramp?
- A. 0.8 m/s
- B. 1.6 m/s
- C. 3.5 m/s
- D. 4.8 m/s
- E. 6.4 m/s

35. A force of 16.88 N is applied tangentially to a wheel of radius 0.340 m and gives rise to angular acceleration of 1.20 rad/s². What is the rotational inertia of the wheel?
- A. 1.48 kg·m²
- B. 2.26 kg·m²
- C. 3.48 kg·m²
- D. 4.78 kg·m²
- E. 6.42 kg·m²

36. A machine does 3.9 kJ of work on a spinning flywheel to bring it from 500.0 rpm to rest. This flywheel is in the shape of a solid uniform disk of a radius of 1.2 m. What is the mass of this flywheel?
- A. 2.6 kg
- B. 4.0 kg
- C. 5.2 kg
- D. 6.4 kg
- E. 8.6 kg

37. A rectangular billboard with $h = 20.0$ cm high and $w = 11.0$ cm wide loses three of its four support bolts and rotates into the position as shown, with P_1 directly over P_3. It is supported by P_2, which is so tight that it holds the billboard from further rotation. What is the gravitational torque about P_2 if the mass of the billboard is 5.0 kg?
- A. 1.2 Nm
- B. 2.5 Nm
- C. 4.7 Nm
- D. 6.8 Nm
- E. 8.2 Nm

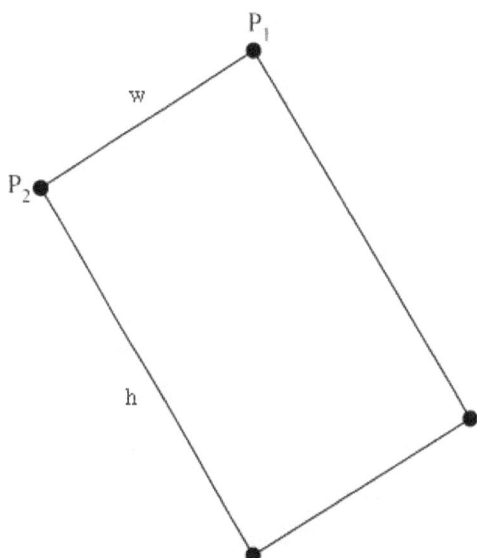

38. A disk, a hoop, and a solid sphere are released simultaneously at the top of an inclined plane. In which order do they reach the bottom if each is uniform and rolls without slipping?

A. sphere, hoop, disk
B. sphere, disk, hoop
C. hoop, sphere, disk
D. disk, hoop, sphere
E. hoop, disk, sphere

39. A uniform disk is attached at the rim to a vertical shaft and is used as a cam. A side view and top view of the disk and shaft are shown.

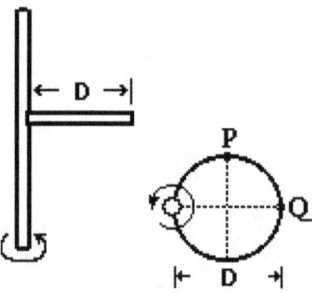

The disk has a diameter of 80.0 cm. The moment of inertia of the disk about the axis of the shaft is 6.0×10^{-3} kg·m². What is the kinetic energy of the disk as the shaft rotates uniformly about its axis at 96.0 rpm?

A. 0.18 J
B. 0.30 J
C. 0.49 J
D. 0.57 J
E. 1.0 J

40. A scooter has wheels with a diameter of 240.0 mm. What is the angular speed of the wheels when the scooter is moving forward at 6.00 m/s?

A. 128.6 rpm
B. 248.2 rpm
C. 472.0 rpm
D. 478.0 rpm
E. 1,260.6 rpm

41. A spinning ice skater on frictionless ice can control the rate at which she rotates by pulling in her arms. Which of the following statements are true about the skater during this process?

  I. Her kinetic energy remains constant
  II. Her moment of inertia remains constant
  III. Her angular momentum remains constant

A. I only
B. II only
C. III only
D. I and II only
E. I, II and III

42. Tanya is riding a merry-go-round that has an instantaneous angular speed of 1.25 rad/s and angular acceleration of 0.745 rad/s². Tanya is standing 4.65 m from the center of the merry-go-round. What is the magnitude of the linear acceleration of Tanya?

- A. 2.45 m/s²
- B. 4.20 m/s²
- C. 6.82 m/s²
- D. 8.05 m/s²
- E. 12.10 m/s²

43. Through how many degrees does a 33.0 rpm turntable rotate in 0.32 s?

- A. 31°
- B. 42°
- C. 63°
- D. 76°
- E. 85°

44. A 50.0 kg uniform ladder, length $L = 5.00$ m long, is against a smooth wall height of $h = 3.70$ m. The base of the ladder rests on a rough horizontal surface whose coefficient of static friction $\mu = 0.750$. An 80.0 kg block is suspended from the top rung of the ladder, just at the wall. What is the approximate magnitude of the force exerted on the ladder base due to contact with the rough horizontal surface?

- A. 1,370 N
- B. 1,460 N
- C. 1,580 N
- D. 1,640 N
- E. 1,760 N

45. At time $t = 0$ s, a wheel has an angular displacement of zero radians and an angular velocity of +29.0 rad/s. The wheel has a constant acceleration of –0.52 rad/s². In this situation, what is the maximum value of the angular displacement?

- A. +467 rad
- B. +809 rad
- C. +1,110 rad
- D. +1,460 rad
- E. +1,840 rad

46. A disk lies in the xz-plane with its center at the origin. When viewed from the positive y-axis (i.e., above the disk), the direction of rotation appears clockwise. In what direction does the angular velocity of the disk point?

- A. to her right
- B. to her left
- C. down
- D. up
- E. forwards

Notes for active learning

Work and Energy

1. Consider the following ways that a girl might throw a stone from a bridge. The speed of the stone as it leaves her hand is the same in each of the three cases.

  I. Thrown straight up
  II. Thrown straight down
  III. Thrown straight out horizontally

Ignoring air resistance, in which case is the vertical speed of the stone the greatest when it hits the water below?

A. I only
B. II only
C. III only
D. I and II only
E. II and III only

2. A package is pulled along the ground by a 5 N force F directed 45° above the horizontal. Approximately how much work is done by the force when it pulls the package 10 m?

A. 14 J
B. 35 J
C. 70 J
D. 46 J
E. 64 J

3. Which quantity has the greatest influence on the amount of kinetic energy that a large truck has while moving down the highway?

A. Velocity
B. Mass
C. Density
D. Direction
E. Acceleration

4. No work is done by gravity on a bowling ball that rolls along the floor of a bowling alley because:

A. no potential energy is converted to kinetic energy
B. the force on the ball is at a right angle to the ball's motion
C. its velocity is constant
D. the total force on the ball is zero
E. its kinetic energy remains constant

5. A 5 kg toy car is moving along the level ground. At a given time, it travels at a speed of 2 m/s and accelerates at 3 m/s². What is the cart's kinetic energy?

A. 20 J
B. 8 J
C. 12 J
D. 4 J
E. 10 J

6. A treehouse is 8 m above the ground. If Peter does 360 J of work while pulling a box from the ground up to his treehouse with a rope, what is the mass of a box? (Use the acceleration due to gravity $g = 10$ m/s^2)

   A. 4.5 kg
   B. 3.5 kg
   C. 5.8 kg
   D. 2.5 kg
   E. 1.4 kg

7. For an ideal elastic spring, what is the slope of the curve for a displacement (x) vs. applied force (F) graph?

   A. The acceleration of gravity
   B. The square root of the spring constant
   C. The spring constant
   D. The reciprocal of the spring constant
   E. The square of the spring constant

8. A spring with a spring constant of 22 N/m is stretched from equilibrium to 3 m. How much work is done in the process?

   A. 33 J
   B. 66 J
   C. 99 J
   D. 198 J
   E. 242 J

9. A baseball is thrown straight up. Compare the sign of the work done by gravity while the ball goes up with the sign of the work done by gravity while it goes down:

   A. negative on the way up and positive on the way down
   B. negative on the way up and negative on the way down
   C. positive on the way up and positive on the way down
   D. positive on the way up and negative on the way down
   E. requires information about the mass of the baseball

10. Let A_1 represent the magnitude of the work done by gravity as mass A's gravitational energy increases by 400 J. Let B_1 represent the total amount of work necessary to increase mass B's kinetic energy by 400 J. How do A_1 and B_1 compare?

   A. $A_1 > B_1$
   B. $A_1 = B_1$
   C. $A_1 < B_1$
   D. $A_1 = 400\ B_1$
   E. $400\ A_1 = B_1$

11. According to the definition of work, pushing on a rock accomplishes no work unless there is:

   A. an applied force equal to the rock's weight
   B. movement perpendicular to the force
   C. an applied force greater than the rock's weight
   D. movement parallel to the force
   E. force perpendicular to the movement

12. A job is done slowly, while an identical job is done quickly. Both jobs require the same amount of work, but different amounts of:

   I. energy   II. power   III. torque

A. I only
B. II only
C. I and II only
D. I and III only
E. none are true

13. On a force (F) vs. distance (d) graph, what represents the work done by the force F?

A. The area under the curve
B. A line connecting two points on the curve
C. The slope of the curve
D. The length of the curve
E. The maximum F × the maximum d

14. A 3 kg cat leaps from a tree to the ground, a distance of 4 m. What is its kinetic energy just before the cat reaches the ground? (Use acceleration due to gravity $g = 10$ m/s^2)

A. 0 J
B. 9 J
C. 120 J
D. 60 J
E. 36 J

15. A book is resting on a plank of wood. Jackie pushes the plank and accelerates it so that the book is stationary with respect to the plank. The work done by static friction is:

A. zero
B. positive
C. negative
D. parallel to the surface
E. perpendicular to the surface

16. 350 J of work is required to drive a stake into the ground thoroughly. If the average resistive force on the stake by the ground is 900 N, how long is the stake?

A. 2.3 m
B. 0.23 m
C. 3 m
D. 0.39 m
E. 0.46 m

17. A lightweight object and a very heavy object are sliding with equal speeds along a level, frictionless surface. They slide up the same frictionless hill with no air resistance. Which object rises to a greater height?

A. The lightweight object because the force of gravity on it is less
B. The heavy object because it has more kinetic energy to carry it up the hill
C. The heavy object because it has greater potential energy
D. The lightweight object because it has more kinetic energy to carry it up the hill
E. They both slide to the same height

18. If Investigator II does 3 times the work of Investigator I in one third the time, the power output of Investigator II is:

A. 9 times greater
B. 3 times greater
C. 1/3 times greater
D. the same
E. √3 times greater

19. A diver who weighs 450 N steps off a diving board 9 m above the water. What is the kinetic energy when the diver strikes the water?

A. 160 J
B. 540 J
C. 45 J
D. 4,050 J
E. 5,400 J

20. A vertical, hanging spring stretches by 23 cm when a 160 N object is attached. What is the weight of a hanging plant that stretches the spring by 34 cm?

A. 237 N
B. 167 N
C. 158 N
D. 309 N
E. 249 N

21. A mule pulled with a horizontal force F on a covered wagon of mass M. The mule and covered wagon travel at a constant speed v on level ground. How much work is done by the mule on the covered wagon during time Δt? (Use acceleration due to gravity $g = 10$ m/s^2)

A. $-Fv\Delta t$
B. $Fv\Delta t$
C. 0 J
D. $-F\sqrt{v}\Delta t$
E. $-Fv/\Delta t$

22. Jane pulls on a sled strap at an angle of 32° above the horizontal. If 540 J of work is done by the strap while moving the sled at a horizontal distance of 18 m, what is the tension in the strap?

A. 86 N
B. 112 N
C. 24 N
D. 35 N
E. 69 N

23. A vertical spring stretches 6 cm from equilibrium when a 120 g mass is attached to the bottom. If an additional 120 g mass is added to the spring, how does the potential energy of the spring change?

A. the same
B. 4 times greater
C. 2 times greater
D. √2 times greater
E. 3 times greater

24. A Ferrari, Maserati, and Lamborghini are moving at the same speed, and each driver slams on his brakes and brings the car to a stop. The most massive is the Ferrari, and the least massive is the Lamborghini. If the tires of the three cars have identical coefficients of friction with the road surface, which car experiences the highest amount of work done by friction?

- **A.** Maserati
- **B.** Lamborghini
- **C.** Ferrari
- **D.** The amount is the same
- **E.** Requires more information

25. A hammer does the work of driving a nail into a wooden board. Compared to the moment before the hammer strikes the nail after it impacts the nail, the hammer's mechanical energy is:

- **A.** the same
- **B.** less, because work has been done on the hammer
- **C.** greater, because the hammer has done work
- **D.** greater, because work has been done on the hammer
- **E.** less, because the hammer has done work

26. A 1,500 kg car travels at 25 m/s on a level road and the driver slams on the brakes. The skid marks are 10 m long. What is the work done by the road on the car?

- **A.** -4.7×10^5 J
- **B.** 0 J
- **C.** 2.0×10^5 J
- **D.** 3.5×10^5 J
- **E.** -3.5×10^5 J

27. A 1,000 kg car is traveling at 4.72 m/s. If a 2,000 kg truck has 20 times the kinetic energy of the car, how fast is the truck traveling?

- **A.** 23.6 m/s
- **B.** 47.2 m/s
- **C.** 94.4 m/s
- **D.** 14.9 m/s
- **E.** 9.71 m/s

28. A 1,500 kg car travels at 25 m/s on a level road, and the driver slams on the brakes. The skid marks are 30 m long. What forces are acting on the car while it is coming to a stop?

- **A.** Gravity down, normal force up, and a frictional force forwards
- **B.** Gravity down, normal force up, and the engine force forwards
- **C.** Gravity down, normal force forward, and a frictional force backward
- **D.** Gravity down, normal force forward, and the engine force backward
- **E.** Gravity down, normal force up, and a frictional force backward

29. A 6,000 N piano is raised via a pulley. For every 1 m that the rope is pulled down, the piano rises 0.15 m. In this pulley system, what is the force needed to lift the piano?

A. 60 N
B. 900 N
C. 600 N
D. 300 N
E. 6 N

30. What does the area under the curve on a force *vs.* position graph represent?

A. Kinetic energy
B. Momentum
C. Work
D. Displacement
E. Friction

31. What is the form in which most energy comes to and leaves the Earth?

A. Kinetic
B. Radiant
C. Chemical
D. Light
E. Heat

32. A driver abruptly slams on the brakes in her car, and the car skids a certain distance on a straight level road. If she had been traveling twice as fast, what distance would the car have skid under the same conditions?

A. 1.4 times farther
B. ½ as far
C. 4 times farther
D. 2 times farther
E. 8 times farther

33. A crane hoists an object weighing 2,000 N to the top of a building. The crane raises the object straight upward at a constant rate. Ignoring the forces of friction, at what rate is energy consumed by the electric motor of the crane if it takes 60 s to lift the mass 320 m?

A. 2.5 kW
B. 6.9 kW
C. 3.50 kW
D. 10.7 kW
E. 16.3 kW

34. A barbell with a mass of 25 kg is raised 3.0 m in 3.0 s before it reaches constant velocity. What is the net power expended by all forces in raising the barbell? (Use acceleration due to gravity $g = 9.8$ m/s^2 and the acceleration of the barbell is constant)

A. 138 W
B. 34 W
C. 67 W
D. 98 W
E. 17 W

35. Susan carried a 6.5 kg bag of groceries 1.4 m above the ground at a constant velocity for 2.4 m across the kitchen. How much work did Susan do on the bag in the process? (Use acceleration due to gravity $g = 10$ m/s^2)

  A. 52 J
  B. 0 J
  C. 164 J
  D. 138 J
  E. 172 J

36. A 1,000 kg car experiences a net force of 9,600 N while decelerating from 30 m/s to 22 m/s. How far does it travel while slowing down?

  A. 17 m
  B. 22 m
  C. 12 m
  D. 34 m
  E. 26 m

37. What is the power output in relation to the work W if a person exerts 100 J in 50 s?

  A. ¼ W
  B. ½ W
  C. 2 W
  D. 4 W
  E. √2 W

38. If a ball is released from a cliff ledge 58 m above the ground, how fast is the ball traveling when it reaches the ground? (Use the acceleration due to gravity $g = 10$ m/s^2)

  A. 68 m/s
  B. 16 m/s
  C. 44 m/s
  D. 34 m/s
  E. 53 m/s

39. A stone is held at a height *h* above the ground, and a second stone with four times the mass is held at the same height. What is the gravitational potential energy of the second stone compared to that of the first stone?

  A. Twice as much
  B. The same
  C. One fourth as much
  D. One half as much
  E. Four times as much

40. A 1.3 kg coconut falls off a coconut tree, landing on the ground 600 cm below. How much work is done on the coconut by the gravitational force? (Use the acceleration due to gravity $g = 10$ m/s^2)

  A. 6 J
  B. 78 J
  C. 168 J
  D. 340 J
  E. 236 J

41. The potential energy of a pair of interacting objects is related to their:

A. relative position
B. momentum
C. acceleration
D. kinetic energy
E. velocity

42. A spring has a spring constant of 65 N/m. One end of the spring is fixed at point P, while the other is connected to a 7 kg mass *m*. The fixed end and the mass sit on a horizontal, frictionless surface so that the mass and the spring can rotate about P. The mass moves in a circle of radius $r = 4$ m, and the centripetal force of the mass is 15 N. What is the potential energy stored in the spring?

A. 1.7 J
B. 2.8 J
C. 3.7 J
D. 7.5 J
E. 11.2 J

43. If electricity costs 8.16 cents/kW·h, how much would it cost to run a 120 W stereo system 3.5 hours per day for 5 weeks?

A. $1.11
B. $1.46
C. $1.20
D. $0.34
E. $0.49

44. A boy does 120 J of work to pull his sister back on a swing with a 5.1 m chain until the swing makes an angle of 32° with the vertical. What is the mass of his sister? (Use the acceleration due to gravity $g = 9.8$ m/s^2)

A. 18 kg
B. 16.4 kg
C. 13.6 kg
D. 11.8 kg
E. 15.8 kg

45. What is the spring constant value if 111 J of work is needed to stretch a spring from 1.4 m to 2.9 m if the spring's equilibrium position is at 0.0 m?

A. 58 N/m
B. 53 N/m
C. 67 N/m
D. 34 N/m
E. 41 N/m

46. The metric unit of a joule (J) is a unit of:

I. potential energy
II. kinetic energy
III. work

A. I only
B. II only
C. III only
D. I and III only
E. I, II and III

47. A horizontal spring-mass system oscillates on a frictionless table. Find the maximum extension of the spring if the ratio of the mass to the spring constant is 0.038 kg·m/N, and the maximum speed of the mass is 18 m/s?

A. 3.5 m
B. 0.67 m
C. 3.4 cm
D. 67 cm
E. 34 cm

48. A truck weighs twice as much as a car and is moving at twice the speed of the car. Which statement is true about the truck's kinetic energy compared to that of the car?

A. The truck has 8 times the KE
B. The truck has twice the KE
C. The truck has √2 times the KE
D. The truck has 4 times the KE
E. The truck has √8 times the KE

49. When a car brakes to a stop, its kinetic energy is transformed into:

A. energy of rest
B. energy of momentum
C. potential energy
D. stopping energy
E. heat

50. A 30 kg block hangs from a spring with a spring constant of 900 N/m. How far does the spring stretch from its equilibrium position? (Use acceleration due to gravity $g = 10$ m/s^2)

A. 12 cm
B. 33 cm
C. 50 cm
D. 0.5 cm
E. 5 cm

51. What is the kinetic energy of a 0.33 kg baseball thrown at a velocity of 40 m/s?

A. 426 J
B. 574 J
C. 318 J
D. 264 J
E. 138 J

52. An object is acted upon by a force as represented by force *vs.* position graph below. What is the work done as the object moves from 0 m to 4 m?

A. 10 J
B. 50 J
C. 20 J
D. 30 J
E. 60 J

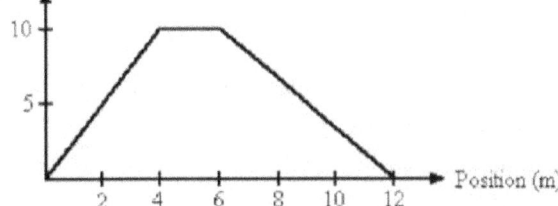

53. James and Bob throw identical balls vertically upward. James throws his ball with an initial speed twice that of Bob's. Assuming no air resistance, what is the maximum height of James's ball compared with that of Bob's ball?

    A. Equal
    B. Eight times
    C. Four times
    D. Two times
    E. √2 times

54. The graphs show the magnitude of the force (F) exerted by a spring as a function of the distance (x) the spring has been stretched. Which graph shows a spring that obeys Hooke's Law?

A.
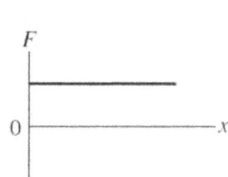

C.
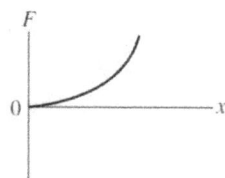

B.
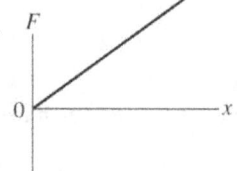

D.
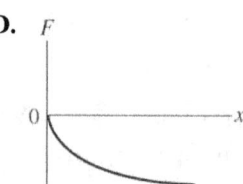

E.
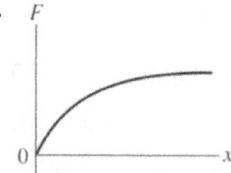

55. If a rocket travels through the air, it loses some kinetic energy due to air resistance. Some of this transferred energy:

    A. decreases the temperature of the air around the rocket
    B. is found in increased KE of the rocket
    C. is found in increased KE of the air molecules
    D. decreases the temperature of the rocket
    E. is found in increased PE of the air molecules

Notes for active learning

Notes for active learning

Waves and Periodic Motion

1. A simple harmonic oscillator oscillates with frequency f when its amplitude is A. What is the new frequency if the amplitude is doubled to 2A?

- **A.** $f/2$
- **B.** f
- **C.** $4f$
- **D.** $2f$
- **E.** $f/4$

2. Springs A and B are attached in series, with the free end of spring B attached to a wall. The free end of spring A is pulled, and both springs expand from their equilibrium lengths. The length of spring A increases by L_A, and the length of spring B increases by L_B. What is the expression for the spring constant k_B of spring B?

- **A.** L_B/k_A
- **B.** k_A^2
- **C.** $k_A L_B$
- **D.** $(k_A L_A)/L_B$
- **E.** $2k_A$

3. Particles of the material moving back and forth in the same direction the wave moves are what type of wave?

- **A.** Standing
- **B.** Torsional
- **C.** Transverse
- **D.** Longitudinal
- **E.** Diffusion

4. If a wave has a wavelength of 25 cm and a frequency of 1.68 kHz, what is its speed?

- **A.** 44 m/s
- **B.** 160 m/s
- **C.** 420 m/s
- **D.** 314 m/s
- **E.** 16 m/s

5. The total stored energy in a system undergoing simple harmonic motion (SHM) is proportional to:

- **A.** (amplitude)^2
- **B.** wavelength
- **C.** (spring constant)^2
- **D.** amplitude
- **E.** $\sqrt{\lambda}$

6. An 11 kg mass m is attached to a spring and allowed to hang in the Earth's gravitational field. The spring stretches 3 cm before reaching its equilibrium position. If the spring were allowed to oscillate, what would be its frequency? (Use the acceleration due to gravity $g = 9.8$ m/s^2)

- **A.** 0.7 Hz
- **B.** 1.8 Hz
- **C.** 4.1 Hz
- **D.** 0.6×10^{-3} Hz
- **E.** 2.9 Hz

7. The time required for one cycle of a repeating event is the:

   A. amplitude
   B. frequency
   C. period
   D. rotation
   E. second

8. A pendulum of length L is suspended from the ceiling of an elevator. When the elevator is at rest, the period of the pendulum is T. How does T change when the elevator moves upward with a constant velocity?

   A. Decreases only if the upward acceleration is less than ½g
   B. Decreases
   C. Increases
   D. Remains the same
   E. Increases only if the upward acceleration is more than ½g

9. What is the period of a transverse wave with a frequency of 100 Hz?

   A. 0.01 s
   B. 0.05 s
   C. 0.2 s
   D. 20 s
   E. 50 s

10. Two radio antennae are located on a seacoast 10 km apart on a North-South axis. The antennas broadcast identical in-phase AM radio waves at a frequency of 4.7 MHz. 200 km offshore, a steamship travels North at 15 km/h, passing East of the antennae with a radio tuned to the broadcast frequency. From the moment of the maximum reception of the radio signal on the ship, what is the time interval until the next occurrence of maximum reception? (Use the speed of radio waves equals the speed of light $c = 3 \times 10^8$ m/s and the path difference = 1 λ)

    A. 7.7 min
    B. 6.4 min
    C. 3.8 min
    D. 8.9 min
    E. 5.1 min

11. A 2.31 kg rope is stretched between supports 10.4 m apart. If one end of the rope is tweaked, how long will it take for the resulting disturbance to reach the other end? Assume that the tension in the rope is 74.4 N.

    A. 0.33 s
    B. 0.74 s
    C. 0.65 s
    D. 0.57 s
    E. 0.42 s

12. Simple pendulum A swings back and forth twice the frequency of simple pendulum B. Which statement is correct?

- A. Pendulum A is ¼ as long as B
- B. Pendulum A is twice as massive as B
- C. Pendulum A is ½ as long as B
- D. Pendulum B is twice as massive as A
- E. Pendulum B is ¼ as long as A

13. A weight attached to the free end of an anchored spring is allowed to slide back and forth in simple harmonic motion on a frictionless table. How many times greater is the spring's restoring force at $x = 5$ cm compared to $x = 1$ cm (measured from equilibrium)?

- A. 2.5
- B. 5
- C. 7.5
- D. 15
- E. $\sqrt{2.5}$

14. A massless, ideal spring projects horizontally from a wall and is connected to a 1 kg mass. The mass is oscillating in one dimension, such that it moves 0.5 m from one end of its oscillation to the other. It undergoes 10 complete oscillations in 60 s. What is the period of oscillation?

- A. 9 s
- B. 3 s
- C. 6 s
- D. 12 s
- E. 0.6 s

15. The total mechanical energy of a simple harmonic oscillating system is:

- A. always zero, which is why it is oscillating
- B. maximum when it reaches the maximum displacement
- C. zero when it reaches the maximum displacement
- D. zero as it passes the equilibrium point
- E. a nonzero constant

16. What is the frequency of the oscillations when a vibrating spring moves from its maximum elongation position to its maximum compression position in 1 s?

- A. 0.75 Hz
- B. 0.5 Hz
- C. 1 Hz
- D. 2.5 Hz
- E. 4 Hz

17. Which of the following is not a transverse wave?

    I. Radio   II. Light   III. Sound

- A. I only
- B. II only
- C. III only
- D. I and II only
- E. I and III only

18. If a wave has a speed of 362 m/s and a period of 4 ms, its wavelength is closest to:

    A. 8.6 m
    B. 5.2 m
    C. 0.86 m
    D. 15 m
    E. 1.5 m

19. Simple harmonic motion is characterized by:

    A. acceleration that is proportional to the negative displacement
    B. acceleration that is proportional to the velocity
    C. constant positive acceleration
    D. acceleration that is inversely proportional to the negative displacement
    E. acceleration that is inversely proportional to the velocity

20. If the frequency of the harmonic oscillator doubles, by what factor does the maximum value of acceleration change?

    A. $2/\pi$
    B. $\sqrt{2}$
    C. 2
    D. 4
    E. ½

21. An object hangs from the ceiling of a stationary elevator by an ideal spring oscillates with a period T. If the elevator were to accelerate upwards with an acceleration of 2g, what is the period of oscillation of the object?

    A. T/2
    B. T
    C. 2T
    D. 4T
    E. T/4

22. Which of the following changes made to a transverse wave must increase wavelength?

    A. An increase in frequency and a decrease in speed
    B. The wavelength is only affected by a change in amplitude
    C. An increase in frequency and an increase in speed
    D. A decrease in frequency and a decrease in speed
    E. A decrease in frequency and an increase in speed

23. If a wave travels 30 m in 1 s, making 60 vibrations per second, what are its frequency and speed, respectively?

    A. 30 Hz and 60 m/s
    B. 60 Hz and 30 m/s
    C. 30 Hz and 30 m/s
    D. 60 Hz and 15 m/s
    E. 15 Hz and 30 m/s

24. Transverse waves propagate at 40 m/s in a string subjected to a tension of 60 N. If the string is 16 m long, what is its mass?

A. 0.6 kg
B. 0.9 kg
C. 0.2 kg
D. 9 kg
E. 2 kg

25. Doubling only the amplitude of a vibrating mass-on-spring system changes the system frequency by what factor?

A. Increases by 3
B. Increases by 2
C. Increases by 5
D. Increases by 4
E. Remains the same

26. A leaky faucet drips 60 times in 40 s. What is the frequency of the dripping?

A. 0.75 Hz
B. 0.67 Hz
C. 1.5 Hz
D. 12 Hz
E. 0.3 Hz

27. Particles of material move up and down perpendicular to the direction the wave is moving are in what type of wave?

A. torsional
B. mechanical
C. longitudinal
D. transverse
E. surface

28. The figure shows a graph of the velocity v as a function of time t for a system undergoing simple harmonic motion. Which one of the following graphs represents the acceleration of this system as a function of time?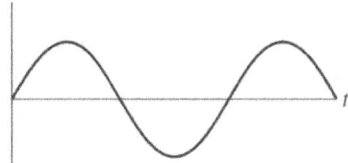

A.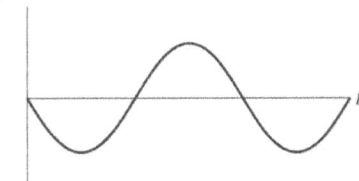

C.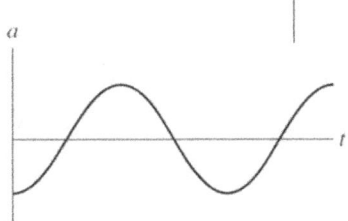

B.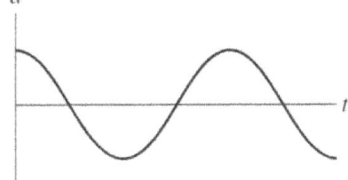

D.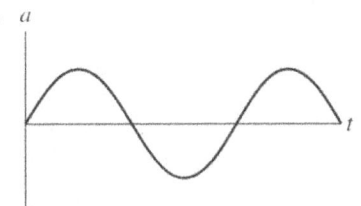

E. None of these

29. When compared, a transverse wave and a longitudinal wave have amplitudes of equal magnitude. Which statement is true about their speeds?

A. The waves have the same speeds
B. The transverse wave has exactly twice the speed of the longitudinal wave
C. The transverse wave has a slower speed
D. The longitudinal wave has a slower speed
E. The speeds of the two waves are unrelated to their amplitudes

30. What is the frequency when the weight on the end of a spring bobs up and down and completes one cycle every 2 s?

A. 0.5 Hz
B. 1 Hz
C. 2 Hz
D. 2.5 Hz
E. Depends on the mass

31. The velocity of a given longitudinal sound wave in an ideal gas is $v = 340$ m/s at constant pressure and constant volume. Assuming an ideal gas, what is the wavelength for a 2,100 Hz sound wave?

A. 0.08 m
B. 0.16 m
C. 1.6 m
D. 7.3 m
E. 0.73 m

32. When the mass of a simple pendulum is quadrupled, how does the time t required for one complete oscillation change?

A. Decreases to $¼t$
B. Decreases to $¾t$
C. Increases to $4t$
D. Remains the same
E. Decreases to $½t$

33. An object undergoing simple harmonic motion has an amplitude of 2.5 m. If the maximum velocity of the object is 15 m/s, what is the object's angular frequency (ω)?

A. 0.17 rad/s
B. 3.6 rad/s
C. 37.5 rad/s
D. 8.8 rad/s
E. 6.0 rad/s

34. Unpolarized light is incident upon two polarization filters that do not have their transmission axes aligned. If 14% of the light passes through, what is the angle between the transmission axes of the filters?

A. 73°
B. 81°
C. 43°
D. 58°
E. 64°

35. A mass on a spring undergoes simple harmonic motion. Which of the statements is true when the mass is at its maximum distance from the equilibrium position?

- **A.** KE is nonzero
- **B.** Acceleration is at a minimum
- **C.** Speed is zero
- **D.** Speed is maximum
- **E.** Total mechanical energy = KE

36. What is the frequency if the speed of a sound wave is 240 m/s and its wavelength is 10 cm?

- **A.** 2.4 Hz
- **B.** 24 Hz
- **C.** 240 Hz
- **D.** 2,400 Hz
- **E.** 0.24 Hz

37. Unlike a transverse wave, a longitudinal wave has no:

- **A.** wavelength
- **B.** crests or troughs
- **C.** amplitude
- **D.** frequency
- **E.** all the above

38. The density of aluminum is 2,700 kg/m^3. If transverse waves propagate at 36 m/s in a 9.2 mm diameter aluminum wire, what is the tension in the wire?

- **A.** 43 N
- **B.** 68 N
- **C.** 233 N
- **D.** 350 N
- **E.** 72 N

39. When a wave obliquely crosses a boundary into another medium, it is:

- **A.** always slowed down
- **B.** reflected
- **C.** diffracted
- **D.** refracted
- **E.** always sped up

40. A floating leaf oscillates up and down two complete cycles each second as a water wave passes. What is the wave's frequency?

- **A.** 0.5 Hz
- **B.** 1 Hz
- **C.** 2 Hz
- **D.** 3 Hz
- **E.** 6 Hz

41. A higher pitch for a sound wave means the wave has a greater:

- **A.** frequency
- **B.** wavelength
- **C.** amplitude
- **D.** period
- **E.** acceleration

42. An object is attached to a vertical spring and bobs up and down between points A and B. Where is the object located when its kinetic energy is at a maximum?

A. One-fourth of the way between A and B
B. One-third of the way between A and B
C. Midway between A and B
D. At either A or B
E. At none of the above points

43. A pendulum consists of a 0.5 kg mass attached to the end of a 1 m rod of negligible mass. What is the magnitude of the torque τ about the pivot when the rod makes an angle θ of 60° with the vertical? (Use the acceleration due to gravity $g = 10$ m/s²)

A. 2.7 N·m
B. 4.4 N·m
C. 5.2 N·m
D. 10.6 N·m
E. 12.7 N·m

44. The Doppler effect is characteristic of:

I. light waves   II. sound waves   III. water waves

A. I only
B. II only
C. III only
D. I and III only
E. I, II and III

45. A crane lifts a 2,500 kg cement block using a steel cable with a mass per unit length of 0.65 kg/m. What is the speed of the transverse waves on this cable? (Use the acceleration due to gravity $g = 10$ m/s²)

A. 196 m/s
B. 1,162 m/s
C. 322 m/s
D. 558 m/s
E. 1,420 m/s

46. A simple pendulum consists of a mass M attached to a weightless string of length L. Which statement about the frequency f is accurate for this system when it experiences small oscillations?

A. The f is directly proportional to the period
B. The f is independent of the mass M
C. The f is inversely proportional to the amplitude
D. The f is independent of the length L
E. The f is dependent on the mass M

47. A child on a swing set swings back and forth. If the length of the supporting cables for the swing is 3.3 m, what is the period of oscillation? (Use acceleration due to gravity $g = 10$ m/s²)

A. 3.6 s
B. 5.9 s
C. 4.3 s
D. 2.7 s
E. 5 s

48. A massless, ideal spring projects horizontally from a wall and is connected to a 0.3 kg mass. The mass is oscillating in one dimension; it moves 0.4 m from one end of its oscillation to the other. It undergoes 15 complete oscillations in 60 s. How does the frequency change if the spring constant increases by a factor of 2?

- **A.** Increases by 200%
- **B.** Decreases by 59%
- **C.** Increases by 41%
- **D.** Decreases by 41%
- **E.** Increases by 59%

49. A ball swinging at the end of a massless string undergoes simple harmonic motion. At what point(s) is the instantaneous acceleration of the ball the greatest?

- **A.** A
- **B.** B
- **C.** C
- **D.** A and D
- **E.** B and C

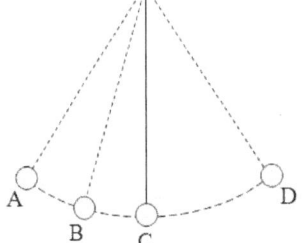

50. A simple pendulum, consisting of a 2 kg weight connected to a 10 m massless rod, is brought to an angle of 90° from the vertical and then released. What is the speed of the weight at its lowest point? (Use the acceleration due to gravity $g = 10$ m/s²)

- **A.** 14 m/s
- **B.** 10 m/s
- **C.** 20 m/s
- **D.** 25 m/s
- **E.** 7 m/s

51. A sound source of high pitch emits a wave with a high:

   I. frequency II. amplitude III. speed

- **A.** I only
- **B.** II only
- **C.** III only
- **D.** I, II and III
- **E.** I and III only

52. Find the wavelength of a train whistle that a stationary observer hears as the train moves toward him with a velocity of 50 m/s. The wind blows at 5 m/s from the observer to the train. The whistle has a natural frequency of 500 Hz. (Use the speed of sound $v = 340$ m/s)

- **A.** 0.75 m
- **B.** 0.43 m
- **C.** 0.58 m
- **D.** 7.5 m
- **E.** 5.5 m

53. Considering a vibrating mass on a spring, what effect on the system's mechanical energy is caused by doubling the amplitude only?

- **A.** Increases by a factor of two
- **B.** Increases by a factor of four
- **C.** Increases by a factor of three
- **D.** Produces no change
- **E.** Increases by a factor of $\sqrt{2}$

54. Which of the following is an accurate statement?

- **A.** Tensile stress is measured in N·m
- **B.** Stress is a measure of external forces on a body
- **C.** Stress is inversely proportional to strain
- **D.** Tensile strain is measured in meters
- **E.** The ratio stress/strain is called the elastic modulus

55. The efficient transfer of energy taking place at a natural frequency occurs in a phenomenon called:

- **A.** reverberation
- **B.** the Doppler effect
- **C.** beats
- **D.** resonance
- **E.** the standing wave phenomenon

Notes for active learning

Notes for active learning

Sound

1. A 20 decibel (dB) noise is heard from a cricket at 30 my. How loud would it sound if the cricket were a 3 m?

A. 30 dB
B. 40 dB
C. $20 \times \sqrt{2}$ dB
D. 80 dB
E. 60 dB

2. A thunderclap occurs at 6 km from a stationary person. How soon does the person hear it? (Use the speed of sound in air $v = 340$ m/s)

A. 18 s
B. 30 s
C. 48 s
D. 56 s
E. 96 s

3. Enrico Caruso, a famous opera singer, made a crystal chandelier shatter with his voice. This is a demonstration of:

A. ideal frequency
B. resonance
C. a standing wave
D. sound refraction
E. interference

4. A taut 2 m string is fixed at both ends and plucked. What is the wavelength corresponding to the third harmonic?

A. 2/3 m
B. 1 m
C. 4/3 m
D. 3 m
E. 4 m

5. High-pitched sound has a high:

    I. number of partial tones   II. frequency   III. speed

A. I only
B. II only
C. III only
D. I and II only
E. I and III only

6. A light ray in air strikes a medium whose index of refraction is 1.5. If the angle of incidence is 60°, which of the following expressions gives the angle of refraction? (Use $n_{air} = 1$)

A. $\sin^{-1}(1.5 \sin 60°)$
B. $\sin^{-1}(1.5 \cos 60°)$
C. $\sin^{-1}(1.5 \sin 30°)$
D. $\sin^{-1}(0.67 \sin 30°)$
E. $\sin^{-1}(0.67 \sin 60°)$

7. A string, 2 m in length, is fixed at both ends and tightened until the wave speed is 92 m/s. What is the frequency of the standing wave shown?

A. 46 Hz
B. 33 Hz
C. 240 Hz
D. 138 Hz
E. 184 Hz

8. A 0.6 m uniform bar of metal with a diameter of 2 cm has a mass of 2.5 kg. A 1.5 MHz longitudinal wave is propagated along the length of the bar. A wave compression traverses the length of the bar in 0.14 ms. What is the wavelength of the longitudinal wave in the metal?

A. 2.9 mm
B. 1.8 mm
C. 3.2 mm
D. 4.6 mm
E. 3.8 mm

Questions **9-12** are based on the following:

The velocity of a wave on a wire or string is not dependent (to a close approximation) on frequency or amplitude and is given by $v^2 = T / \rho_L$. T is the tension in the wire. The linear mass density ρ_L (rho) is the mass per unit length of wire. Therefore, ρ_L is the product of the mass density and the cross-sectional area (A).

A sine wave is traveling to the right with a frequency of 250 Hz. Wire A is composed of steel and has a circular cross-section diameter of 0.6 mm, and a tension of 2,000 N. Wire B is under the same tension and is made of the same material as wire A but has a circular cross-section diameter of 0.3 mm. Wire C has the same tension as wire A and is made of a composite material. (Use the density of steel wire $\rho = 7$ g/cm^3 and the density of the composite material $\rho = 3$ g/cm^3)

9. How much does the tension need to be increased to increase the wave velocity on a wire by 30%?

A. 37%
B. 60%
C. 69%
D. 81%
E. 74%

10. What is the linear mass density of wire B compared to wire A?

A. √2 times
B. 2 times
C. 1/8
D. 1/4
E. 4 times

11. What must the diameter of wire C be to have the same wave velocity as wire A?

A. 0.41 mm
B. 0.92 mm
C. 0.83 mm
D. 3.2 mm
E. 0.2 mm

12. How does the cross-sectional area change if the diameter increases by a factor of 4?

   A. Increases by a factor of 16
   B. Increases by a factor of 4
   C. Increases by a factor of 2
   D. Decreases by a factor of 4
   E. Increases by a factor of √2

13. A bird, emitting sounds with a frequency of 60 kHz, is moving at a speed of 10 m/s toward a stationary observer. What is the frequency of the sound waves detected by the observer? (Use the speed of sound in air $v = 340$ m/s)

   A. 55 kHz
   B. 46 kHz
   C. 68 kHz
   D. 76 kHz
   E. 62 kHz

14. What is observed for a frequency heard by a stationary person when a sound source is approaching?

   A. Equal to zero
   B. The same as the source
   C. Higher than the source
   D. Lower than the source
   E. Requires more information

15. Which of the following is a false statement?

   A. The transverse waves on a vibrating string are different from sound waves
   B. Sound travels much slower than light
   C. Sound waves are longitudinal pressure waves
   D. Sound can travel through a vacuum
   E. Perceived musical pitch is correlated with the frequency

16. Which of the following is a real-life example of the Doppler effect?

   A. London police whistle, which uses two short pipes to produce a three-note sound
   B. Radio signal transmission
   C. Sound becomes quieter as the observer moves away from the source
   D. Human hearing is most acute at 2,500 Hz
   E. Changing the pitch of the siren as an ambulance passes by the observer

17. Two sound waves have the same frequency and amplitudes of 0.4 Pa and 0.6 Pa, respectively. When they arrive at point X, what is the range of possible amplitudes for sound at point X?

   A. 0 – 0.4 Pa
   B. 0.4 – 0.6 Pa
   C. 0.2 – 1.0 Pa
   D. 0.4 – 0.8 Pa
   E. 0.2 – 0.6 Pa

18. The intensity of the waves from a point source at a distance d from the source is I. What is the intensity at a distance $2d$ from the source?

- **A.** I/2
- **B.** I/4
- **C.** 4I
- **D.** 2I
- **E.** I/√2

19. The sound would be expected to travel most slowly in a medium that exhibited:

- **A.** low resistance to compression and high density
- **B.** high resistance to compression and low density
- **C.** low resistance to compression and low density
- **D.** high resistance to compression and high density
- **E.** equal resistance to compression and density

20. Which is valid for a resonating pipe open at both ends?

- **A.** Displacement node at one end and a displacement antinode at the other end
- **B.** Displacement antinodes at each end
- **C.** Displacement nodes at each end
- **D.** Displacement node at one end and a one-fourth antinode at the other end
- **E.** Displacement antinode at one end and a one-fourth node at the other end

21. In a pipe of length L open at both ends, the lowest tone to resonate is 200 Hz. Which of the following frequencies does not resonate in this pipe?

- **A.** 400 Hz
- **B.** 600 Hz
- **C.** 1,200 Hz
- **D.** 800 Hz
- **E.** 500 Hz

22. In general, the sound is conducted fastest through:

- **A.** vacuum
- **B.** gases
- **C.** liquids
- **D.** solids
- **E.** warm air

23. If an electric charge is shaken up and down:

- **A.** electron excitation occurs
- **B.** a magnetic field is created
- **C.** sound is emitted
- **D.** its charge changes
- **E.** its mass decreases

24. What is the wavelength of a sound wave of frequency 620 Hz in steel, given that the speed of sound in steel is 5,000 m/s?

- A. 1.8 m
- B. 6.2 m
- C. 8.1 m
- D. 2.6 m
- E. 5.7 m

25. If the sound from a constant sound source radiates equally in all directions, as the distance doubles, by what amount is the intensity of the sound reduced?

- A. 1/8
- B. 1/16
- C. $1/\sqrt{2}$
- D. ½
- E. ¼

26. Why does the intensity of waves from a sound source decrease with the square of the distance from the source?

- A. The medium through which the waves travel absorbs the energy of the waves
- B. The waves speed up as they travel away from the source
- C. The waves lose energy as they travel
- D. The waves spread out as they travel
- E. The frequency of the waves decreases as they get farther from the source

Questions **27-30** are based on the following:

Steven is preparing a mailing tube 1.5 m long and 4 cm in diameter. The tube is open at one end and sealed at the other. Before he inserted his documents, the mailing tube fell to the floor and produced a note. (Use the speed of sound in air $v = 340$ m/s)

27. What is the wavelength of the fundamental?

- A. 0.04 m
- B. 6 m
- C. 0.75 m
- D. 1.5 m
- E. 9 m

28. If the tube was filled with helium, in which sound travels at 960 m/s, what is the fundamental frequency?

- A. 160 Hz
- B. 320 Hz
- C. 80 Hz
- D. 640 Hz
- E. 590 Hz

29. What is the wavelength of the fifth harmonic?

A. 3.2 m
B. 0.6 m
C. 2.4 m
D. 1.5 m
E. 1.2 m

30. What is the frequency of the note Steven heard?

A. 57 Hz
B. 85 Hz
C. 30 Hz
D. 120 Hz
E. 25 Hz

31. A 4 g string, 0.34 m long, is under tension. The string vibrates in the third harmonic. What is the wavelength of the standing wave in the string? (Use the speed of sound in air $v = 344$ m/s)

A. 0.56 m
B. 0.33 m
C. 0.23 m
D. 0.61 m
E. 0.87 m

32. Two pure tones are sounded together, and a particular beat frequency is heard. What happens to the beat frequency if the frequency of one of the tones is increased?

A. Increases
B. Decreases
C. Remains the same
D. Either increase or decrease
E. Increases logarithmically

33. Consider a closed pipe of length L. What are the wavelengths of the three lowest tones produced by this pipe?

A. $4L, 4/3L, 4/5L$
B. $2L, L, 2/3L$
C. $2L, L, ½L$
D. $4L, 2L, L$
E. $4L, 4/3L, L$

34. Mary hears the barely audible buzz of a mosquito one meter from her ear in a quiet room. How much energy does a mosquito produce in 200 s? (Note: an almost inaudible sound has a threshold value of 9.8×10^{-12} W/m²)

A. 6.1×10^{-8} J
B. 1.3×10^{-8} J
C. 6.4×10^{-10} J
D. 3.6×10^{-10} J
E. 2.5×10^{-8} J

35. How long does it take for a light wave to travel 1 km through the water with a refractive index of 1.33? (Use the speed of light $c = 3 \times 10^8$ m/s)

A. 4.4×10^{-6} s
B. 4.4×10^{-9} s
C. 2.8×10^{-9} s
D. 2.8×10^{-12} s
E. 3.4×10^{-9} s

36. In designing a music hall, an acoustical engineer deals mainly with:

A. beats
B. resonance
C. forced vibrations
D. modulation
E. wave interference

37. Which curve in the figure represents wave speed variation (v) as a function of tension (T) for transverse waves on a stretched string?

A. A
B. B
C. C
D. D
E. E

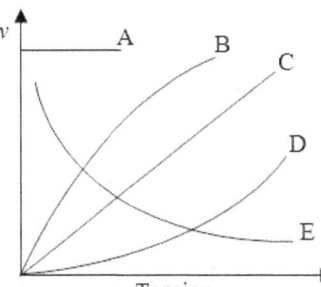

38. A string, 4 meters long, is fixed at both ends and tightened until the wave speed is 20 m/s. What is the frequency of the standing wave shown?

A. 13 Hz
B. 8.1 Hz
C. 5.4 Hz
D. 15.4 Hz
E. 7.8 Hz

39. Compared to the velocity of a 600 Hz sound, the velocity of a 300 Hz sound through air is:

A. one-half as great
B. one-fourth as great
C. twice as great
D. four times as great
E. the same

40. Consider a string with a linear mass density of 0.40 g/m stretched to a length of 0.50 m by the tension of 75 N, vibrating at the 6th harmonic. It excites an open pipe into the second overtone. What is the length of the pipe?

A. 0.25 m
B. 0.1 m
C. 0.20 m
D. 0.6 m
E. 0.32 m

41. A string of length L is under tension, and the speed of a wave in the string is v. What is the speed of a wave in a string of the same mass under the same tension but twice as long?

A. $v\sqrt{2}$
B. $2v$
C. $v/2$
D. $v/\sqrt{2}$
E. $4v$

42. If a guitar string has a fundamental frequency of 500 Hz, which one of the following frequencies can set the string into resonant vibration?

A. 450 Hz
B. 760 Hz
C. 1,500 Hz
D. 2,250 Hz
E. 1,250 Hz

43. When a light wave passes from a lower refractive index to a medium with a higher refractive index, some of the incident light is refracted, while some are reflected. What is the angle of refraction?

A. Greater than the angle of incidence and less than the angle of reflection
B. Less than the angle of incidence and greater than the angle of reflection
C. Greater than the angles of incidence and reflection
D. Less than the angles of incidence and reflection
E. Equal to the angles of incidence and reflection

44. The speed of a sound wave in air depends on:

I. air temperature   II. wavelength   III. frequency

A. I only
B. II only
C. III only
D. I and II only
E. I and III only

45. Which of the following statements is false?

A. The speed of a wave and the speed of the vibrating particles that constitute the wave are different entities
B. Waves transport energy and matter from one region to another
C. In a transverse wave, the particle motion is perpendicular to the velocity vector of the wave
D. Not all waves are mechanical
E. A wave in which particles move back and forth in the same direction the wave moves is a longitudinal wave

46. A 2.5 g string, 0.75 m long, is under tension. The string produces a 700 Hz tone when it vibrates in the third harmonic. What is the wavelength of the tone in the air? (Use the speed of sound in air $v = 344$ m/s)

- A. 0.65 m
- B. 0.57 m
- C. 0.33 m
- D. 0.4 m
- E. 0.5 m

47. Suppose a source of sound is emitting waves uniformly in all directions. If an observer moves to a point twice as far from the source, what is the frequency of the sound?

- A. $\sqrt{2}$ as large
- B. Twice as large
- C. Unchanged
- D. Half as large
- E. One-fourth as large

48. A 2.5 kg rope is stretched between supports 8 m apart. If one end of the rope is tweaked, how long will it take for the resulting disturbance to reach the other end? Assume the tension in the rope is 40 N.

- A. 0.71 s
- B. 0.62 s
- C. 0.58 s
- D. 0.47 s
- E. 0.84 s

49. An office machine makes a rattling sound with an intensity of 10^{-5} W/m² when perceived by an office worker sitting 3 m away. What is the sound level in decibels for the sound of the machine? (Use threshold of hearing $I_0 = 10^{-12}$ W/m²)

- A. 10 dB
- B. 35 dB
- C. 70 dB
- D. 95 dB
- E. 45 dB

50. A taut 1 m string is plucked. Point B is midway between both ends, and a finger is placed on point B such that a waveform exists with a node at B. What is the lowest frequency that can be heard? (Use the speed of waves on the string $v = 3.8 \times 10^4$ m/s)

- A. 4.8×10^5 Hz
- B. 2.3×10^4 Hz
- C. 9.7×10^3 Hz
- D. 7.4×10^3 Hz
- E. 3.8×10^4 Hz

51. For a light wave traveling in a vacuum, which of the following properties is true?

- A. Increased f results in increased amplitude
- B. Increased f results in decreased speed
- C. Increased f results in an increased wavelength
- D. Increased f results in a decreased wavelength
- E. Increased f results in decreased amplitude

52. Which wave is different from the others (i.e., does not belong to the same grouping)?

- **A.** Pressure wave
- **B.** Radio wave
- **C.** Ultrasonic wave
- **D.** Infrasonic wave
- **E.** Acoustic wave

53. Two speakers are 2 m apart, and both produce a sound wave (in-phase) with a wavelength of 0.8 m. A microphone is at an equal distance from both speakers to determine the intensity of the sound at various points. What point is precisely halfway between the two speakers? (Use the speed of sound $v = 340$ m/s)

- **A.** An antinode and a node
- **B.** Neither an antinode nor a node
- **C.** A node
- **D.** An antinode
- **E.** Need information about the frequency

54. The siren of an ambulance blares at 1,200 Hz when the ambulance is stationary. What frequency does a stationary observer hear after this ambulance passes her while traveling at 30 m/s? (Use the speed of sound $v = 342$ m/s)

- **A.** 1,240 Hz
- **B.** 1,128 Hz
- **C.** 1,103 Hz
- **D.** 1,427 Hz
- **E.** 1,182 Hz

55. Compared to the wavelength of a 600 Hz sound, the wavelength of a 300 Hz sound in air is:

- **A.** one-half as long
- **B.** the same
- **C.** one-fourth as long
- **D.** four times as long
- **E.** twice as long

Notes for active learning

Notes for active learning

Light and Optics

1. What is the minimum thickness of a soap film that reflects a given wavelength of light?

- A. ¼ the wavelength
- B. ½ the wavelength
- C. One wavelength
- D. Two wavelengths
- E. There is no minimum thickness

2. As the angle of an incident ray of light increases, the angle of the reflected ray:

- A. increases
- B. decreases
- C. stays the same
- D. increases or decreases
- E. requires more information

3. At what distance from a concave spherical mirror (with a focal length of 100 cm) must a woman stand to see an upright image of herself twice her actual height?

- A. 100 cm
- B. 50 cm
- C. 300 cm
- D. 25 cm
- E. 150 cm

4. If a person's eyeball is too long from front to back, what is the condition that the person likely suffers?

- A. Hyperopia
- B. Astigmatism
- C. Presbyopia
- D. Myopia
- E. Diplopia

5. According to the relationship between frequency and energy of light ($E = hf$), which color of light has more energy?

- A. Red
- B. Yellow
- C. Green
- D. Orange
- E. Blue

6. A candle 18 cm tall sits 4 m from a diverging lens with a focal length of 3 m. What is the size of the image?

- A. 6.3 cm
- B. 7.7 cm
- C. 2.9 cm
- D. 13.5 cm
- E. 18 cm

Questions **7-8** are based on the following:

A tank holds a layer of oil 1.58 m thick that floats on a layer of syrup 0.66 m thick. Both liquids are clear and do not intermix. A ray, which originates at the bottom of the tank on a vertical axis (see figure), crosses the oil-syrup interface at a point 0.9 m to the right of the vertical axis. The ray continues and arrives at the oil-air interface, 2 m from the axis and at the critical angle. (Use the refractive index $n = 1$ for air)

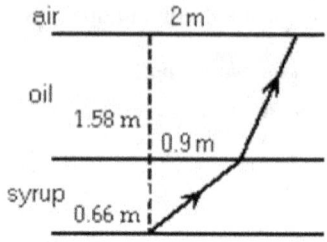

7. The index of refraction of the oil is closest to:

   A. 1.39
   B. 1.56
   C. 1.75
   D. 1.82
   E. 1.94

8. What is the index of refraction of the syrup?

   A. 1.53
   B. 1.46
   C. 1.17
   D. 1.24
   E. 1.33

9. Which of the following cannot be explained with the wave theory of light?

   A. Photoelectric effect
   B. Interference
   C. Polarization
   D. Diffusion
   E. All the above

10. The use of wavefronts and rays to describe optical phenomena is:

    A. dispersive optics
    B. reflector optics
    C. wave optics
    D. geometrical optics
    E. array optics

11. In investigating a new type of optical fiber (index of refraction $n = 1.26$), a laser beam is incident on the flat end of a straight fiber in air, as shown in the figure. What is the maximum angle of incidence (θ_1) if the beam does not escape from the fiber?

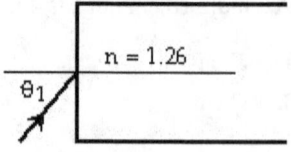

   A. 36°
   B. 43°
   C. 58°
   D. 50°
   E. 28°

12. An object is 0.5 m from a converging lens with a power of 10 diopters. At what distance from the lens does the image appear?

- **A.** 0.13 m
- **B.** 0.47 m
- **C.** 0.7 m
- **D.** 1.5 m
- **E.** 1.8 m

13. A virtual image is:

  I. produced by light rays
  II. the brain's interpretations of light rays
  III. found only on a concave mirror

- **A.** I only
- **B.** II only
- **C.** III only
- **D.** I and II only
- **E.** I and III only

14. If Karen stands in front of a convex mirror, at the same distance from it as its radius of curvature:

- **A.** Karen does not see her image because it is focused at a different distance
- **B.** Karen sees her image, and she appears the same size
- **C.** Karen does not see her image, and she is not within its range
- **D.** Karen sees her image, and she appears larger
- **E.** Karen sees her image, and she appears smaller

15. An object is viewed at various distances using a mirror with a focal length of 10 m. If the object is 20 m from the mirror, what best characterizes the image?

- **A.** Inverted and real
- **B.** Inverted and virtual
- **C.** Upright and real
- **D.** Upright and virtual
- **E.** Real, but it cannot be determined if it is inverted or upright

16. If an object is placed at a position beyond $2f$ of the focal point of a converging lens, the image is:

- **A.** real, upright and enlarged
- **B.** virtual, inverted and enlarged
- **C.** virtual, upright and reduced
- **D.** real, inverted and enlarged
- **E.** real, inverted and reduced

17. Which form of electromagnetic radiation has photons with the lowest energy?

- **A.** X-rays
- **B.** Ultraviolet radiation
- **C.** Radio waves
- **D.** Microwaves
- **E.** Infrared radiation

18. If the index of refraction of a diamond is 2.43, a given wavelength of light travels:

A. 2.43 times faster in diamond than it does in air
B. 2.43 times faster in a vacuum than it does in diamond
C. 2.43 times faster in diamond than it does in a vacuum
D. 2.43 times faster in a vacuum than it does in air
E. 2.43 times faster in the air than it does in a vacuum

19. An object is 15 cm to the left of a double-convex lens of focal length 20 cm. Where is the image of this object located?

A. 15 cm to the left of the lens
B. 30 cm to the left of the lens
C. 60 cm to the right of the lens
D. 60 cm to the left of the lens
E. 30 cm to the right of the lens

20. A sheet of red paper appears black when it is illuminated with:

A. orange light
B. cyan light
C. red light
D. yellow light
E. violet light

21. Where is an object if the image produced by a lens appears close to its focal point?

A. near the center of curvature of the lens
B. far from the lens
C. near the lens
D. near the focal point
E. requires more information

22. A light with the frequency 4.9×10^{14} Hz is produced by a source located 6 m from a converging lens with a focal length of 3 m. For a different frequency of light, the focal length of the lens is different than 3 m. This phenomenon is:

A. Diffusion
B. Incidence
C. Interference
D. Refraction
E. Dispersion

23. If an image appears at the same distance from a mirror as the object, the size of the image is:

A. exactly quadruple the size of the object
B. exactly ¼ the size of the object
C. the same size as the object
D. exactly twice the size of the object
E. exactly ½ the size of the object

24. When viewed straight down (90° to the surface), an incident light ray moving from water to air is refracted:

    A. 37° away from the normal
    B. 37° toward the normal
    C. 28° toward the normal
    D. 28° away from the normal
    E. 0°

25. Suppose a beachgoer uses two lenses from a pair of disassembled polarized sunglasses and places one on top of the other. What would be observed if he rotates one lens 90° with respect to the normal position of the other lens and looks directly at the sun overhead?

    A. Light with an intensity reduced to about 50% of what it would be with one lens
    B. Light with an intensity the same as what it would be with one lens
    C. Complete darkness, since no light would be transmitted
    D. Light with an intensity reduced to about 25% of what it would be with one lens
    E. Light with intensity increased to about 150% of what it would be with one lens

26. A glass plate with an index of refraction of 1.45 is immersed in a liquid. The liquid is an oil with an index of refraction of 1.35. The surface of the glass is inclined at an angle of 54° with the vertical. A horizontal ray in the glass is incident on the interface of glass and liquid, and the horizontal incident ray refracts at the interface. The angle that the refracted ray in the oil makes with the horizontal is closest to:

    A. 8.3°
    B. 14°
    C. 6°
    D. 12°
    E. 17°

27. Two plane mirrors make an angle of 30°. A light ray enters the system and is reflected once off each mirror. Through what angle is the ray turned?

    A. 60°
    B. 90°
    C. 120°
    D. 160°
    E. 180°

28. Which of the following statements about light is TRUE?

    A. A packet of light energy is known as a photon
    B. Color can be used to determine the approximate energy of visible light
    C. Light travels through space at a speed of 3.0×10^8 m/s
    D. Ultraviolet light cannot be seen with the unaided eye
    E. All of the above

29. The angle of incidence:

  A. may be greater than, less than, or equal to the angle of refraction
  B. is always less than the angle of refraction
  C. must equal the angle of refraction
  D. is always greater than the angle of refraction
  E. is independent of the angle of refraction

30. As a person walks away from a plane mirror on a wall, her image:

  A. is a real image, no matter how far she is from the mirror
  B. changes from being upright to being inverted as she passes the focal point
  C. gets smaller
  D. may or may not get smaller, depending on where she is positioned
  E. is always the same size

31. If a spherical concave mirror has a radius of curvature R, its focal length is:

  A. $2R$
  B. R
  C. $R/2$
  D. $R/4$
  E. $4R$

32. Let n_1 be the index of refraction of the incident medium and let n_2 be the index of refraction of the refracting medium. Which of the following must be true if the angle that the refracted ray makes with the boundary (not with the normal) is less than the angle that the incident ray makes with the boundary?

  A. $n_1 < n_2$
  B. $n_1 > n_2$
  C. $n_1 < 1$
  D. $n_2 < 1$
  E. $n_1 = n_2$

33. If a person's eyeball is too short from front to back, the person is likely to suffer from:

  A. nearsightedness
  B. farsightedness
  C. presbyopia
  D. astigmatism
  E. diplopia

34. The shimmering observed over a hot surface is:

  A. changing refraction from the mixing of warm and cool air
  B. a mirage
  C. heat rays
  D. reflections from evaporating water vapor
  E. reflections from condensing water vapor

35. When two parallel white rays pass through the outer edges of a converging glass lens, chromatic aberrations cause colors to appear on the screen in what order, from the top down?

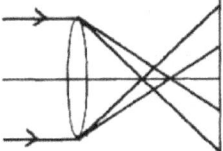

- A. blue, blue, red, red
- B. red, blue, blue, red
- C. blue, red, blue, red
- D. red, red, blue, blue
- E. blue, red, red, blue

36. Two thin converging lenses are near each other so that the lens on the left has a focal length of 2 m and the one on the right has a focal length of 4 m. What is the focal length of the combination?

- A. 1/4 m
- B. 4/3 m
- C. 3/4 m
- D. 4 m
- E. 8 m

37. A cylindrical tank is 50 ft. deep, 37.5 ft. in diameter, and filled to the top with water. A flashlight shines into the tank from above. What is the minimum angle θ that its beam can make with the water surface if the beam is to illuminate part of the bottom? (Use the index of refraction $n = 1.33$ for water)

- A. 25°
- B. 31°
- C. 37°
- D. 53°
- E. 18°

38. Which color of the visible spectrum has the shortest wavelength (400 nm)?

- A. Violet
- B. Green
- C. Orange
- D. Blue
- E. Yellow

39. An object is placed at a distance d in front of a plane mirror. The size of the image is:

- A. dependent on where the observer is positioned when looking at the image
- B. twice the size of the object
- C. half the size of the object
- D. dependent on the distance d
- E. the same as the object, independent of the position of the observer or distance d

40. If a single lens forms a virtual image of an object, then the:

    I. image must be upright
    II. lens must be a converging lens
    III. lens could be diverging or converging

- A. I only
- B. I and III only
- C. III only
- D. I and II only
- E. II only

41. When neon light passes through a prism, what is observed?

- **A.** White light
- **B.** Bright spots or lines
- **C.** The same neon light
- **D.** Continuous spectrum
- **E.** Both A and B

42. The law of reflection holds for:

    I. plane mirrors   II. curved mirrors   III. spherical mirrors

- **A.** I only
- **B.** II only
- **C.** III only
- **D.** I and III only
- **E.** I, II and III

43. The image formed by a single concave lens:

- **A.** can be real or virtual but is real when the object is placed at the focal point
- **B.** can be real or virtual, depending on the object's distance compared to the focal length
- **C.** is always virtual
- **D.** is always real
- **E.** is always inverted

44. A lens forms a virtual image of an object. Which of the following must be true of the image?

- **A.** It is inverted
- **B.** It is upright
- **C.** It is larger than the object and upright
- **D.** It is smaller than the object and inverted
- **E.** It is the same size as the object and upright

45. Light with the lowest frequency (longest wavelength) detected by your eyes is perceived as:

- **A.** violet
- **B.** green
- **C.** yellow
- **D.** orange
- **E.** red

46. A 0.1 m tall candle is observed through a converging lens 3 m away and has a focal length of 6 m. The resulting image is:

- **A.** 3 m from the lens on the opposite side of the object
- **B.** 6 m from the lens on the opposite side of the object
- **C.** 3 m from the lens on the same side as the object
- **D.** 6 m from the lens on the same side as the object
- **E.** 0.5 m from the lens on the opposite side of the object

47. Which statement about thin lenses is correct when considering only a single lens?

A. A diverging lens always produces a virtual, erect image
B. A diverging lens always produces a real, erect image
C. A diverging lens always produces a virtual, inverted image
D. A diverging lens always produces a real, inverted image
E. A converging lens always produces a real, inverted image

48. A double-concave lens has equal radii of curvature of 15 cm. An object 14 cm from the lens forms a virtual image 5 cm from the lens. What is the index of refraction of the lens material?

A. 0.8
B. 1.4
C. 2
D. 2.6
E. 2.8

49. The magnification m for an object reflected from a mirror is the ratio of what characteristic of the image to the object?

A. Center of curvature
B. Focal distances
C. Orientation
D. Angular size
E. Distance

50. Suppose Mike places his face in front of a concave mirror. Which of the following statements is correct?

A. Mike's image is diminished in size
B. Mike's image is always inverted
C. No matter where Mike places himself, a virtual image is formed
D. If Mike positions himself between the center of curvature and the focal point of the mirror, he will not be able to see his image
E. Mike's image is enlarged in size

51. Single-concave spherical mirrors produce images that:

A. are always smaller than the actual object
B. are always the same size as the actual object
C. are always larger than the actual object
D. could be smaller than, larger, or the same size as the object, depending on the placement of the object
E. are always upright

52. When two converging lenses of equal focal lengths are used together, the effective combined focal length is less than the individual lenses' focal length. The combined power of the two lenses together is:

   A. greater than the power of either individual lens
   B. the same as the power of either individual lens
   C. less than the power of either individual lens
   D. greater than the sum of the powers of both individual lens
   E. exactly the sum of the powers of both individual lens

53. The index of refraction is based on the ratio of the speed of light in:

   A. water to the speed of light in the transparent material
   B. a vacuum to the speed of light in the transparent material
   C. two different transparent materials
   D. air to the speed of light in the transparent material
   E. a solid to the speed of light in the transparent material

54. An object is located 2.2 m in front of a plane mirror. The image formed by the mirror appears:

   A. 4.4 m behind the mirror's surface
   B. 2.2 m in front of the mirror's surface
   C. 4.4 m in front of the mirror's surface
   D. on the mirror's surface
   E. 2.2 m behind the mirror's surface

55. An upright object is 40 cm from a concave mirror with a radius of 50 cm. The image is:

   A. virtual and inverted
   B. virtual and upright
   C. real and inverted
   D. real and upright
   E. real or virtual

Notes for active learning

Notes for active learning

Diagnostic Tests

Diagnostic Test 1

This Diagnostic Test is designed to assess your proficiency on each topic and NOT to mimic the test. Use your test results and identify areas of strength and weakness to adjust your study plan and enhance your fundamental knowledge. The length of the Diagnostic Tests is optimal for a single study session.

#	Answer:					Review	#	Answer:					Review
1:	A	B	C	D	E	___	26:	A	B	C	D	E	___
2:	A	B	C	D	E	___	27:	A	B	C	D	E	___
3:	A	B	C	D	E	___	28:	A	B	C	D	E	___
4:	A	B	C	D	E	___	29:	A	B	C	D	E	___
5:	A	B	C	D	E	___	30:	A	B	C	D	E	___
6:	A	B	C	D	E	___	31:	A	B	C	D	E	___
7:	A	B	C	D	E	___	32:	A	B	C	D	E	___
8:	A	B	C	D	E	___	33:	A	B	C	D	E	___
9:	A	B	C	D	E	___	34:	A	B	C	D	E	___
10:	A	B	C	D	E	___	35:	A	B	C	D	E	___
11:	A	B	C	D	E	___	36:	A	B	C	D	E	___
12:	A	B	C	D	E	___	37:	A	B	C	D	E	___
13:	A	B	C	D	E	___	38:	A	B	C	D	E	___
14:	A	B	C	D	E	___	39:	A	B	C	D	E	___
15:	A	B	C	D	E	___	40:	A	B	C	D	E	___
16:	A	B	C	D	E	___	41:	A	B	C	D	E	___
17:	A	B	C	D	E	___	42:	A	B	C	D	E	___
18:	A	B	C	D	E	___	43:	A	B	C	D	E	___
19:	A	B	C	D	E	___	44:	A	B	C	D	E	___
20:	A	B	C	D	E	___	45:	A	B	C	D	E	___
21:	A	B	C	D	E	___	46:	A	B	C	D	E	___
22:	A	B	C	D	E	___	47:	A	B	C	D	E	___
23:	A	B	C	D	E	___	48:	A	B	C	D	E	___
24:	A	B	C	D	E	___	49:	A	B	C	D	E	___
25:	A	B	C	D	E	___	50:	A	B	C	D	E	___

#	Answer:					Review	#	Answer:					Review
51:	A	B	C	D	E	___	76:	A	B	C	D	E	___
52:	A	B	C	D	E	___	77:	A	B	C	D	E	___
53:	A	B	C	D	E	___	78:	A	B	C	D	E	___
54:	A	B	C	D	E	___	79:	A	B	C	D	E	___
55:	A	B	C	D	E	___	80:	A	B	C	D	E	___
56:	A	B	C	D	E	___	81:	A	B	C	D	E	___
57:	A	B	C	D	E	___	82:	A	B	C	D	E	___
58:	A	B	C	D	E	___	83:	A	B	C	D	E	___
59:	A	B	C	D	E	___	84:	A	B	C	D	E	___
60:	A	B	C	D	E	___	85:	A	B	C	D	E	___
61:	A	B	C	D	E	___	86:	A	B	C	D	E	___
62:	A	B	C	D	E	___	87:	A	B	C	D	E	___
63:	A	B	C	D	E	___	88:	A	B	C	D	E	___
64:	A	B	C	D	E	___	89:	A	B	C	D	E	___
65:	A	B	C	D	E	___	90:	A	B	C	D	E	___
66:	A	B	C	D	E	___	91:	A	B	C	D	E	___
67:	A	B	C	D	E	___	92:	A	B	C	D	E	___
68:	A	B	C	D	E	___	93:	A	B	C	D	E	___
69:	A	B	C	D	E	___	94:	A	B	C	D	E	___
70:	A	B	C	D	E	___	95:	A	B	C	D	E	___
71:	A	B	C	D	E	___	96:	A	B	C	D	E	___
72:	A	B	C	D	E	___	97:	A	B	C	D	E	___
73:	A	B	C	D	E	___	98:	A	B	C	D	E	___
74:	A	B	C	D	E	___	99:	A	B	C	D	E	___
75:	A	B	C	D	E	___	100:	A	B	C	D	E	___

1. What property of matter determines an object's resistance to change in its state of motion?

   I. mass   II. density   III. volume

   A. I only
   B. II only
   C. III only
   D. I and II only
   E. I and III only

2. Two forces of equal magnitude act on an object. If each force is 4.6 N and the angle between them is 40°, what is the magnitude and direction of a third force for the object to be in equilibrium?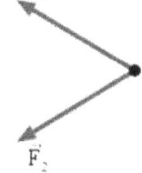

   A. 2.3 N, to the right
   B. 4.3 N, to the right
   C. 6.5 N, to the right
   D. 0.6 N, to the right
   E. 8.6 N, to the right

3. A thermally-isolated system comprises a hot piece of aluminum and a cold piece of copper, with the aluminum and copper in thermal contact. The specific heat capacity of aluminum is more than double that of copper. Which object experiences the greater magnitude of gain or loss of heat during the time the system takes to reach thermal equilibrium?

   A. Aluminum
   B. Copper
   C. Neither, because both undergo the same magnitude of gain or loss of heat
   D. Requires knowing the masses
   E. Requires knowing the volumes

4. In the absence of friction, how much work would a boy do while pulling a 10 kg sled for 3.5 m with a 20 N force?

   A. 57 J
   B. 70 J
   C. 1.8 J
   D. 85 J
   E. 280 J

5. Total constructive interference is observed when two waves with the same frequency and wavelength are at:

   A. 45° phase difference
   B. 90° phase difference
   C. 180° phase difference
   D. −90° phase difference
   E. 0° phase difference

6. The Doppler shift occurs when the source of waves and a detector move relative. There is an increase in the detected frequency when the source and detector are approaching each other and a decrease in the detected frequency when moving away. A commuter train is moving rapidly at 50 m/s towards Kevin, who is standing still. The train sounds its horn at 420 Hz, and the speed of sound is 350 m/s at 29 °C. What frequency does Kevin hear after the train passes?

- **A.** 335 Hz
- **B.** 368 Hz
- **C.** 424 Hz
- **D.** 446 Hz
- **E.** 295 Hz

7. As a lead weight drops into the water of uniform density and continues to sink deeper, what happens to the buoyant force on the lead weight from its origin above the surface of the water?

- **A.** First increases, then remain constant
- **B.** First decreases, then remain constant
- **C.** Increases steadily
- **D.** First decreases, then increase steadily
- **E.** First increases, then decrease steadily

8. A charged particle moving in a uniform static magnetic field:

- **A.** may experience a magnetic force, but its speed does not change
- **B.** may experience a magnetic force, but its direction of motion does not change
- **C.** experiences a magnetic force, and its direction of motion does change
- **D.** experiences a magnetic force, and its speed does not change
- **E.** experiences a magnetic force, and its direction of motion does not change

9. Two objects, I and II, have equal charge and mass. Because of equal gravitational and electrostatic forces between them, neither body is in motion. If the mass of object I is halved, equilibrium is maintained if which change occurs for object II:

- **A.** mass is quadrupled
- **B.** mass is halved
- **C.** charge is doubled
- **D.** charge is halved
- **E.** charge is increased by $\sqrt{2}$

10. What is the lens's focal length if a candle is 4 m from the lens, and the image is 2 m from the other side of the lens?

- **A.** −2 m
- **B.** −4/3 m
- **C.** 3/4 m
- **D.** 2 m
- **E.** 4/3 m

11. When a nucleus captures a β⁻ particle, the atomic number of the nucleus:

    A. increases by two
    B. decreases by one
    C. increases by one
    D. remains the same
    E. decreases by two

12. How far from the heavier end must the fulcrum of a massless 10 m seesaw be if an 800 N father on one side is to balance his 200 N son at the other end?

    A. 0.5 m
    B. 2 m
    C. 1 m
    D. 8 m
    E. 6 m

13. A ball bounces on the floor three times, whereby it loses 20% of its energy with each bounce due to heating. How high is the third bounce, provided the ball was released 250 cm from the floor?

    A. 115 cm
    B. 150 cm
    C. 75 cm
    D. 180 cm
    E. 128 cm

14. What is the period of a wave if its frequency is 10 Hz?

    A. 0.1 s
    B. 1 s
    C. 100 s
    D. 10 s
    E. 0.01 s

15. What is the frequency of a pressure wave with a wavelength of 2.5 m traveling at 1,600 m/s?

    A. 640 Hz
    B. 5.6 kHz
    C. 0.64 Hz
    D. 4 kHz
    E. 64 Hz

16. How long does it take for a rotating object to speed up from 15.0 rad/s to 33.3 rad/s if it has a uniform angular acceleration of 3.45 rad/s²?

    A. 3.45 s
    B. 5.30 s
    C. 8.35 s
    D. 14.60 s
    E. 20.80 s

17. Which characteristic is required for a mass spectrometer?

   A. Perpendicular electric and gravitational fields
   B. Perpendicular gravitational and magnetic fields
   C. Perpendicular magnetic and electric fields
   D. Collinear magnetic and electric fields
   E. Collinear gravitational and magnetic fields

18. What is the RMS current for a 26 µF capacitor connected across a 120 V_{rms} 60 Hz source?

   A. 1.2 A
   B. 7.3 A
   C. 2.7 A
   D. 0 A
   E. 0.13 A

19. A light ray in glass arrives at the glass-water interface at an angle of $\theta = 48°$ with respect to the normal. The refracted ray in the water makes an angle of $\phi = 61°$ with respect to the normal. If the angle of incidence changes to $\theta = 25°$, what is the new angle of refraction ϕ in the water? (Use index of refraction of water = 1.33)

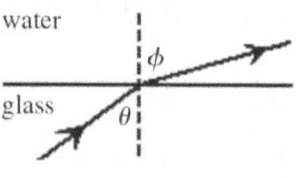

   A. 16°
   B. 54°
   C. 30°
   D. 24°
   E. 38°

20. An organ pipe is a cylindrical tube open at both ends. The air column is set to vibrate by air flowing through the lower portion of the pipe. The length of the pipe is 0.1 m, and the diameter is 0.2 m. What is the frequency of the fundamental? (Use the velocity of sound $v = 340$ m/s)

   A. 300 Hz
   B. 600 Hz
   C. 1,700 Hz
   D. 800 Hz
   E. 1,200 Hz

21. As a solid goes through a phase change to a liquid, heat is absorbed, and the temperature:

   A. fluctuates
   B. remains the same
   C. decreases
   D. increases
   E. depends on the heat absorbed

22. Which of the following statements is TRUE regarding the acceleration experienced by a block moving down a frictionless plane inclined at a 20° angle?

    A. It decreases as the block moves down the plane
    B. It increases as the block moves down the plane
    C. It increases at a rate proportional to the incline
    D. It decreases at a rate proportional to the incline
    E. It remains constant

23. Which of the following statements is FALSE?

    A. Waves from a vibrating string are transverse waves
    B. Sound travels much slower than light
    C. Sound waves are longitudinal pressure waves
    D. Sound can travel through a vacuum
    E. In music, pitch, and frequency have approximately the same meaning

24. In a given medium with fixed boundaries, the longest wavelength that produces a standing wave is 4 m. What is the lowest possible frequency associated with a standing wave within this medium if waves propagate through the medium at 8 m/s?

    A. 0.5 Hz
    B. 1 Hz
    C. 2 Hz
    D. 6 Hz
    E. 8 Hz

25. Two charges, $Q_1 = 3.4 \times 10^{-10}$ C and $Q_2 = 6.8 \times 10^{-9}$ C, are separated by 1 cm. Let F_1 be the magnitude of the electrostatic force felt by Q_1 due to Q_2 and let F_2 be the magnitude of the electrostatic force felt by Q_2 due to Q_1. What is the ratio of F_1 / F_2?

    A. 2
    B. 1
    C. 16
    D. 8
    E. 4

26. The electric power of a lamp that carries 2 A at 120 V is:

    A. 24 W
    B. 2 W
    C. 60 W
    D. 120 W
    E. 240 W

27. What happens to an atom when it absorbs energy?

  A. The atom re-emits the energy as light
  B. The atom stores the energy as potential energy
  C. The average distance between the electron and nucleus is reduced
  D. The atom stores the energy as kinetic energy
  E. The atom re-emits the energy as alpha particles

28. What is the de Broglie wavelength of a 1.30 kg missile moving at 28.10 m/s? (Use $h = 6.626 \times 10^{-34}$ J·s)

  A. 1.85×10^{-37} m
  B. 2.40×10^{-36} m
  C. 1.81×10^{-35} m
  D. 3.37×10^{-35} m
  E. 6.43×10^{-35} m

29. Ignoring air resistance, how long does a coin take to reach the ground when dropped from a 42 m building? (Use the acceleration due to gravity $g = 10$ m/s²)

  A. 1.4 s
  B. 2.9 s
  C. 3.6 s
  D. 5.4 s
  E. 4.7 s

30. A box that weighs 40 N is on a rough horizontal surface. An external force F is applied horizontally to the box. A normal force and a friction force are present. When force F equals 8.8 N, the box is in motion at a constant velocity. The box decelerates when force F is removed. What is the magnitude of the acceleration of the box? (Use acceleration due to gravity $g = 10$ m/s²)

  A. 0.55 m/s²
  B. 1.1 m/s²
  C. 4.4 m/s²
  D. 2.2 m/s²
  E. 0 m/s²

31. A bullet shot from a gun with a longer barrel has a higher muzzle velocity because the bullet receives a greater:

  I. force   II. impulse   III. acceleration

  A. I only
  B. II only
  C. III only
  D. I and II only
  E. II and III only

32. A 1,000 kg car travels at 30 m/s on a level road when the driver slams the brakes, bringing the car to a stop. What is the change in kinetic energy during the braking if the skid marks are 35 m long?

  A. -4.5×10^5 J
  B. 0 J
  C. -9×10^{10} J
  D. 4.2×10^5 J
  E. 9×10^{10} J

33. The graph shows the position (x) as a function of time (t) for a system undergoing simple harmonic motion. Which graph represents the acceleration of this system as a function of time?

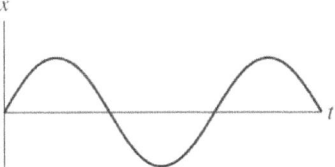

A.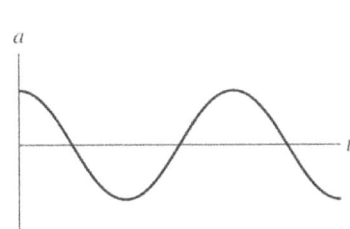

C.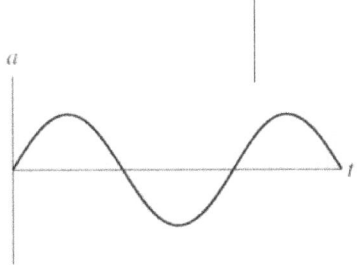

B.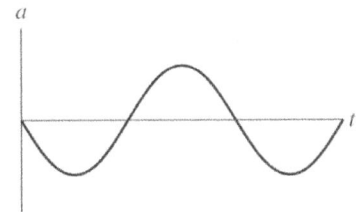

D.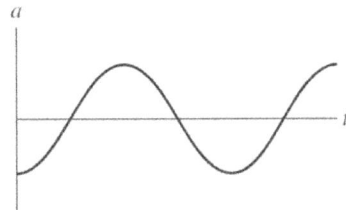

E. None of the above

34. What is the average kinetic energy of a molecule in an ideal gas at 740 K? (Use Boltzmann's constant $k = 1.38 \times 10^{-23}$ J/K)

A. 3.9×10^{-19} J
B. 2.4×10^{-17} J
C. 5.8×10^{-21} J
D. 1.5×10^{-20} J
E. 4.5×10^{-22} J

35. If two converging lenses with focal lengths of 10 cm and 20 cm are in contact, what is the power of the combination?

A. 10 D
B. 15 D
C. 20 D
D. 30 D
E. 25 D

36. What is the capacitance of a capacitor having an impedance of 4 kΩ when operating at 0.6 kHz?

A. 96 µF
B. 2.4 µF
C. 0.15 µF
D. 0.024 µF
E. 0.066 µF

37. A 3 Ω and a 1.5 Ω resistor are connected in parallel within a circuit. If the voltage drop across the 3 Ω resistor is 2 V, what is the sum of the currents through these two resistors?

- **A.** 4/3 amps
- **B.** 3/2 amps
- **C.** 2 amps
- **D.** 2/3 amps
- **E.** 3/4 amps

38. The density of the material at the center of a neutron star is about 1×10^{18} kg/m³. Calculate the approximate mass of a cube of this material that is 1.76 microns on each side. (Use the conversion of 1 micron = 1×10^{-6} m)

- **A.** 5.5 kg
- **B.** 4.8 kg
- **C.** 7.8 kg
- **D.** 6.4 kg
- **E.** 3.6 kg

39. An adiabatic and isothermal process are shown on the pressure *vs.* volume diagram. Which is the isothermal process?

I. Process A
II. Process B
III. Requires knowing if the gas is monatomic or diatomic

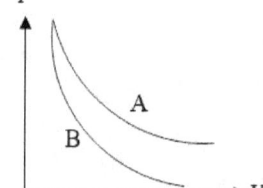

- **A.** I only
- **B.** II only
- **C.** III only
- **D.** I and II only
- **E.** None are correct

40. Crests of an ocean wave pass a pier every 10 s. What is the wavelength of the ocean waves if the waves are moving at 4.6 m/s?

- **A.** 4.4 m
- **B.** 0.46 m
- **C.** 4.6 m
- **D.** 2.2 m
- **E.** 46 m

41. How much work is done on a crate if pushed 2 m with a force of 20 N?

- **A.** 10 J
- **B.** 20 J
- **C.** 30 J
- **D.** 40 J
- **E.** 50 J

42. Two friends are standing on opposite ends of a canoe, which is initially at rest with respect to the lake. Steve is on the right when he throws a very massive ball to the left, and Mike, on the left, catches it. Ignoring friction between the canoe and the water, after the ball is caught, the canoe:

- **A.** moves to the right before reversing direction
- **B.** moves to the left before reversing direction
- **C.** remains stationary
- **D.** moves to the right
- **E.** moves to the left

43. At age 21, Joseph sets out for a star 50 light-years from Earth. How fast would his spaceship have to travel to reach that star when he is 61 years old? One light-year is the distance light travels in one year.

- **A.** $0.58c$
- **B.** $0.68c$
- **C.** $0.78c$
- **D.** $0.88c$
- **E.** $0.96c$

44. A massless, ideal spring is projected horizontally from a wall and is connected to a 0.3 kg mass. The mass is oscillating in one dimension, such that it moves 0.5 m from one end of its oscillation to the other. What is the frequency of the oscillation if it undergoes 10 complete oscillations in 60 s?

- **A.** 0.17 Hz
- **B.** 3.9 Hz
- **C.** 3.1 Hz
- **D.** 11.2 Hz
- **E.** 17.6 Hz

45. Aluminum has a positive coefficient of thermal expansion. A round hole has been drilled in a large sheet of aluminum. As the temperature increases and the surrounding metal expands, the diameter of the hole:

- **A.** increases or decreases, depending on how much metal surrounds the hole
- **B.** remains constant
- **C.** decreases
- **D.** increases
- **E.** increases or decreases, depending on the total change in temperature

46. Which statement is correct when a flowerpot of mass m falls from rest to the ground for a distance h below?

- **A.** The speed of the flowerpot when it hits the ground is proportional to m
- **B.** The KE of the flowerpot when it hits the ground does not depend on h
- **C.** The KE of the flowerpot when it hits the ground is proportional to h
- **D.** The speed of the flowerpot when it hits the ground is proportional to h
- **E.** The speed of the flowerpot when it hits the ground is inversely proportional to h

47. Two speakers 3 m apart produce in-phase sound waves with a wavelength of 1 m. A microphone is placed between the speakers to determine the intensity of the sound at various points. What kind of point exists exactly 0.5 m to the left of the speaker on the right? (Use the speed of sound $v = 340$ m/s)

- A. Node
- B. Antinode
- C. Node and antinode
- D. Destructive interference
- E. None of the above

48. What causes an object to become electrostatically charged?

- A. Charge is created
- B. Protons are transferred
- C. Electrons are transferred
- D. Protons and electrons are transferred
- E. Charge is destroyed

49. A 12-liter volume of oil is subjected to pressure, which produces a volume strain of -3×10^{-4}. The bulk modulus of the oil is 6.3×10^9 Pa and is independent of the pressure. What is the change in the pressure of the oil? (Use the conversion of 1 atm = 10^5 Pa)

- A. 26 atm
- B. 4 atm
- C. 7 atm
- D. 19 atm
- E. 13 atm

50. A small boat is moving at a velocity of 3.35 m/s when it is accelerated by a river current perpendicular to the initial direction of motion. Relative to the initial direction of motion, what is the new velocity of the boat after 33.5 s if the current acceleration is 0.75 m/s^2?

- A. 62 m/s at 7.6°
- B. 62 m/s at 82.4°
- C. 25 m/s at 7.6°
- D. 25 m/s at 82.4°
- E. 40 m/s at 82.4°

51. A machinist turns the power on for a stationary grinding wheel at time $t = 0$ s. The wheel accelerates uniformly for 10 s and reaches the operating angular velocity of 58 radians/s. The wheel is run at that angular velocity for 30 s before the power is shut off. The wheel slows down uniformly at 1.4 radians/s^2 until it stops. What is the approximate total number of revolutions for the wheel?

- A. 460
- B. 320
- C. 380
- D. 510
- E. 720

52. A light ray in glass arrives at the glass-water interface at an angle of $\theta = 48°$ with the normal. The refracted ray in water makes an angle of $\phi = 72°$ with respect to the normal. What is the new angle of refraction ϕ in the water if the angle of incidence is changed to $\theta = 37°$? (Use index of refraction for water n = 1.33)

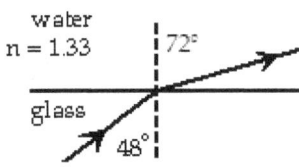

A. 43°
B. 55°
C. 37°
D. 59°
E. 50°

53. With a total of four tuning forks, what is the highest number of different beat frequencies heard by striking the forks one pair at a time?

A. 2
B. 4
C. 6
D. 8
E. 10

54. Initially, the switch S is open for the circuit shown, and the capacitor voltage is 295 V. The switch S is closed at time $t = 0$. What is the charge on the capacitor when the current in the circuit is 33 μA?

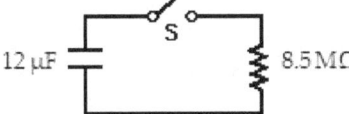

A. 2,200 μC
B. 2,600 μC
C. 3,000 μC
D. 1,800 μC
E. 3,400 μC

55. A solid uniform sphere of mass 120.0 kg and radius 1.7 m starts from rest and rolls without slipping down an inclined plane of vertical height 5.3 m; the sphere starts at the top of the ramp. What is the angular speed of the sphere at the bottom of the inclined plane? The moment of inertia of a solid sphere is $(2/5)mR^2$.

A. 0.81 rad/s
B. 1.7 rad/s
C. 2.9 rad/s
D. 4.3 rad/s
E. 5.1 rad/s

56. Which of the following is a TRUE statement?

A. It is impossible to convert work entirely into heat
B. It is impossible to transfer heat from a cooler to a hotter body
C. The second law of thermodynamics is a consequence of the first law of thermodynamics
D. Heat can flow spontaneously from a hot body to a cold one or vice versa, depending on whether the process is reversible or irreversible
E. All statements are false

57. Ignoring air resistance, what is the speed of a rock as it hits the ground if it was dropped from a 50 m cliff? (Use acceleration due to gravity $g = 10$ m/s^2)

A. 21 m/s
B. 14 m/s
C. 32 m/s
D. 42 m/s
E. 9 m/s

58. A string connects a 15 kg block on a table to a 60 kg mass hanging over the edge of the table. Ignoring the frictional force, what is the acceleration of the 15 kg block when the 60 kg block is released? (Use acceleration due to gravity $g = 10$ m/s^2)

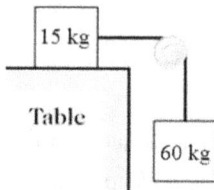

A. 9.5 m/s^2
B. 7.5 m/s^2
C. 10.5 m/s^2
D. 8 m/s^2
E. 6 m/s^2

59. Suppose a van de Graaff generator builds a negative static charge, and a grounded conductor is near enough to it so 9 µC of negative charge arcs to the conductor. What is the number of electrons transferred? (Use Coulomb's constant $k = 9 \times 10^9$ N·m^2/C^2 and the charge of an electron = 1.6×10^{-19} C)

A. 1.6×10^{18} electrons
B. 43.8×10^{12} electrons
C. 5.6×10^{13} electrons
D. 7 electrons
E. 4.1×10^{20} electrons

60. What type of radioactive decay produces a daughter nuclide with the same element as the parent nuclide?

I. Alpha   II. Gamma   III. Beta

A. I only
B. II only
C. III only
D. I and II only
E. I and III only

61. Two charged objects attract each other with a specific force. If the charges on both objects are doubled with no change in separation, what is the force between them?

A. Quadruples
B. Halves
C. Doubles
D. Becomes zero
E. Requires knowing the distance between them

62. A potted plant of mass M is resting on a flat board, and one end of the board is lifted slowly until the potted plant begins to slide. What does the angle θ that the board must make for sliding to occur?

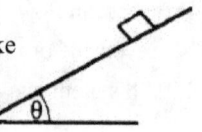

A. M
B. μ_s, static friction
C. μ_k, kinetic friction
D. g, acceleration due to gravity
E. all the above

63. Objects 1 and 2 are heated from the same initial temperature (T_i) to the same final temperature (T_f). Object 1 has three times the specific heat capacity of Object 2 and four times the mass. If Object 1 absorbs heat Q during this process, what is the amount of heat absorbed by Object 2?

- **A.** $(4/3)Q$
- **B.** $(3/4)Q$
- **C.** $6Q$
- **D.** $12Q$
- **E.** $(1/12)Q$

64. A 20 kg object is dropped from a height of 100 m. Ignoring air resistance, how much gravitational PE has the object lost when its speed is 30 m/s?

- **A.** 2,050 J
- **B.** 2,850 J
- **C.** 9,000 J
- **D.** 5,550 J
- **E.** 6,750 J

65. What is the approximate wavelength of a wave with a speed of 360 m/s and a period of 4.2 s?

- **A.** 85.7 m
- **B.** 1.86 m
- **C.** 1,512 m
- **D.** 288.6 m
- **E.** 422.1 m

66. What is the effect on a system's mechanical energy if only the amplitude of a vibrating mass-and-spring system is doubled?

- **A.** Increases by a factor of 2
- **B.** Increases by a factor of 4
- **C.** Increases by a factor of 3
- **D.** Remains the same
- **E.** Increases by a factor of $\sqrt{2}$

67. When an 8.8 kg mass is suspended from a 4.4 m long wire with a 1.6 mm diameter, the wire stretches by 3.3 mm. What is Young's modulus for the wire? (Use the acceleration due to gravity $g = 9.8$ m/s^2)

- **A.** 2.4×10^{10} N/m^2
- **B.** 3.6×10^{11} N/m^2
- **C.** 5.7×10^{10} N/m^2
- **D.** 7.1×10^{12} N/m^2
- **E.** 6.9×10^{11} N/m^2

68. If the distance between two electrostatic charges is doubled, how is the force between them affected?

- **A.** Increases by a factor of 2
- **B.** Increases by a factor of 4
- **C.** Decreases by a factor of $\sqrt{2}$
- **D.** Decreases by a factor of 4
- **E.** Remains the same

69. When fully charged, a particular battery provides 1 mW of power at 9 V. What is the current it delivers?

A. 0.13 kA
B. 9 kA
C. 0.11 mA
D. 18 mA
E. 0.55 mA

70. A girl of height *h* stands in front of a plane mirror. What must the minimum length of the mirror be so that she can view her entire body?

A. ¼*h*
B. 2*h*
C. ½*h*
D. *h*
E. Depends on her distance from the mirror

71. The fission of an atom that has a larger atomic number (e.g., uranium) can be induced by bombarding the atom with:

A. electrons
B. positrons
C. neutrons
D. protons
E. gamma rays

72. A 1 kg chunk of putty moving at 1 m/s collides and sticks to a stationary 6 kg box. What is the total momentum of the box and putty? (Assume the box rests on a frictionless surface)

A. 0 kg·m/s
B. 1 kg·m/s
C. 2 kg·m/s
D. 3 kg·m/s
E. 5 kg·m/s

73. Which statement correctly describes the situation when a 6 kg mass moves at 2 m/s and a 3 kg mass moving at 4 m/s glides over a horizontal frictionless surface? A horizontal force *F*, which directly opposes their motion, results in the objects coming to rest.

A. The 6 kg mass travels twice the distance of the 3 kg mass before stopping
B. The 3 kg mass travels farther, but less than twice the distance of the 6 kg mass before stopping
C. The 3 kg mass travels twice the distance of the 6 kg mass before stopping
D. The 6 kg mass loses four times more KE than the 3 kg mass before stopping
E. The 6 kg mass loses two times more KE than the 3 kg mass before stopping

74. Which statement is correct for separating adjacent maxima in a double-slit interference pattern for monochromatic light?

A. Greatest for red light
B. Greatest for violet light
C. Greatest for yellow light
D. Greatest for blue light
E. The same for all colors of light

75. If a 25 cm violin string vibrates at its fundamental frequency of 860 Hz, what is the speed of transverse waves on the string?

- A. 220 m/s
- B. 430 m/s
- C. 880 m/s
- D. 1,680 m/s
- E. 2,260 m/s

76. A pipe with a 3 cm radius carries water at a velocity of 4 m/s. What is the volume flow rate?

- A. 1.1×10^{-2} m^3/s
- B. 48 m^3/s
- C. 7.5×10^{-3} m^3/s
- D. 2.7×10^{2} m^3/s
- E. 4.3 m^3/s

77. A conductor differs from an insulator in that a conductor has:

- A. slower-moving molecules
- B. tightly-bound outer electrons
- C. more protons than electrons
- D. more electrons than protons
- E. none of the above

78. Which statement is valid for two conductors joined by a long copper wire?

- A. One conductor must have a lower potential than the other conductor
- B. Shortening the wire increases the potential of both conductors
- C. Each conductor must have the same potential
- D. The potential on the wire is the sum of the potentials of each conductor
- E. The potential on the wire is the average of the potentials of each conductor

79. Two antennas 130 m apart on a North-South line radiate in phase at a frequency of 3.6 MHz, and radio measurements are recorded far away from the antennas. What is the smallest angle, East of North from the antennas, for constructive interference of the two radio waves? (Use the speed of light $c = 3 \times 10^8$ m/s)

- A. 45°
- B. 60°
- C. 50°
- D. 30°
- E. 90°

80. According to the quantum mechanical model of the He atom, if the orbital angular momentum quantum number is ℓ, how many magnetic quantum numbers are possible?

- A. $2\ell + 1$
- B. $2\ell - 1$
- C. 2ℓ
- D. $\ell/2$
- E. 3ℓ

81. The closest star to our solar system is Alpha Centauri, which is 4.367 light-years away. A spaceship with a constant speed of $0.800c$ relative to Earth travels toward the star. How many years would elapse on a clock on the Earth before the spaceship reaches Alpha Centauri?

- **A.** 0.32
- **B.** 0.97
- **C.** 3.56
- **D.** 5.49
- **E.** 7.32

82. A stable nucleus has many positively-charged protons packed very close to each other. Why do the protons not move apart due to mutual Coulomb repulsion?

- **A.** The neutrons in the nucleus shield the protons from each other's positive charge
- **B.** The Coulomb force cannot operate within small nuclei
- **C.** An attractive nuclear force in the nucleus counteracts the Coulomb force
- **D.** There are an equal number of electrons in the nucleus, which neutralize the protons
- **E.** The gravitational force on the protons and neutrons overcome their repulsion at such close distances

83. The intensity of a sound wave is directly proportional to the:

- **A.** Doppler shift
- **B.** power
- **C.** decibel level
- **D.** wavelength
- **E.** frequency

84. What force needs to be applied to a 6 cm diameter piston to lift a 12,000 N container with a hydraulic piston that has a diameter of 25 cm?

- **A.** 26 N
- **B.** 360 N
- **C.** 1,040 N
- **D.** 2,080 N
- **E.** 691 N

85. Two parallel metal plates, separated by a 0.05 m distance, are charged to produce a uniform electric field between them that points down. What is the magnitude of the force experienced by a proton between the two plates? (Use the acceleration due to gravity $g = 10$ m/s², the charge of a proton = 1.6×10^{-19} C and uniform electric field = 4×10^4 N/C)

- **A.** 6.4×10^{-10} N
- **B.** 3.2×10^{-10} N
- **C.** 3.2×10^{-15} N
- **D.** 6.4×10^{-15} N
- **E.** 2.5×10^{-5} N

86. 200 kcal of heat raises the temperature of 3 kg of material by 90 °C. What is the material's specific heat capacity?

   A. 0.74 kcal/kg·°C
   B. 0.33 kcal/kg·°C
   C. 1.42 kcal/kg·°C
   D. 1.13 kcal/kg·°C
   E. 0.14 kcal/kg·°C

87. Optical density is proportional to:

   A. index of refraction
   B. index of reflection
   C. mass density
   D. light speed
   E. wavelength

88. Which radiation type penetrates about 1 cm of human tissue and requires a minimum protective shielding made of wood or aluminum?

   A. Gamma
   B. Beta
   C. Alpha
   D. Nuclide
   E. None of the above

89. Reynold's number is given by Re = $L\rho v / \mu$, where L is some characteristic length related to the path that the fluid travels (e.g., the length or diameter of the pipe), ρ is the density of the fluid, v is the velocity of the fluid, and μ is the viscosity. A larger Reynold's number for a given flow indicates the increased likelihood of developing turbulence. For water flowing in a pipe, which might reduce the likelihood of turbulent flow?

   A. Raising the temperature
   B. Increasing the radius of the pipe
   C. Increasing the flow rate
   D. Increasing the viscosity
   E. None of the above

90. What is the tension in a cable that pulls a 900 kg object straight upward at an acceleration of 0.6 m/s^2? (Use the acceleration due to gravity g = 9.8 m/s^2)

   A. 8,280 N
   B. 980 N
   C. 9,800 N
   D. 9,360 N
   E. 930 N

91. The masses of the blocks and the velocities before and after a collision are:

   In this example, the collision is:

   A. completely inelastic
   B. completely elastic
   C. characterized by an increase in KE
   D. characterized by a decrease in momentum
   E. characterized by an increase in PE

92. What is Amanda's mass if Steve does 174 J of work while pulling her backward on a swing with a 5.1 m chain until the swing makes an angle of 32° with the vertical? (Use the acceleration due to gravity $g = 9.8$ m/s^2)

   A. 17.4 kg
   B. 22.9 kg
   C. 29.8 kg
   D. 37.8 kg
   E. 14.4 kg

93. For an object that exhibits simple harmonic motion (SHM), what is the frequency of the motion if the shortest interval between the two extremes of the object's displacement from its equilibrium position is 2 s?

   A. 0.25 cycle/s
   B. 0.5 cycle/s
   C. 2.5 cycles/s
   D. 5 cycles/s
   E. 1 cycle/s

94. A few 10 cm long aluminum rods and 8 cm long steel rods are at 5 °C and are joined to form a 60 cm long rod. What is the increase in the length of the joined rod when the temperature is raised to 80 °C? (Use the coefficient of linear expansion for aluminum = 2.4×10^{-5} K^{-1} and the coefficient of linear expansion for steel = 1.2×10^{-5} K^{-1})

   A. 0.3 mm
   B. 0.5 mm
   C. 1.8 mm
   D. 0.72 mm
   E. 1.4 mm

95. Relative to the mirror, where is the resulting image when a light source is 12 m in front of a diverging mirror with a focal length of 6 m?

   A. 2 m in front
   B. 2 m behind
   C. 4 m in front
   D. 4 m behind
   E. 0.5 m in front

96. A beam of light falling on a metal surface is causing electrons to be ejected from the surface. If the frequency of the light now doubles, which of the following statements is always true?

   A. The number of electrons ejected per second doubles
   B. Twice as many photons hit the metal surface as before
   C. The kinetic energy of the ejected electrons doubles
   D. The speed of the ejected electrons doubles
   E. None of the above statements are always true

97. What is the efficiency of a heat engine that receives 8,500 J of heat and loses 4,500 J in each cycle?

   A. 57%
   B. 47%
   C. 33%
   D. 16%
   E. 29%

98. A marble cube is lowered at a steady rate into the ocean by a crane, while its top and bottom surfaces remain parallel with the water's surface. Which graph describes the buoyant force (B) on this cube as a function of time (t) if the cube enters the water at time $t = 0$ s and is lowered until its top surface is well below the water?

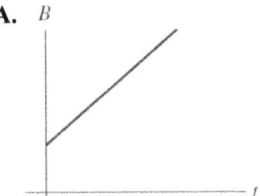
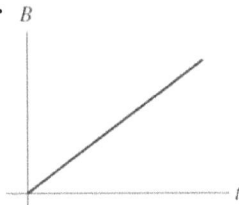
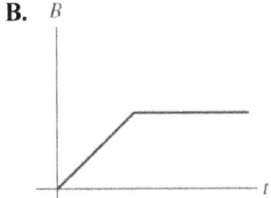
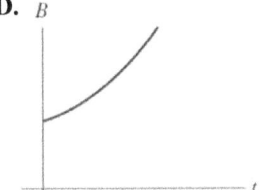
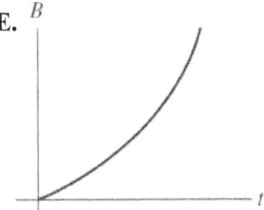

99. A commuter train moves at 50 m/s directly toward Daud, who is whistling at 420 Hz while standing. If a passenger on the train hears him, what frequency would she hear? (Use the speed of sound $v = 350$ m/s)

A. 300 Hz
B. 360 Hz
C. 480 Hz
D. 500 Hz
E. 520 Hz

100. Both constructive and destructive interference are necessary to produce the sound phenomena of:

    I. beats   II. resonance   III. refraction

A. I only
B. II only
C. III only
D. I and II only
E. I and III only

Notes for active learning

Diagnostic Test 2

This Diagnostic Test is designed to assess your proficiency on each topic and NOT to mimic the test. Use your test results and identify areas of strength and weakness to adjust your study plan and enhance your fundamental knowledge. The length of the Diagnostic Tests is optimal for a single study session.

#	Answer:					Review	#	Answer:					Review
1:	A	B	C	D	E	___	26:	A	B	C	D	E	___
2:	A	B	C	D	E	___	27:	A	B	C	D	E	___
3:	A	B	C	D	E	___	28:	A	B	C	D	E	___
4:	A	B	C	D	E	___	29:	A	B	C	D	E	___
5:	A	B	C	D	E	___	30:	A	B	C	D	E	___
6:	A	B	C	D	E	___	31:	A	B	C	D	E	___
7:	A	B	C	D	E	___	32:	A	B	C	D	E	___
8:	A	B	C	D	E	___	33:	A	B	C	D	E	___
9:	A	B	C	D	E	___	34:	A	B	C	D	E	___
10:	A	B	C	D	E	___	35:	A	B	C	D	E	___
11:	A	B	C	D	E	___	36:	A	B	C	D	E	___
12:	A	B	C	D	E	___	37:	A	B	C	D	E	___
13:	A	B	C	D	E	___	38:	A	B	C	D	E	___
14:	A	B	C	D	E	___	39:	A	B	C	D	E	___
15:	A	B	C	D	E	___	40:	A	B	C	D	E	___
16:	A	B	C	D	E	___	41:	A	B	C	D	E	___
17:	A	B	C	D	E	___	42:	A	B	C	D	E	___
18:	A	B	C	D	E	___	43:	A	B	C	D	E	___
19:	A	B	C	D	E	___	44:	A	B	C	D	E	___
20:	A	B	C	D	E	___	45:	A	B	C	D	E	___
21:	A	B	C	D	E	___	46:	A	B	C	D	E	___
22:	A	B	C	D	E	___	47:	A	B	C	D	E	___
23:	A	B	C	D	E	___	48:	A	B	C	D	E	___
24:	A	B	C	D	E	___	49:	A	B	C	D	E	___
25:	A	B	C	D	E	___	50:	A	B	C	D	E	___

Any duplication (copies, uploads, PDFs) is illegal.

#	Answer:					Review	#	Answer:					Review
51:	A	B	C	D	E	___	76:	A	B	C	D	E	___
52:	A	B	C	D	E	___	77:	A	B	C	D	E	___
53:	A	B	C	D	E	___	78:	A	B	C	D	E	___
54:	A	B	C	D	E	___	79:	A	B	C	D	E	___
55:	A	B	C	D	E	___	80:	A	B	C	D	E	___
56:	A	B	C	D	E	___	81:	A	B	C	D	E	___
57:	A	B	C	D	E	___	82:	A	B	C	D	E	___
58:	A	B	C	D	E	___	83:	A	B	C	D	E	___
59:	A	B	C	D	E	___	84:	A	B	C	D	E	___
60:	A	B	C	D	E	___	85:	A	B	C	D	E	___
61:	A	B	C	D	E	___	86:	A	B	C	D	E	___
62:	A	B	C	D	E	___	87:	A	B	C	D	E	___
63:	A	B	C	D	E	___	88:	A	B	C	D	E	___
64:	A	B	C	D	E	___	89:	A	B	C	D	E	___
65:	A	B	C	D	E	___	90:	A	B	C	D	E	___
66:	A	B	C	D	E	___	91:	A	B	C	D	E	___
67:	A	B	C	D	E	___	92:	A	B	C	D	E	___
68:	A	B	C	D	E	___	93:	A	B	C	D	E	___
69:	A	B	C	D	E	___	94:	A	B	C	D	E	___
70:	A	B	C	D	E	___	95:	A	B	C	D	E	___
71:	A	B	C	D	E	___	96:	A	B	C	D	E	___
72:	A	B	C	D	E	___	97:	A	B	C	D	E	___
73:	A	B	C	D	E	___	98:	A	B	C	D	E	___
74:	A	B	C	D	E	___	99:	A	B	C	D	E	___
75:	A	B	C	D	E	___	100:	A	B	C	D	E	___

1. Determine the resting length of a spring if one end (spring constant $k = 40$ N/m) is fixed at point P, while the other end is connected to a 7 kg mass. The fixed end and the mass sit on a horizontal frictionless surface, and the mass and the spring can rotate about P. The mass moves in a circle with $r = 2$ m, and the force on the mass is 12 N.

   A. 0.1 m
   B. 1.7 m
   C. 0.8 m
   D. 2.3 m
   E. 3.8 m

2. An electrical motor spins at a constant of 2,640 rpm. What is the acceleration of the edge of the motor if the armature radius is 7.2 cm?

   A. 87 m/s^2
   B. 1,690 m/s^2
   C. 8,432 m/s^2
   D. 2,420 m/s^2
   E. 5,495 m/s^2

3. As in the figure, when a 100 kg block is released from rest from a height of 1 m, it takes 0.51 s to hit the floor. Assuming no friction and that the pulley is massless, what is the mass of the block on the other end? (Use the acceleration due to gravity $g = 9.8$ m/s^2)

   A. 9 kg
   B. 23 kg
   C. 16 kg
   D. 12 kg
   E. 19 kg

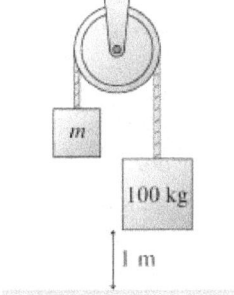

4. A rock from a volcanic eruption is propelled straight up into the air. Ignoring air resistance, which of the statements is correct about the rock in the air? (Take the positive axis to point up)

   A. Throughout its motion, the acceleration is always negative
   B. Throughout its motion, the acceleration is always negative, and the velocity is continually increasing
   C. On the way up, its velocity is increasing, and its acceleration is positive
   D. On the way down, its velocity is increasing, and its acceleration is negative
   E. At the highest point, its velocity and acceleration are zero

5. Three solid, uniform, cylindrically shaped flywheels, each of mass 65.0 kg and radius 1.47 m, rotate independently around a common axis. Two flywheels rotate in one direction at 3.83 rad/s; the other rotates in the opposite direction at 3.42 rad/s. What is the magnitude of the net angular momentum of the system?

   A. 168.0 kg·m^2/s
   B. 298.0 kg·m^2/s
   C. 456.0 kg·m^2/s
   D. 622.0 kg·m^2/s
   E. 882.0 kg·m^2/s

6. An object weighing 50 N is traveling vertically upward from the Earth without air resistance at a constant velocity of 10 m/s. What is the power required to keep the object in motion?

A. 0 W
B. 10 W
C. 50 W
D. 100 W
E. 500 W

7. Which statement is correct for a pipe with a length of L, closed at one end, and is resonating at its fundamental frequency?

A. The wavelength is $2L$, and there is a displacement node at the pipe's closed end
B. The wavelength is $2L$, and there is a displacement antinode at the pipe's open end
C. The wavelength is $4L$, and there is a displacement antinode at the pipe's closed end
D. The wavelength is $4L$, and there is a displacement antinode at the pipe's open end
E. The wavelength is $4L$, and there is a displacement node at the pipe's open end

8. Which statement regarding the electric charge is NOT correct if the electric charge is conserved?

A. Will not interact with neighboring electric charges
B. Can neither be created nor destroyed
C. Is a whole-number multiple of the charge of one electron
D. May occur in an infinite variety of quantities
E. Can only occur in restricted (i.e., allowable) quantities

9. Which of the graphs illustrates Hooke's Law?

A.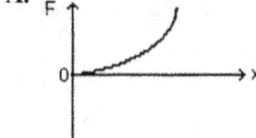

C.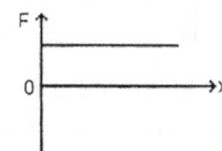

B.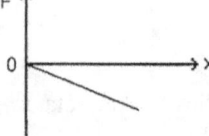

D.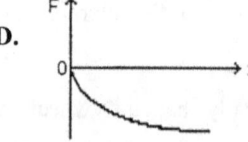

E. None of the above

10. The area under the curve in a velocity *vs.* time graph determines:

A. position
B. displacement
C. velocity
D. acceleration
E. time

11. Which statement is correct if two hockey pucks, each with a nonzero velocity, undergo an elastic collision as they slide toward each other on a surface of frictionless ice and collide head-on?

   A. Momentum and KE are doubled
   B. Neither momentum nor KE is conserved
   C. Momentum is conserved, but KE is not conserved
   D. Momentum is not conserved, but KE is conserved
   E. Momentum and KE are conserved

12. A gamma-ray is a pulse of electromagnetic energy with a frequency of 2.4×10^{20} Hz. What is the ratio of its wavelength to the radius of the nucleus which produced it? (Use the speed of light $c = 3 \times 10^8$ m/s and the radius of the nucleus = 5×10^{-13} cm)

   A. 0.028
   B. 1.33×10^7
   C. 38.2
   D. 250
   E. 2.58×10^{-7}

13. A water tank open to the atmosphere is elevated above the ground by 25 m and is filled to a depth of 12 m. What is the approximate water pressure in a hose with a 2 cm diameter at ground level? (Use the acceleration due to gravity $g = 9.8$ m/s^2, the density of water $\rho = 1{,}000$ kg/m^3 and the atmospheric pressure P = 1 atm or 101,325 N/m^2)

   A. 8.4 N/m^2
   B. 3.2×10^5 N/m^2
   C. 4.6×10^5 N/m^2
   D. 5.6 N/m^2
   E. Requires the cross-sectional area of the tank

14. Which of the following diagrams is correct for a circuit with a battery connected to four resistors, R_1, R_2, R_3, and R_4? Resistors R_1 and R_2 are connected in parallel, resistors R_3 and R_4 are connected in parallel, and parallel sets of resistors are connected in series across the battery.

A.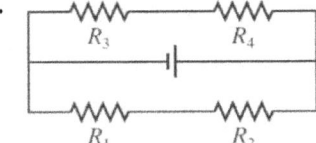

C.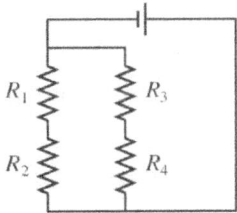

B.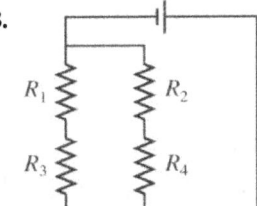

D.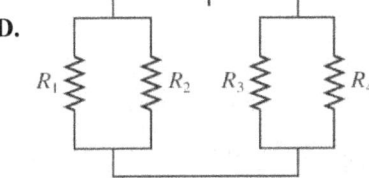

E. None of the above

15. Monochromatic light is incident on a metal surface, and electrons are ejected. How do the ejection rate and the maximum energy of the electrons change if the intensity of the light is increased?

    A. Same rate; same maximum energy
    B. Greater rate; higher maximum energy
    C. Same rate; lower maximum energy
    D. Greater rate; same maximum energy
    E. Greater rate; lower maximum energy

16. Which is an important feature of the Carnot cycle?

    A. Efficiency is determined only by the properties of the working substance used
    B. It is an irreversible process that can be analyzed precisely without approximations
    C. Efficiency can be 100%
    D. Efficiency depends only on the absolute temperature of the hot reservoir used
    E. It is the most efficient engine operating between two temperatures

17. An object travels along the x-axis at a constant speed of 3 m/s in the $-x$-direction. If the object is on $x = 4$ m at $t = 0$, where is it at time $t = 4$ s?

    A. $x = -16$ m
    B. $x = -12$ m
    C. $x = -8$ m
    D. $x = -6$ m
    E. $x = -2$ m

18. In an air-free chamber, a pebble is thrown horizontally, and at the same instant, a second pebble is dropped from the same height. Compare the time it took for the two pebbles to hit the ground:

    A. They hit at the same time
    B. Requires values for the initial velocities of both pebbles
    C. The thrown pebble hits first
    D. The dropped pebble hits first
    E. Requires the height from which they were released

19. Cart 1 (2 kg) and Cart 2 (2.5 kg) run along a frictionless, level, one-dimensional track. Cart 2 is initially at rest, and Cart 1 travels 0.6 m/s toward the right when it encounters Cart 2. After the collision, Cart 1 is at rest. What is the efficiency of the collision for kinetic energy?

    A. 16%
    B. 65%
    C. 80%
    D. 25%
    E. 53%

20. The two strongest forces that act between protons in a nucleus are the:

    A. electrostatic and gravitational forces
    B. strong nuclear and electrostatic forces
    C. weak nuclear and electrostatic forces
    D. strong nuclear and gravitational forces
    E. weak nuclear and gravitational forces

21. A projectile is fired at time $t = 0$ s from point O of a ledge. It has initial velocity components of $v_{ox} = 30$ m/s and $v_{oy} = 300$ m/s with time in flight of 75 s. The projectile lands at point P. What is the horizontal distance that the projectile travels?

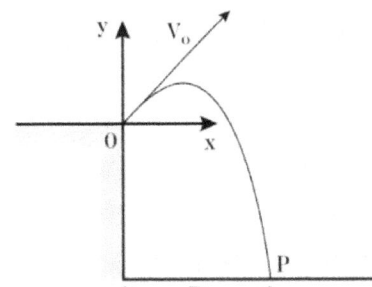

A. 3,020 m
B. 2,880 m
C. 2,420 m
D. 2,250 m
E. 3,360 m

22. A car of mass m is traveling along the roadway up a slight incline of angle θ to the horizontal when the driver sees a deer and suddenly applies the brakes. The car skids before coming to rest. Which expression gives the force of friction on the car if the coefficient of static friction between the tires and the road is μ_s and the coefficient of kinetic friction is μ_k?

A. $\mu_k N$
B. $\mu_s N$
C. mg
D. $mg \sin \theta$
E. $(\mu_s - \mu_k)N$

23. Once a steady-state heat flow is established, the thickness of a wall built from solid uniform material is doubled. Relative to the original value, what is the result of the rate of heat loss for a given temperature difference across the wall?

A. $1/\sqrt{2}$
B. ¼
C. 2 times
D. ½
E. 4 times

24. Energy is the:

I. ability to do work
II. work that can be done by an object with potential or kinetic energy
III. work needed to generate potential or kinetic energy

A. I only
B. II only
C. III only
D. I and III only
E. I, II and III

25. The displacement of a vibrating tuning fork and the resulting sound wave is related to:

A. period
B. wavelength
C. resonance
D. frequency
E. amplitude

26. The decibel level of sound is related to its:

A. velocity
B. frequency
C. wavelength
D. intensity
E. pitch

27. A tank of water has a hose filled with water projecting from the top. The system acts as a siphon as the other end of the hose is below the tank. The end of the hose outside the tank is at height $h = 0$ m. The bottom of the tank is at height h_1, the end of the hose inside the tank is at height h_2, and the top of the water is at height h_3. Assuming the flow is without viscosity, which is the best expression for the pressure at the bottom of the tank?

A. $P_{atm} + \rho g(h_3 + h_1)$
B. $P_{atm} - \rho g(h_3 + h_1)$
C. $P_{atm} + \rho g(h_3 - h_1)$
D. $P_{atm} - \rho g(h_3 - h_1)$
E. $P_{atm} \times \rho g(h_3 - h_1)$

28. Which statement(s) is/are correct?

I. Current results in a voltage
II. Current flows through a circuit
III. Voltage flows through a circuit

A. I only
B. II only
C. III only
D. I and II only
E. I and III only

29. At what speed is a moving clock traveling if a stationary observer observes it as running at one-half its regular rate?

A. $0.402c$
B. $0.536c$
C. $0.682c$
D. $0.866c$
E. $1.090c$

30. Which of the following types of electromagnetic radiation has the highest energy per photon?

I. Microwave
II. Infrared
III. Ultraviolet

A. I only
B. II only
C. III only
D. I and II only
E. I and III only

31. This is an example of what type of nuclear reaction: $^{126}_{50}Sn \rightarrow \, ^{126}_{51}Sb$?

A. Transmutation
B. Gamma particle
C. Fusion
D. Fission
E. Beta emission

32. A torque of 14 N·m is applied to a solid, uniform disk with a radius of 0.6 m. What is the mass of the disk if it accelerates at 5.3 rad/s²?

A. 7.6 kg
B. 4.2 kg
C. 14.7 kg
D. 21.4 kg
E. 13.8 kg

33. Marshall drops a water balloon from the top of a building onto Peter on the sidewalk below. Ignoring air resistance, how tall is the building if the balloon travels at 29 m/s when it strikes Peter's head? (Use the acceleration due to gravity $g = 10$ m/s² and the distance of Peter's head above the ground = 1 m)

A. 50.5 m
B. 37.5 m
C. 43 m
D. 26 m
E. 33 m

34. Assuming no change in the system's mass m, increasing the spring constant k of a spring system causes what kind of change in the resonant frequency of the system?

A. No change
B. Increase
C. Decrease only if the ratio k/m is > 1
D. Increase only if the ratio k/m is ≥ 1
E. Decrease

35. Assuming that other factors remain constant, what happens to the velocity of sound as the temperature of the air increases?

A. Does not change because it is dependent only on the state of the substance
B. Increases when atmospheric pressure is high and decreases when the pressure is low
C. Increases
D. Decreases
E. Decreases when atmospheric pressure is high and increases when the pressure is low

36. A 0.1 m cube consists of six aluminum plates insulated from each other. Plates A and D are opposite and maintained at 750 V. Plates B and E are opposite and maintained at 0 V. Plates C and F are opposite and maintained at –750 V. A force moves a charge of 10^{-14} C at constant speed straight across from the center of plate A to the center of plate D. What is the total work done by this force? (Use the elementary charge = 1.6×10^{-19} C)

A. 7.5×10^{-12} J
B. 0 J
C. 5.1×10^{-14} J
D. 7.5×10^{-14} J
E. 2.3×10^{5} J

37. A likely cause for the existence of Earth's magnetic field is:

   I. moving charges in the liquid part of Earth's core
   II. convection currents in the liquid part of Earth's core
   III. great numbers of very slow-moving charges in the Earth

A. I only
B. II only
C. III only
D. I and III only
E. I, II and III

38. How often does the polarity of the voltage reverse in a 60 Hz circuit?

A. 60 times/s
B. 120 times/s
C. 90 times/s
D. 1/60 times/s
E. 30 times/s

39. A blue object appears black when illuminated with which color of light?

A. Green
B. Yellow
C. Cyan
D. Blue
E. None of the above

40. What type of radiation is released when $^{220}_{86}Rn \rightarrow {}^{216}_{84}Po$?

   I. Gamma   II. Beta   III. Alpha

A. I only
B. II only
C. III only
D. I and II only
E. I and III only

41. According to the laws of thermodynamics:

A. entropy decreases as more energy is consumed
B. heat flows naturally from a region of lower to a region of higher temperature
C. mechanical energy cannot be converted entirely into heat
D. at a constant temperature, entropy increases as heat is extracted from a system
E. heat energy cannot be converted entirely into mechanical energy

42. What is the net force on a 1,200 kg Alfa Romeo moving at a constant speed of 3.5 m/s and turning to the left on a curve of the road with an effective radius of 4 m?

A. 1,550 N
B. 2,160 N
C. 3,675 N
D. 8,465 N
E. 5,830 N

43. A guitar has a 14 cm string and sounds a 440 Hz musical note when played without fingering. How far from the end of the string should Samantha place her fingers to play a 520 Hz note?

A. 5.8 cm
B. 0.8 cm
C. 1.6 cm
D. 2.2 cm
E. 3.4 cm

44. In Egypt, the Aswan Dam on the Nile River is 110 m high. Assuming the density of water is 1,000 kg/m^3, what is the gauge pressure of the water at the foot of the dam? (Use the acceleration due to gravity $g = 10$ m/s^2)

A. 2.1×10^5 Pa
B. 2.9×10^3 Pa
C. 1.8×10^7 Pa
D. 1.1×10^6 Pa
E. 0.8×10^2 Pa

45. Consider two current-carrying circular loops. Both are made from one strand of wire each and carry the same amount of current, but one has double the radius. Compared to the magnetic moment of the smaller loop, the magnetic moment of the larger loop is:

A. √2 times stronger
B. 4 times stronger
C. √2 times weaker
D. 2 times stronger
E. 3 times stronger

46. What is the equivalent resistance of the circuits if each has a resistance of 600 Ω?

A. 60 Ω
B. 1,200 Ω
C. 600 Ω
D. 175 Ω
E. 350 Ω

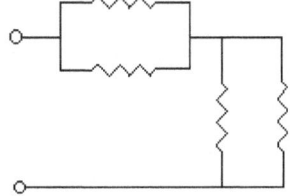

47. A simple compound microscope commonly uses a:

A. long focal length objective and a longer focal length eyepiece
B. long focal length objective and a short focal length eyepiece
C. focal length objective and focal length eyepiece of the same length
D. short focal length objective and a shorter focal length eyepiece
E. short focal length objective and a long focal length eyepiece

48. What happens to the de Broglie wavelength for a particle as it increases its velocity?

A. Increases
B. Decreases
C. Remains constant
D. Increases by $\sqrt{\Delta v}$
E. Increases by Δv^2

49. What is the longest wavelength of a photon emitted by a hydrogen atom, for which the initial state is n = 3?

A. 486 nm
B. 510 nm
C. 540 nm
D. 610 nm
E. 656 nm

50. What is the reaction force if, as a ball falls, the action force pulls the Earth's mass on the ball?

A. None present
B. The pull of the ball's mass on Earth
C. The downward acceleration due to gravity
D. The air resistance acting against the ball
E. Less than the action force

51. An irregularly-shaped object 10 m long is placed with each end on two nearby scales. If the scale on the right reads 94 N and the left reads 69 N, how far from the left is the object's center of gravity? (Use the acceleration due to gravity $g = 9.8$ m/s^2)

A. 6.8 m
B. 6.3 m
C. 7.7 m
D. 8.1 m
E. 5.8 m

52. Is it possible for a system to have negative potential energy?

A. Yes, because the choice of the zero for potential energy is arbitrary
B. No, because this has no physical meaning
C. Yes, if the kinetic energy is positive
D. Yes, if the total energy is positive
E. No, because the kinetic energy of a system must be equal to its potential energy

53. What is the speed of 2 m long water waves as they pass by a floating piece of cork that bobs up and down for one complete cycle each second?

A. 8 m/s
B. 0.5 m/s
C. 1 m/s
D. 2 m/s
E. 4 m/s

54. When 110 J of heat is added to a system performing 40 J of work, the total thermal energy change of the system is:

A. 2.8 J
B. 40 J
C. 70 J
D. 0 J
E. 150 J

55. The image of a real object from a plane mirror has the following characteristics:

   A. real, erect, with magnification = 1
   B. real, inverted, with magnification = 1
   C. real, erect, with magnification > 1
   D. virtual, erect, with magnification = 1
   E. virtual, erect, with magnification < 1

56. What is the current through the 2 Ω resistor if the current through the 8 Ω resistor is 0.8 A?

   A. 15.2 A
   B. 18.7 A
   C. 1.5 A
   D. 6.6 A
   E. 8.8 A

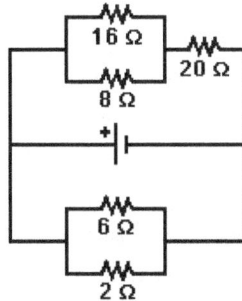

57. A proton is traveling to the right and encounters region Y contains an electric field where the proton speeds up. In what direction does the electric field in region Y point?

   A. To the left
   B. To the right
   C. Down into the page
   D. Up from the page
   E. To the left and into the page

58. An over-taut violin string was tuned with a tuning fork that produced an accurate pitch of 340 Hz. What is the period of vibration of the violin string if a beat frequency of 4 Hz is produced when the string and the fork are sounded together?

   A. 1/336 sec
   B. 1/321 sec
   C. 1/340 sec
   D. 1/327 sec
   E. 1/344 sec

59. Simple harmonic motion (SMH) is characterized by an acceleration that:

   A. is proportional to the displacement
   B. is proportional to the velocity
   C. decreases linearly
   D. is inversely proportional to the displacement
   E. is inversely proportional to the velocity

60. A simple pendulum has a bob of mass M and a period T. If M is doubled, what is the new period?

   A. $T/\sqrt{2}$
   B. T
   C. $T\sqrt{2}$
   D. 2T
   E. T/2

61. Susan pulls on a wagon with a force of 70 N. What is the average power generated by Susan if the wagon moves a total of 45 m in 3 min?

A. 18 W
B. 27 W
C. 14 W
D. 21 W
E. 28 W

62. Shawn, with a mass of 105 kg, sits 5.5 m to the left of the center of a seesaw. Mark and John, each with a mass of 20 kg, are seated on the right side of the seesaw. If Mark sits 10 m to the right of the center, how far to the right from the center should John sit to balance the seesaw? (Use the acceleration due to gravity $g = 10$ m/s^2)

A. 5 m
B. 10 m
C. 19 m
D. 20 m
E. 25 m

63. A cylinder and a sphere are released simultaneously at the top of an inclined plane. Which reaches the bottom first if they roll down the inclined plane without slipping?

A. The one of smallest diameter
B. The one of highest mass
C. The disk
D. The sphere
E. They reach the bottom at the same time

64. How long does it take for a rock to reach the maximum height of its trajectory if a boy throws it with an initial velocity of 3.13 m/s at 30° above the horizontal? (Use the acceleration due to gravity $g = 9.8$ m/s^2)

A. 0.16 s
B. 0.28 s
C. 0.333 s
D. 0.446 s
E. 0.84 s

65. If Susan stands in front of a concave mirror, at the same distance from it as its focal length:

A. her image appears larger and upright
B. no image is formed
C. her image appears larger, but upside down
D. her image appears the same size as her and upright
E. her image appears the same size as her, but upside down

66. A hammer of mass *m* is dropped from a roof and falls a distance *h* before striking the ground. How does the maximum velocity of the hammer, just before it hits the ground, change if *h* is doubled? Assume no air resistance.

A. It is multiplied by √2
B. It is multiplied by 2
C. It is increased by 200%
D. It is multiplied by 4
E. It remains constant

67. Color always depends on what characteristics of light?

    I. frequency   II. wavelength   III. amplitude

A. I only
B. II only
C. III only
D. I and II only
E. I and III only

68. The water fountain pump recirculates water from a pool and pumps it up to a trough, where it flows along the trough and passes through a hole in the bottom of it. As the water falls back into the pool, it turns a water wheel. What aspect of this water fountain is analogous to an electric current within an electric circuit?

A. Volume flow rate
B. Height of water
C. Density of water
D. Flow velocity
E. Volume of the trough

69. A 25-year-old astronaut goes off on a long-term mission in a spacecraft travels at speeds close to that of light. The mission lasts exactly 15 years as measured on Earth. With respect to biology, at the end of the mission, the astronaut's age would be:

A. exactly 25 years
B. less than 40 years
C. exactly 40 years
D. more than 40 years
E. exactly 10 years

70. What is the shape of the line on a position *vs.* time graph for constant linear acceleration?

A. curve
B. sloped line
C. sinusoidal graph
D. horizontal line
E. vertical line

71. A steel ball A is thrown in the air with a speed of 4 m/s at an angle of 60° from the horizontal. It drops onto steel ball B, which is 1.4 times the mass of A. If ball A comes to rest after the collision and ball B bounces, what is the horizontal component of ball B's velocity?

A. 0.4 m/s
B. 0.6 m/s
C. 1.4 m/s
D. 1.8 m/s
E. 1.1 m/s

72. Which of the following increases if the brightness of a beam of light is increased without changing color?

   I. the speed of the photons
   II. the average energy of each photon
   III. the number of photons

A. I only
B. II only
C. III only
D. I and II only
E. I, II and III

73. A pump uses a piston with a 20 cm diameter moving at 3 cm/s to push a liquid through a pipe. Assuming the liquid is ideal and incompressible, what is the speed of the liquid when it enters a portion of the pipe 4 mm in diameter?

A. 5 m/s
B. 60 cm/s
C. 22 m/s
D. 38 cm/s
E. 75 m/s

74. Which statement is accurate?

A. The magnetic force on a moving charge does not change its energy
B. The magnetic force on a current-carrying wire is minimal when the wire is perpendicular to magnetic field
C. All magnetic fields originate from the North and South poles
D. By definition, a magnetic field line is a tangent to the direction of the magnetic force on a moving charge at a given point in space
E. A current-carrying loop of wire tends to line up with its plane parallel to an external magnetic field in which it is positioned

75. In β^- decay, the number of protons in the nucleus:

A. increases by 2
B. increases by 1
C. decreases by 2
D. decreases by 1
E. remains unchanged

76. If 60 g of material at 100 °C is mixed with 200 g of water at 0 °C, the final temperature is 40 °C. What is the specific heat of the material?

A. 2.2 kcal/kg·°C
B. 6.3 kcal/kg·°C
C. 0.4 kcal/kg·°C
D. 4.6 kcal/kg·°C
E. 1.6 kcal/kg·°C

77. A ball is projected horizontally with an initial speed of 5 m/s from an initial height of 50 m. Ignoring air resistance, how far has the ball traveled horizontally from its original position when it lands? (Use the acceleration due to gravity $g = 10$ m/s^2)

- A. 11 m
- B. 16 m
- C. 20 m
- D. 7 m
- E. 27 m

78. A 200 g hockey puck slides up a metal ramp inclined at a 30° angle. The coefficients of static and kinetic friction between the hockey puck and the metal ramp are $\mu_s = 0.4$ and $\mu_k = 0.3$, respectively. The initial speed of the hockey puck is 14 m/s. What vertical height does the puck reach above its starting point? (Use the acceleration due to gravity $g = 9.8$ m/s^2)

- A. 11 m
- B. 4.8 m
- C. 6.6 m
- D. 14 m
- E. 14.3 m

79. A 6.5 g bullet was fired horizontally into a 2 kg wooden block suspended on a 1.5 m string. The bullet becomes embedded in the woodblock, and immediately after, the block and the bullet move at 2 m/s. The suspended wooden block with embedded bullet swings upward by height h. How high does the block with bullet swing before it comes to rest? (Use the acceleration due to gravity $g = 9.8$ m/s^2)

- A. 5.5 cm
- B. 20 cm
- C. 12 cm
- D. 44 cm
- E. 56 cm

80. In a transition from one vibrational state to another, a molecule emits a photon of wavelength 6.5 μm. What is the energy difference between these two states? (Use the speed of light $c = 3 \times 10^8$ m/s and Planck's constant $h = 4.136 \times 10^{-15}$ eV·s)

- A. 11.1 eV
- B. 11.1 MeV
- C. 0.28 MeV
- D. 2.6 MeV
- E. 0.19 eV

81. An object starting from rest accelerates uniformly along a straight line until its final velocity is v while traveling a distance d. What would be the distance traveled if the object accelerated uniformly from rest until its final velocity was $4v$?

- A. $2d$
- B. $4d$
- C. $6d$
- D. $12d$
- E. $16d$

82. Satellite #1 has mass M, which takes time T to orbit Earth. If satellite #2 has twice the mass, how long does it take for satellite #2 to orbit Earth?

A. T/2
B. T
C. 2T
D. 4T
E. T/4

83. An engineer is studying the rate of heat loss, $\Delta Q / \Delta t$, through a sheet of insulating material as a function of the thickness of the sheet. Assuming fixed temperatures on the two faces of the sheet and steady-state heat flow, which of the graphs best represents the heat transfer rate as a function of the thickness of the insulating sheet?

A.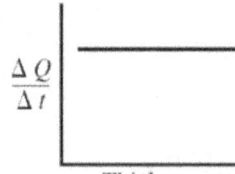

C.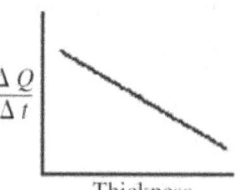

B.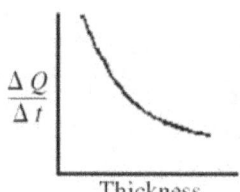

D.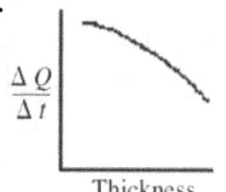

E.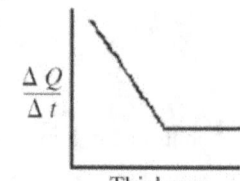

84. Two identical arrows, one with twice the kinetic energy, are fired into a hay bale. Compared to the slower arrow, the faster arrow penetrates:

A. the same distance
B. twice as far
C. four times as far
D. more than four times as far
E. less than twice as far

85. On the Moon, the acceleration of gravity is $g/6$. If a pendulum has a period T on Earth, what will be the period on the Moon?

A. 6T
B. T/6
C. T/√6
D. T/3
E. T√6

86. What is the decibel level of a sound with an intensity of 10^{-7} W/m^2?

   A. 10 dB
   B. 20 dB
   C. 30 dB
   D. 50 dB
   E. 70 dB

87. What would be the apparent mass of a 2 in × 4 in × 6 in lead brick if it were placed in oil? (Use acceleration due to gravity $g = 9.8$ m/s^2, density of oil $\rho = 0.92$ g/cm^3, density of lead $\rho = 11.4$ g/cm^3 and conversion of 1 in^3 = 16.4 cm^3)

   A. 8.2 kg
   B. 6 kg
   C. 0.3 kg
   D. 1.8 kg
   E. 3.4 kg

88. A positive charge Q is held fixed at the origin. A positive charge z is let go from point p on the positive x-axis. Ignoring friction, which statement describes the velocity of z after it is released?

   A. Increases indefinitely
   B. Decrease to zero
   C. Increases, then decreases, but never reaches zero
   D. Increases, but never exceeds a specific limit
   E. Increases, then decrease forever to zero

89. A 9 V battery is connected to two resistors in a series. One resistance is 5 ohms, and the other is 10 ohms. Which is true about the current for the locations (A, B, C, D) marked along the circuit?

   A. Current at A > current at B > current at C > current at D
   B. Current at A > current at B = current at C = current at D
   C. Current at A = current at B = current at C = current at D
   D. Current at A = current at B = current at C > current at D
   E. Current at A = current at B > current at C = current at D

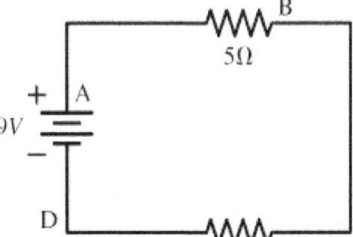

90. Which expression describes the critical angle for the interface of water with air? (Use the index of refraction for water n = 1.33 and the index of refraction for air n = 1)

   A. $\sin^{-1}(1/3)$
   B. $\sin^{-1}(3/4)$
   C. $\sin^{-1}(2/3)$
   D. $\sin^{-1}(4/3)$
   E. $\sin^{-1}(3/2)$

91. What is the amount of energy required to ionize a hydrogen atom from the ground state? (Use Rydberg formula where $E_0 = -13.6$ eV)

- **A.** 4.1 eV
- **B.** 9.8 eV
- **C.** 13.6 eV
- **D.** 22.3 eV
- **E.** ∞

92. Sonja is sitting on the outer edge of a carousel 18 m in diameter. What is the velocity of Sonja in m/s if the carousel makes 5 rev/min?

- **A.** 3.3 m/s
- **B.** 0.8 m/s
- **C.** 8.8 m/s
- **D.** 4.7 m/s
- **E.** 3.2 m/s

93. A projectile weighing 120 N is traveling horizontally to the surface of the Earth at a constant velocity of 6 m/s. Ignoring air resistance, what is the power required to maintain this motion?

- **A.** 0 W
- **B.** 20 W
- **C.** 120 W
- **D.** 2 W
- **E.** 12 W

94. A 0.4 kg mass is attached to a massless spring. The mass oscillates and has a total energy of 10 J. What is the oscillation frequency if the oscillation amplitude is 20 cm? (Use 1 J = 1 N·m)

- **A.** 3 Hz
- **B.** 4.3 Hz
- **C.** 2.1 Hz
- **D.** 5.6 Hz
- **E.** 9.9 Hz

95. Electromagnetic waves consist of:

- **A.** particles of heat energy
- **B.** high-frequency gravitational waves
- **C.** compressions and rarefactions of electromagnetic pulses
- **D.** low-frequency gravitational waves
- **E.** oscillating electric and magnetic fields

96. Jack is breathing through a snorkel as he swims in the Caribbean Sea. He has trouble breathing when his chest is submerged about 1 meter underwater. Which expression gives the force that his muscles must exert to expand his chest?

- **A.** (atmospheric pressure) × (area of his chest)
- **B.** (atmospheric pressure) × (area of snorkel hole + area of his chest)
- **C.** (gauge pressure of the water) × (area of his chest)
- **D.** (gauge pressure of the water) × (area of snorkel hole)
- **E.** (gauge pressure of the water) × (area of his chest + area of snorkel hole)

97. Two solenoids are close to each other with the switch S open. In which direction does the induced current flow through the galvanometer in the left-hand solenoid when the switch is closed?

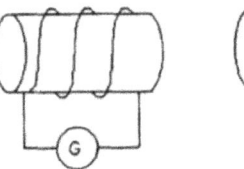
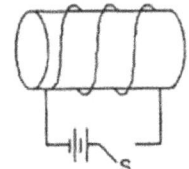

A. From left to right
B. From right to left
C. There will be no induced current through the galvanometer
D. It depends on the amount of the induced current
E. Closing the switch does not affect the left-hand solenoid because they are independent

98. If the length and the cross-sectional diameter of a wire are doubled, the resistance is:

A. halved
B. increased fourfold
C. doubled
D. decreased by one-fourth
E. unchanged

99. Using a mirror with a focal length of 10 m, an object is viewed at various distances. What is its magnification and orientation when the object is 5 m in front of the mirror?

A. Twice as large and upright
B. Twice as large and inverted
C. Half as large and upright
D. Same size and inverted
E. Same size and upright

100. Which energy source provides most of a person's annual exposure to radiation?

A. Cell phones and hand-held electronic devices
B. Televisions (i.e., cathode ray tubes)
C. Background radiation
D. Sunlight and UV rays
E. Dental and medical X rays

Notes for active learning

Diagnostic Test 3

This Diagnostic Test is designed to assess your proficiency on each topic and NOT to mimic the test. Use your test results and identify areas of strength and weakness to adjust your study plan and enhance your fundamental knowledge. The length of the Diagnostic Tests is optimal for a single study session.

#	Answer:					Review	#	Answer:					Review
1:	A	B	C	D	E	___	26:	A	B	C	D	E	___
2:	A	B	C	D	E	___	27:	A	B	C	D	E	___
3:	A	B	C	D	E	___	28:	A	B	C	D	E	___
4:	A	B	C	D	E	___	29:	A	B	C	D	E	___
5:	A	B	C	D	E	___	30:	A	B	C	D	E	___
6:	A	B	C	D	E	___	31:	A	B	C	D	E	___
7:	A	B	C	D	E	___	32:	A	B	C	D	E	___
8:	A	B	C	D	E	___	33:	A	B	C	D	E	___
9:	A	B	C	D	E	___	34:	A	B	C	D	E	___
10:	A	B	C	D	E	___	35:	A	B	C	D	E	___
11:	A	B	C	D	E	___	36:	A	B	C	D	E	___
12:	A	B	C	D	E	___	37:	A	B	C	D	E	___
13:	A	B	C	D	E	___	38:	A	B	C	D	E	___
14:	A	B	C	D	E	___	39:	A	B	C	D	E	___
15:	A	B	C	D	E	___	40:	A	B	C	D	E	___
16:	A	B	C	D	E	___	41:	A	B	C	D	E	___
17:	A	B	C	D	E	___	42:	A	B	C	D	E	___
18:	A	B	C	D	E	___	43:	A	B	C	D	E	___
19:	A	B	C	D	E	___	44:	A	B	C	D	E	___
20:	A	B	C	D	E	___	45:	A	B	C	D	E	___
21:	A	B	C	D	E	___	46:	A	B	C	D	E	___
22:	A	B	C	D	E	___	47:	A	B	C	D	E	___
23:	A	B	C	D	E	___	48:	A	B	C	D	E	___
24:	A	B	C	D	E	___	49:	A	B	C	D	E	___
25:	A	B	C	D	E	___	50:	A	B	C	D	E	___

Any duplication (copies, uploads, PDFs) is illegal.

#	Answer:					Review	#	Answer:					Review
51:	A	B	C	D	E	___	76:	A	B	C	D	E	___
52:	A	B	C	D	E	___	77:	A	B	C	D	E	___
53:	A	B	C	D	E	___	78:	A	B	C	D	E	___
54:	A	B	C	D	E	___	79:	A	B	C	D	E	___
55:	A	B	C	D	E	___	80:	A	B	C	D	E	___
56:	A	B	C	D	E	___	81:	A	B	C	D	E	___
57:	A	B	C	D	E	___	82:	A	B	C	D	E	___
58:	A	B	C	D	E	___	83:	A	B	C	D	E	___
59:	A	B	C	D	E	___	84:	A	B	C	D	E	___
60:	A	B	C	D	E	___	85:	A	B	C	D	E	___
61:	A	B	C	D	E	___	86:	A	B	C	D	E	___
62:	A	B	C	D	E	___	87:	A	B	C	D	E	___
63:	A	B	C	D	E	___	88:	A	B	C	D	E	___
64:	A	B	C	D	E	___	89:	A	B	C	D	E	___
65:	A	B	C	D	E	___	90:	A	B	C	D	E	___
66:	A	B	C	D	E	___	91:	A	B	C	D	E	___
67:	A	B	C	D	E	___	92:	A	B	C	D	E	___
68:	A	B	C	D	E	___	93:	A	B	C	D	E	___
69:	A	B	C	D	E	___	94:	A	B	C	D	E	___
70:	A	B	C	D	E	___	95:	A	B	C	D	E	___
71:	A	B	C	D	E	___	96:	A	B	C	D	E	___
72:	A	B	C	D	E	___	97:	A	B	C	D	E	___
73:	A	B	C	D	E	___	98:	A	B	C	D	E	___
74:	A	B	C	D	E	___	99:	A	B	C	D	E	___
75:	A	B	C	D	E	___	100:	A	B	C	D	E	___

1. A charged parallel-plate capacitor has an electric field E_0 between its plates. A stationary proton and an electron are between the plates. Ignoring the force of gravity, how does the magnitude of the acceleration of the proton a_p compare with the magnitude of the acceleration of the electron a_e? (Use the mass of an electron = 9×10^{-31} kg, the mass of a proton = 1.67×10^{-27} kg and the charge of a proton = 1.6×10^{-19} C)

   A. $a_p = (1,850)^2 a_e$
   B. $a_p = 1,850 a_e$
   C. $a_p = a_e$
   D. $a_p = (1 / 1,850) a_e$
   E. $a_p = (1 / 1,850)^2 a_e$

2. A 5.5 kg box slides down an inclined plane makes an angle of 40° with the horizontal. At what rate does the box accelerate down the slope if the coefficient of kinetic friction μ_k is 0.19? (Use the acceleration due to gravity $g = 9.8$ m/s²)

   A. 7.5 m/s²
   B. 6.4 m/s²
   C. 4.9 m/s²
   D. 5.9 m/s²
   E. 6.5 m/s²

3. Assume the sound level of a whisper is 20 dB and a shout is 90 dB. How many times greater is the intensity of a shout than a whisper, given the decibel level of a sound wave is related to the intensity I of the wave by:

   $$dB = 10 \log(I / I_0), \text{ where } I_0 = 10^{-12} \text{ W/m}^2$$

   A. Seven
   B. Seventy thousand
   C. Seventy million
   D. Seven million
   E. Ten million

4. What is the mass of a cylindrical rod with a length of 14 cm and a diameter of 2 cm that just barely floats in water? (Use the density of water $\rho = 1,000$ kg/m³)

   A. 44 g
   B. 70 g
   C. 140 g
   D. 28 g
   E. 90 g

5. What is the angular speed of a flywheel turning at 813.0 rpm?

   A. 8.33 rad/s
   B. 56.23 rad/s
   C. 33.84 rad/s
   D. 85.14 rad/s
   E. 116.48 rad/s

6. Which quantity is expressed in units of $\Omega \cdot m$?

   A. Flow
   B. Capacitance
   C. Resistivity
   D. Potential
   E. Current

7. A candle is viewed through a converging lens. What is the magnification of the image when the candle is 6 m from the lens and the image is 3 m from the lens on the other side?

- **A.** Twice as large and upright
- **B.** Same size and inverted
- **C.** Half as large and upright
- **D.** Half as large and inverted
- **E.** Same size and upright

8. Uranium has an atomic number of 92 but often contains 146 or more neutrons and undergoes radioactive decay. Which statement describes why this occurs?

   I. The electromagnetic repulsion overcomes the strong nuclear force
   II. Excess neutrons increase the electromagnetic repulsion
   III. The strong nuclear force has a limited range

- **A.** I only
- **B.** II only
- **C.** III only
- **D.** I and III only
- **E.** I, II and III

9. The captain of a yacht intends to travel due north. He checks his navigation gear and discovers that due to ocean currents, the yacht is traveling NE at a constant 10.7 m/s. To correct the yacht's bearing, the captain turns the vessel to point north-west and accelerates. How long does it take for the yacht to correct the bearing and achieve a due north bearing, given that the engine delivers a constant acceleration of 4.4 m/s²?

- **A.** 1.3 s
- **B.** 1.8 s
- **C.** 3.3 s
- **D.** 3.6 s
- **E.** 2.4 s

10. A 100 kg lion sees an antelope and, from rest, accelerates uniformly to 20 m/s in 10 s. How much distance does the lion cover in 10 s?

- **A.** 100 m
- **B.** 200 m
- **C.** 180 m
- **D.** 50 m
- **E.** 150 m

11. An object is released from rest at a height h above the surface of the Earth, where h is much smaller than the radius of the Earth. The object's speed is v as it strikes the ground. Ignoring air resistance, at what height should the object be released from rest for it to strike the ground with a speed of $2v$? (Use g = the acceleration due to gravity)

- **A.** $4gh$
- **B.** $4h$
- **C.** $2gh$
- **D.** $2h$
- **E.** h

12. A 4 kg ball is attached to one end of a 1.4 m light rod, while the other end is loosely affixed at a frictionless pivot. The rod is raised until it is vertical, with the ball above the pivot. The ball moves in a circle when the rod is released. What is the tension in the rod as the ball moves through the bottom of the circle? (Use the acceleration due to gravity $g = 9.8$ m/s^2)

A. 30.0 N
B. 84.8 N
C. 46.6 N
D. 120.0 N
E. 196.0 N

13. The crests of ocean waves pass a pier every 12 s. What is the wavelength of the ocean waves if the waves are moving at 4.5 m/s?

A. 84 m
B. 66 m
C. 54 m
D. 38 m
E. 47 m

14. Compared to a giant iceberg, a hot cup of coffee has:

A. a higher temperature, but more thermal energy
B. a greater specific heat and more thermal energy
C. a higher temperature, but less thermal energy
D. more thermal energy and lower temperature
E. a higher temperature and the same amount of thermal energy

15. What is the focal length of the mirror if, when an object is 24 cm in front of a concave spherical mirror, the image is formed 3 cm in front of the mirror?

A. 1.5 cm
B. 2.7 cm
C. 5 cm
D. 6.3 cm
E. 7.4 cm

16. Which of the following correctly balances the decay reaction when $^{230}_{90}\text{Th} \rightarrow ^{0}_{-1}\text{e} + \underline{}$?

A. $^{230}_{91}\text{Pa}$
B. $^{233}_{89}\text{Ac}$
C. $^{230}_{89}\text{Ac}$
D. $^{230}_{91}\text{Th}$
E. $^{231}_{90}\text{Th}$

17. By what magnitude does a magnetic field produced by a wire decrease when the distance from a long current-carrying wire is doubled?

A. $1/\sqrt{2}$
B. ½
C. ¼
D. 1/6
E. $\sqrt{(½)}$

18. A pipe with a circular cross-section has water flowing from point I to point II. The pipe radius is 6 cm at point I, while the radius at point II is 3 cm. At the end of point I, the flow rate is 0.04 m³/s. What is the velocity of the water at point I?

A. 3.5 m/s
B. 22 m/s
C. 18.5 m/s
D. 6 m/s
E. 9 m/s

19. If the intensity of sound increases by a factor of 100, the decibel level increases by:

A. 20
B. 1,000
C. $\log_{100}$
D. 100
E. 0.1

20. What is the tension on an aluminum wire with a diameter of 4.4 mm and a density of 2,600 kg/m³ when transverse waves propagate at 42 m/s?

A. 15 N
B. 24 N
C. 46 N
D. 70 N
E. 76 N

21. A 5 kg box of books slides 10 m down a ramp inclined at 30° from the horizontal. What is the work done by gravity if the box slides at a constant velocity of 4 m/s? (Use the acceleration due to gravity $g = 9.8$ m/s²)

A. 0 J
B. −32 J
C. 245 J
D. 32 J
E. 133 J

22. What is the average momentum of a 65 kg runner who travels 400 m in 50 s?

A. 19 kg·m/s
B. 63 kg·m/s
C. 520 kg·m/s
D. 112 kg·m/s
E. 386 kg·m/s

23. Michael is moving at a speed $2/3c$ toward Susan when she shines a light toward Michael. At what speed does Michael see the light approaching him?

A. $1/3c$

B. $3/5c$

C. $4/3c$

D. c

E. $2/3c$

24. An object is traveling uniformly at a v of 5 m/s. What is its final velocity if it experiences a uniform acceleration of 2 m/s² for 6 s?

A. 12 m/s

B. 28 m/s

C. 24 m/s

D. 32 m/s

E. 17 m/s

25. A 830 g meteor impacts the Earth at a speed of 1,250 m/s. If its kinetic energy is entirely converted to the heat of the meteorite, by what temperature does it increase? (Use the specific heat for the meteor = 108 cal/kg·°C and the conversion of 1 cal = 4.186 Joules)

A. 1,728 °C

B. 1,346 °C

C. 2,628 °C

D. 7,142 °C

E. 4,286 °C

26. Mary and Brittany throw identical balls vertically upward. Mary throws her ball with an initial speed of twice Brittany's ball. The maximum height of Mary's ball is:

A. higher than Brittany's ball, but less than two times as high

B. equal to the maximum height of Brittany's ball

C. two times higher than the maximum height of Brittany's ball

D. four times higher than the maximum height of Brittany's ball

E. higher than twice Brittany's ball, but less than four times as high

27. An organ pipe is a cylindrical tube open at both ends. The air column is set to vibrate by air flowing through the pipe. The length of the pipe is 0.2 m, and the diameter is 0.04 m. What is the wavelength of the fundamental? (Use v of sound at 23 °C = 340 m/s)

A. 0.1 m

B. 2 m

C. 1 m

D. 0.4 m

E. 0.8 m

28. A proton, moving in a uniform magnetic field, moves in a circle perpendicular to the field. If the proton's speed is tripled, what happens to the time needed to complete a circular path?

- **A.** Increases
- **B.** Remains constant
- **C.** Decreases
- **D.** Doubles
- **E.** Triples

29. A sodium emission tube produces a light of frequency 4.9×10^{14} Hz. Which is true of the image if it is 6 m from a converging lens of focal length 2 m?

- **A.** Inverted and virtual
- **B.** Inverted and real
- **C.** Upright and virtual
- **D.** Upright and real
- **E.** Same as if it were 1 m away

30. How much heat must be added to a 10 kg block of ice at –8 °C to change it to water at 14 °C? (Use the specific heat of ice = 0.5 kcal/kg·°C, the latent heat of fusion L_f = 80 kcal/kg and the specific heat of water = 1 kcal/kg·°C)

- **A.** 840 kcal
- **B.** 280 kcal
- **C.** 440 kcal
- **D.** 980 kcal
- **E.** 744 kcal

31. A 6 kg ball collides head-on with a stationary 8 kg ball. Which statement is true if the collision between the balls is inelastic? Assume no external forces are acting on the balls.

- **A.** Δp that the 6 kg ball experiences is greater than the Δp of the 8 kg ball
- **B.** Δv that the 6 kg ball experiences is equal to the Δv of the 8 kg ball
- **C.** Δv that the 6 kg ball experiences is greater than the Δv of the 8 kg ball
- **D.** Δv that the 6 kg ball experiences is less than the Δv of the 8 kg ball
- **E.** Δp that the 6 kg ball experiences is less than the Δp of the 8 kg ball

32. A pendulum of length L is suspended from the ceiling of an elevator. When the elevator is at rest, the period of the pendulum is T. How does the period of the pendulum change when the elevator moves upward with constant acceleration?

- **A.** Remains the same
- **B.** Decreases
- **C.** Increases
- **D.** Decreases only if the upward acceleration is less than $g / 2$
- **E.** Increases only if the upward acceleration is greater than $g / 2$

33. A small aircraft is traveling at a constant speed in a circular path with a radius of 200 m parallel to the ground. The center of the circular path is 400 m above an air traffic control tower. Its engine is the source of audible sound waves of a fixed frequency. To a stationary observer in the tower, how would the detected frequency (f_d) of the engine differ from the source (f_s) while the aircraft circled above?

A. Remains constant and equal to f_s
B. Higher than f_s during one orbit and lower during each subsequent orbit
C. Higher than f_s during half the orbit and lower during each subsequent half-orbit
D. Remains constant but is lower than f_s
E. Remains constant but is higher than f_s

34. Which of the following is an accurate statement?

A. The magnetic force of a current-carrying wire is smallest when the wire is perpendicular to magnetic field
B. The magnetic force on a moving charge does not change its energy
C. A magnetic field line is, by definition, tangent to the direction of the magnetic force on a moving charge at a given point in space
D. All magnetic fields have North and South poles as their sources
E. A current-carrying loop of wire tends to line up with its plane parallel to an external magnetic field in which it is positioned

35. The isotope $^{238}_{92}U$ is most likely to emit:

A. a γ ray
B. a β particle
C. an α particle
D. an α and β particle
E. a β particle and a γ ray

36. An auto mechanic needs to remove a tight-fitting pin of material X from a hole in a block made of material Y. The mechanic heats both the pin and the block to the same high temperature and removes the pin easily. What statement relates the coefficient of thermal expansion of material X to that of material Y?

A. Material Y has a negative coefficient of expansion, and material X has a positive coefficient of expansion
B. Material Y has the same coefficient of expansion as material X
C. This not possible; heating material Y shrinks the hole as the material expands with increasing temperature
D. Material Y has a higher coefficient of expansion than material X
E. Material X has a higher coefficient of expansion than material Y

37. A change in the state of motion is evidence of:

   I. a force that is wearing down
   II. an applied force that is unbalanced
   III. an increase in the total force

A. I only
B. II only
C. III only
D. I and II only
E. I and III only

38. The reason an astronaut in one of Earth's satellites feels weightless is that:

A. gravity does not affect the astronaut because there is no atmospheric pressure
B. the forces acting on the astronaut are negative
C. the astronaut is beyond the range of the Earth's gravity
D. the astronaut is in free fall
E. the astronaut's acceleration is zero

39. A child throws a ball over a fence 2 m high at an angle of 40° above the horizontal. The ball leaves her hand 1 m above the ground. How far is the child from the fence if the ball just clears the fence at the peak of its arc and experiences no significant air resistance? (Use the acceleration due to gravity $g = 9.8$ m/s²)

A. 1.6 m
B. 2.4 m
C. 3.9 m
D. 4.6 m
E. 8.8 m

40. A β⁻ decay occurs in an unstable nucleus when a neutron is converted to a:

A. beta particle by the weak force
B. positron by the weak force
C. neutron by the strong force
D. proton by the strong force
E. proton by the weak force

41. In a particular case of Compton scattering, a photon collides with a free electron and scatters backward. The wavelength after the collision is exactly double the wavelength before the collision. What is the wavelength of the incident photon? (Use the $m_{electron} = 9.11 \times 10^{-31}$ kg, $c = 3.00 \times 10^8$ m/s and $h = 6.626 \times 10^{-34}$ J·s)

A. 3.4×10^{-12} m
B. 4.8×10^{-12} m
C. 5.6×10^{-12} m
D. 6.8×10^{-12} m
E. 8.8×10^{-12} m

42. Which statement must be valid for an object moving with constant nonzero velocity?

   A. The net force on the object is zero
   B. The net force on the object is positive
   C. A constant force is being applied to the object in the direction opposite of motion
   D. A constant force is being applied to the object in the direction of motion
   E. Its acceleration is equal to its velocity

43. A solid cylindrical bar conducts heat at a rate of 30 W from a hot to a cold reservoir under steady-state conditions. What is the rate at which it conducts heat between these reservoirs if the diameter and length of the bar are doubled? Assume heat transfer is lengthwise, and the sides of the bar are perfectly insulated.

   A. 30 W
   B. 60 W
   C. 15 W
   D. 120 W
   E. 7.5 W

44. An ideal, massless spring with a spring constant of 3 N/m has a 0.9 kg mass attached to one end, and the other end is attached to a beam. If the system is initially at equilibrium and the mass is then down 18 cm below the equilibrium length and released, what is the magnitude of the net force on the mass just after its release? (Use acceleration due to gravity is $g = 10$ m/s^2)

   A. 0.54 N
   B. 0.75 N
   C. 6 N
   D. 0.35 N
   E. 3.5 N

45. A 30.0 N block is attached to the free end of an anchored spring and is allowed to slide back and forth on a frictionless table. Determine the frequency of motion if the spring constant $k = 40.0$ N/m. (Use the acceleration due to gravity $g = 9.8$ m/s^2)

   A. 0.30 Hz
   B. 0.58 Hz
   C. 2.3 Hz
   D. 3.6 Hz
   E. 5.4 Hz

46. A piano is tuned, so the frequency of the third harmonic of one string is 786.3 Hz. If the fundamental frequency of another string is 785.8 Hz, what is the beat frequency between the notes?

   A. 0 Hz
   B. 1 Hz
   C. 785.8 Hz
   D. 786.3 Hz
   E. 0.5 Hz

47. A 600 N weight sits on the small piston of a hydraulic machine. The small piston has an area of 5 cm². If the large piston has an area of 50 cm², how much force can the large piston support?

A. 200 N
B. 300 N
C. 3,000 N
D. 6,000 N
E. 10,000 N

48. A cube with 0.1 m sides is constructed of six insulated metal plates. Plates I and IV are opposite to each other and maintained at 500 V. Plates II and V are opposite to each other and are maintained at 0 V. Plates III and VI are opposite to each other and are maintained at –500 V. What is the change in potential energy, if an electron is transferred from plate I to plate III? (Use the charge of $e = 1.6 \times 10^{-19}$ C)

A. -3.2×10^{-14} J
B. -1.6×10^{-15} J
C. 3.2×10^{-12} J
D. 1.6×10^{-16} J
E. -1.6×10^{-17} J

49. What quantity does the slope of this graph represent if the graph shows the power dissipated in a resistor as a function of the resistance?

A. Maximum power transferred across the resistor
B. Square of the current across the resistor
C. Current across the resistor
D. The potential difference across the resistor
E. The voltage across the resistor

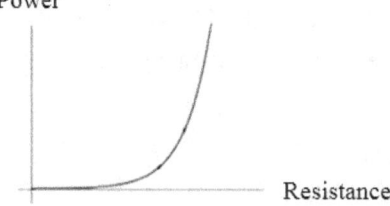

50. What color light allows the investigator to see with the highest resolution in a light microscope?

A. Red light because it is refracted less than other colors by the objective lens
B. Blue light because it has a shorter wavelength
C. Violet light because it has a longer wavelength
D. Blue light because it is brighter
E. The color of light makes no difference for relative resolution

51. A blue laser beam is incident on a metallic surface, causing electrons to be ejected from the metal. What is the effect on the rate of ejected electrons if the frequency of the laser beam is increased while the intensity of the beam is held fixed?

A. Remains the same, but the maximum kinetic energy decreases
B. Decreases and the maximum kinetic energy decreases
C. Decreases, but the maximum kinetic energy remains the same
D. Decreases, but the maximum kinetic energy increases
E. Remains the same, but the maximum kinetic energy increases

52. A 3 kg stone is dropped from a height of 5 m. Ignoring air resistance, what is its momentum on impact? (Use the acceleration due to gravity $g = 10$ m/s^2)

- A. 7.5 kg·m/s
- B. 5 kg·m/s
- C. 30 kg·m/s
- D. 45 kg·m/s
- E. 15 kg·m/s

53. A 1,500 kg car moving at 45 km/h locks its brakes and skids 30 m. How far does the same car skid if it is traveling at 150 km/h?

- A. 230 m
- B. 160 m
- C. 445 m
- D. 90 m
- E. 333 m

54. How does the frequency of vibration relate to the time it takes to complete one cycle?

- A. Inversely with the time
- B. Inversely with the amplitude
- C. Directly with the time
- D. Directly with the amplitude
- E. Directly with the wavelength

55. A 12 L volume of oil is subjected to pressure, producing a volume strain of -3×10^{-4}. The bulk modulus of the oil is 6×10^9 Pa and is independent of the pressure. What is the reduction in the volume of the oil?

- A. 1.8 ml
- B. 2.6 ml
- C. 4.4 ml
- D. 3.1 ml
- E. 3.6 ml

56. How many grams of ethanol should be added to 5 grams of chloroform for the resulting mixture to have a specific gravity of 1.2? (Use the specific gravity of ethanol = 0.8, the specific gravity of chloroform = 1.5 and the conversion factor of 1 mL = 1 g)

- A. 1.5 g
- B. 2.6 g
- C. 3.8 g
- D. 1.0 g
- E. 2.0 g

57. Four 8 V batteries (A + B + C + D) are connected in series to power lights A and B. The resistance of light A is 45 Ω and the resistance of light B is 25 Ω. What is the current through the wire at a point between battery C and D?

- A. 0.46 A
- B. 0.31 A
- C. 0.17 A
- D. 0.84 A
- E. 1.2 A

58. Which statement is correct about the equivalent resistance when four unequal resistors are in parallel?

A. It is the average of the largest and smallest resistance
B. It is less than the smallest resistance
C. It is more than the largest resistance
D. It is the average of the four resistances
E. It is ¼ the largest resistance

59. What is the distance between a lens and a screen, so when the screen and the converging lens of focal length f are arranged, an image of the Moon falls on the screen? Assume the Moon is infinity ∞ away from the lens.

A. $f/2$
B. $2f$
C. f
D. infinity
E. f^2

60. A machinist turns on the power to a grinding wheel at time $t = 0$ s. The wheel accelerates uniformly from rest for 10.0 s and reaches the operating angular speed of 96.0 rad/s. The wheel is run at that angular velocity for 40.0 s, and then power is shut off. The wheel slows down uniformly at 1.5 rad/s² until the wheel stops. For how long after the power is shut off, does it take the wheel to stop?

A. 56.0 s
B. 64.0 s
C. 72.0 s
D. 82.0 s
E. 90.0 s

61. Bombarding ^{23}Na with protons produces nuclide Y and a neutron. What is nuclide Y?

A. ^{24}Na
B. ^{21}Ne
C. ^{24}Mg
D. ^{23}Mg
E. ^{22}Na

62. Why does it take more force to start moving a heavy bookcase across the carpet than to keep it moving?

A. For objects in motion, kinetic friction is a force in the same direction as the motion
B. The coefficient of static friction is higher than the forces of movement
C. The coefficient of static friction is higher than the coefficient of kinetic friction
D. The coefficient of kinetic friction is higher than the coefficient of static friction
E. The cumulative forces acting on it are negative

63. What is the wavelength of the standing wave when a 12 m string, fixed at both ends, is resonating at a frequency that produces 4 nodes?

A. 6 m
B. 8 m
C. 4 m
D. 24 m
E. 12 m

64. A silver necklace with a mass of 60 grams and a volume of 5.7 cm3 is lowered into a container of water and tied to a string connected to a force meter. What is the reading on the force meter? (Use the density of water = 1 g/cm³ and acceleration due to gravity $g = 9.8$ m/s²)

A. 0.53 N
B. 0.22 N
C. 0.62 N
D. 0.38 N
E. Requires more information

65. A charged particle travels in a circular path of radius r in a uniform magnetic field, and the plane of the circular path is perpendicular to the magnetic field. What is the radius of the circular path if the particle travels twice as fast?

A. $\sqrt{2}r$
B. $r/2$
C. $4r$
D. $2r$
E. $r/4$

66. In which direction is the magnetic field if a positive charge moves to the right and experiences a vertical (upward) magnetic force?

A. Out of the page
B. Upward
C. To the left
D. To the right
E. Into the page

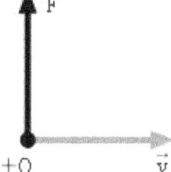

67. What is the result for temperature, measured in Kelvin, if the pressure and volume of a given sample of an ideal gas double?

A. Decreases to one-fourth original value
B. Quadruples original value
C. Decreases to one-half original value
D. Doubles original value
E. Remains the same

68. An electron initially moves to the right when it enters a uniform electric field directed upwards. Which trajectory represents the path of the electron?

A. W
B. X
C. Y
D. Z
E. More than one

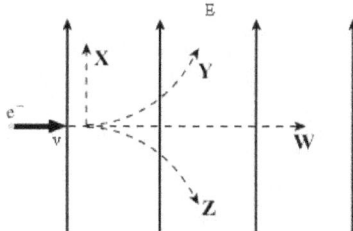

69. The graph shows the position of an object as a function of time. At which moment in time is the speed of the object equal to zero?

A. A
B. B
C. C
D. D
E. E

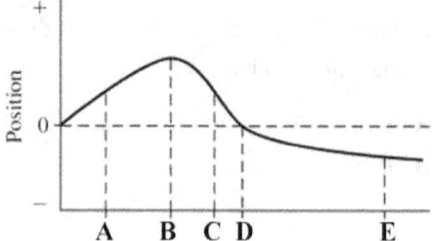

70. Michelle takes off down a 50 m high, 10° slope on her jet-powered skis. The skis have a thrust of 260 N. The combined mass of the skis and Michelle is 50 kg. Michelle's speed at the bottom of the slope is 40 m/s. Assuming the mass of the fuel is negligible, what is the coefficient of kinetic friction of her skis on the snow? (Use the acceleration due to gravity $g = 9.8$ m/s^2)

A. 0.23
B. 0.53
C. 0.68
D. 0.42
E. 0.36

71. A 1 kg chunk of putty moving at 1 m/s collides with and sticks to a 7 kg box initially at rest. What is the speed that the box and putty are then set in motion? (Assume the box rests on a frictionless surface)

A. 1/8 m/s
B. 1/6 m/s
C. 1/4 m/s
D. 1/7 m/s
E. requires more information

72. A 4 kg mass is affixed to the end of a vertical spring with a spring constant of 10 N/m. When the mass comes to rest, how much has the spring stretched?

A. 1 m
B. 4 m
C. 5 m
D. 0.1 m
E. 0.2 m

73. In music, the 3rd harmonic corresponds to which overtone?

A. 1st
B. 2nd
C. 3rd
D. 4th
E. 5th

74. A spaceship traveling at constant velocity passes by Earth and later passes by Mars. In which frame of reference is the amount of time separating these two events the proper time?

A. Any inertial frame of reference
B. The spaceship frame of reference
C. The Mars frame of reference
D. The Earth frame of reference
E. Any frame of reference, inertial or not

75. What is the orientation and magnification of the image of a light bulb if the light bulb is 2 m in front of a mirror, and the image is 6 m behind the mirror?

A. Upright and × 3
B. Inverted and × 3
C. Upright and × 0.5
D. Inverted and × 1.5
E. Upright and × 1

76. A series circuit has a 50 Hz AC source, a 0.4 H inductor, a 50 μF capacitor, and a 30 Ω resistor. If the RMS current in the circuit is 1.8 A, what is the voltage of the source?

A. 142 V
B. 124 V
C. 75.5 V
D. 96.7 V
E. 41.9 V

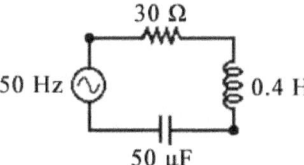

77. A fixed distance separates two particles of like charge and equal mass. What is the effect on the repulsive force between the particles if the mass of one particle is doubled?

A. Doubles
B. Quadruples
C. Increases by ½
D. Remains the same
E. Increases by √2

78. A water tank is filled to a depth of 6 m, and the bottom of the tank is 22 m above ground. A water-filled hose 2 cm in diameter extends from the bottom of the tank to the ground, but no water is flowing in the hose. What is the gauge water pressure at ground level in the hose? (Use the density of water $\rho = 1{,}000$ kg/m^3 and acceleration due to gravity $g = 9.8$ m/s^2)

A. 2.7×10^5 N/m^2
B. 5.3×10^4 N/m^2
C. 8.7 N/m^2
D. Requires the cross-sectional area of the tank
E. Requires the water pressure at the tank level

79. The tension in each of the two strings is adjusted, so each vibrates at precisely 822 Hz. The tension in one string is then increased slightly. Five beats per second are then heard when both strings vibrate. What is the new frequency of the string that was tightened?

A. 824 Hz
B. 816 Hz
C. 827 Hz
D. 818 Hz
E. 837 Hz

80. When a light ray traveling in glass strikes an air boundary, which phase change occurs in the reflected ray?

A. 45° phase change
B. 180° phase change
C. −45° phase change
D. 90° phase change
E. No phase change

81. A stone of mass m is dropped from a height h toward the ground. Ignoring air resistance, which statement is true about the stone as it hits the ground?

A. Its KE is proportional to h
B. Its KE is proportional to h^2
C. Its speed is proportional to h
D. Its speed is inversely proportional to h^2
E. Its speed is inversely proportional to h

82. Johnny is sitting on the outer edge of a carousel 18 m in diameter. What is the velocity of Johnny if the carousel makes 5.3 rev/min?

A. 4.2 m/s
B. 5 m/s
C. 3.1 m/s
D. 9.8 m/s
E. 6.2 m/s

83. A 40 kg runner is running around a track. The curved portions of the track are arcs of a circle that has a radius of 16 m. The runner is running at a constant speed of 4 m/s. What is the net force on the runner on the curved portion of the track?

A. 150 N
B. 5 N
C. 40 N
D. 100 N
E. 10 N

84. If an object is accelerating, which values must change?

    I. Speed   II. Velocity   III. Direction

A. I only
B. II only
C. III only
D. I and II only
E. I and III only

85. A 1,140 g empty iron kettle is on a hot stove. How much heat must it absorb to raise its temperature from 18 °C to 90 °C? (Use specific heat for iron = 113 cal/kg·°C and 1 cal = 4.186 J)

A. 8,230 J
B. 20,340 J
C. 38,825 J
D. 41,650 J
E. 17,300 J

86. A 1.2 kg bowling ball is dropped from a height of 6 m. During its fall, it is constantly acted upon by air resistance, with a force of 3.4 N. Accounting for air resistance, what is the speed of the bowling ball as it hits the ground? (Use acceleration due to gravity $g = 10$ m/s^2)

A. 9.2 m/s
B. 10.6 m/s
C. 11.3 m/s
D. 13.4 m/s
E. 7.6 m/s

87. Sound intensity is defined as:

A. power per unit time
B. power passing through a unit of area per unit time
C. energy passing through a unit of volume per unit time
D. energy passing through a unit of area
E. energy passing through a unit of area per unit time

88. A kilowatt-hour is a unit of:

A. work
B. current
C. power
D. charge
E. force

89. The index of refraction of the core of a piece of fiber optic cable is 1.6. If the index of the surrounding cladding is 1.3, what is the critical angle for total internal reflection of a light ray in the core incident on the core-cladding interface?

A. 82°
B. 40°
C. 34°
D. 54°
E. 69°

90. Consider the group of charges in the figure. All three charges have $Q = 6.2$ nC. What is their electric potential energy? (Use Coulomb's constant $k = 9 \times 10^9$ Nm²/C²).

A. 5.2×10^{-5} J
B. 5.9×10^{-5} J
C. 1.9×10^{-5} J
D. 6.1×10^{-5} J
E. 6.4×10^{-5} J

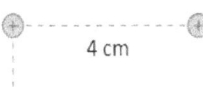

91. A solid cylinder with an 80 cm radius is positioned on a frictionless plane at 30° above the horizontal. A string wrapped around the cylinder exerts a force (F). The center of mass of the cylinder does not move when F has a specific critical value. What is the angular acceleration of the spool when F is at this critical value? (The moment of inertia of a solid cylinder is $I = \frac{1}{2}mr^2$; use acceleration due to gravity $g = 10$ m/s²)

A. 15.8 rad/s²
B. 23 rad/s²
C. 12.5 rad/s²
D. 18.6 rad/s²
E. 8.2 rad/s²

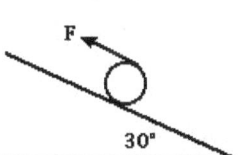

92. As a water wave passes, a floating leaf oscillates up and down completely for two cycles in 1 s. What is the wave's speed if the wave's wavelength is 12 m?

A. 1 m/s
B. 10 m/s
C. 24 m/s
D. 6 m/s
E. 12 m/s

93. When a dam began to leak, Mike placed his finger in the hole to stop the flow. The dam is 20 m high and 100 km long and sits on top of a lake, another 980 m deep, 100 km wide, and 100 km long. The hole that Mike blocked is a square 0.01 m by 0.01 m located 1 m below the surface of the water. Assuming that the viscosity of the water is negligible, what force does Mike have to exert to prevent water from leaking? (Use atmospheric pressure $P_{atm} = 10^5$ Pa, density of water $\rho = 10^3$ kg/m³ and the acceleration due to gravity $g = 10$ m/s²)

A. 10 N
B. 100 N
C. 1,000 N
D. 0.1 N
E. 1 N

94. For a graph of potential *vs.* power, what does the slope represent for a DC circuit?

A. 1 / resistance
B. current
C. 1 / current
D. resistance
E. resistivity

95. What is the amount of energy required to ionize a hydrogen atom from the ground state? (Use Rydberg formula where $E_0 = -13.6$ eV)

A. 4.1 eV
B. 9.8 eV
C. 13.6 eV
D. 22.3 eV
E. ∞

96. How much heat is needed to melt a 70 kg sample of ice at 0 °C? (Use the latent heat of fusion for water $L_f = 334,000$ J/kg and heat of vaporization for water $L_v = 2.3 \times 10^6$ J/kg)

A. 1.3×10^5 kJ
B. 5.7×10^4 kJ
C. 4.0×10^6 kJ
D. 2.3×10^4 kJ
E. 1.4×10^3 kJ

97. A car and a truck are initially alongside each other at time $t = 0$. The velocity *vs.* time graph represents their motions along a straight road. At time T, which statement is true for the vehicles?

A. The car traveled farther than the truck
B. The truck traveled farther than the car
C. They traveled the same distance
D. The truck had a greater acceleration than the car
E. The car's velocity remained constant

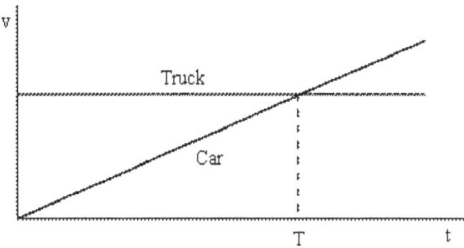

98. Two sources emit beams of microwaves. The microwaves from source A have a frequency of 15 GHz, and the microwaves from source B have a frequency of 30 GHz. This is all the information available for the two beams. Which of the following statements about these microwave beams must be correct?

A. The intensity of beam B is twice as high as the intensity of beam A
B. A photon in beam B has the same energy as a photon in beam A
C. Beam B carries twice as many photons per second as beam A
D. A photon in beam B has twice the energy of a photon in beam A
E. None of the above statements is true

99. A motor can provide a maximum of 120 N·m of torque. If all this torque is used to accelerate a solid, uniform flywheel of mass 12 kg and radius 4 m, what is the time necessary for the flywheel to accelerate from rest to 7.35 rad/s?

A. 4.4 s
B. 3.9 s
C. 1.1 s
D. 1.9 s
E. 5.9 s

100. The image shows three beams of radiation passing between two electrically-charged plates. Which of the beams is due to a high-energy electron?

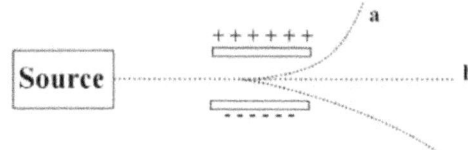

    I. a   II. b   III. C

A. I only
B. II only
C. III only
D. I and II only
E. I and III only

Notes for active learning

Answer Keys
and
Detailed Explanations

Topical Practice Questions

Answer Keys

Kinematics and Dynamics

1: D	11: C	21: E	31: B	41: C	51: A
2: B	12: B	22: A	32: C	42: A	52: A
3: B	13: A	23: B	33: D	43: E	53: E
4: A	14: C	24: D	34: E	44: B	54: D
5: C	15: B	25: B	35: C	45: A	55: A
6: E	16: D	26: E	36: E	46: D	
7: D	17: E	27: C	37: D	47: E	
8: B	18: B	28: D	38: B	48: C	
9: A	19: D	29: D	39: E	49: D	
10: B	20: C	30: E	40: E	50: E	

Force, Motion, Gravitation

1: B	11: A	21: B	31: C	41: A	51: C
2: E	12: E	22: A	32: E	42: C	52: B
3: A	13: B	23: C	33: D	43: B	53: D
4: C	14: C	24: E	34: A	44: C	54: E
5: E	15: E	25: D	35: D	45: A	55: B
6: D	16: A	26: A	36: A	46: C	
7: A	17: D	27: E	37: D	47: E	
8: E	18: C	28: D	38: B	48: E	
9: A	19: E	29: A	39: B	49: C	
10: C	20: B	30: D	40: C	50: A	

Equilibrium and Momentum

1: A	11: E	21: D	31: C	41: B	51: D
2: D	12: C	22: B	32: E	42: A	52: D
3: E	13: D	23: A	33: C	43: C	53: A
4: C	14: B	24: E	34: B	44: B	54: E
5: E	15: E	25: D	35: E	45: D	55: B
6: B	16: D	26: A	36: C	46: C	
7: D	17: B	27: E	37: B	47: B	
8: A	18: E	28: C	38: C	48: A	
9: B	19: B	29: A	39: E	49: D	
10: C	20: C	30: B	40: A	50: C	

Rotational Motion

1: D	11: A	21: B	31: D	41: C
2: B	12: E	22: D	32: D	42: D
3: A	13: C	23: B	33: B	43: C
4: D	14: B	24: A	34: C	44: C
5: C	15: D	25: A	35: D	45: B
6: C	16: D	26: A	36: B	46: C
7: E	17: E	27: A	37: C	
8: A	18: A	28: C	38: B	
9: C	19: D	29: C	39: B	
10: B	20: D	30: C	40: D	

Work and Energy

1: D	11: D	21: B	31: B	41: A	51: D
2: B	12: B	22: D	32: C	42: A	52: C
3: A	13: A	23: B	33: D	43: C	53: C
4: B	14: C	24: C	34: E	44: E	54: B
5: E	15: B	25: E	35: B	45: D	55: C
6: A	16: D	26: A	36: B	46: E	
7: D	17: E	27: D	37: C	47: A	
8: C	18: A	28: E	38: D	48: A	
9: A	19: D	29: B	39: E	49: E	
10: B	20: A	30: C	40: B	50: B	

Waves and Periodic Motion

1: B	11: D	21: B	31: B	41: A	51: A
2: D	12: A	22: E	32: D	42: C	52: C
3: D	13: B	23: B	33: E	43: B	53: B
4: C	14: C	24: A	34: D	44: E	54: E
5: A	15: E	25: E	35: C	45: A	55: D
6: E	16: B	26: C	36: D	46: B	
7: C	17: C	27: D	37: B	47: A	
8: D	18: E	28: B	38: C	48: C	
9: A	19: A	29: E	39: D	49: D	
10: E	20: D	30: A	40: C	50: A	

Answer Keys: Topical Practice Questions

Sound

1: B	11: B	21: E	31: C	41: A	51: D
2: A	12: A	22: D	32: D	42: C	52: B
3: B	13: E	23: B	33: A	43: D	53: D
4: C	14: C	24: C	34: E	44: A	54: C
5: B	15: D	25: E	35: A	45: B	55: E
6: E	16: E	26: D	36: E	46: E	
7: D	17: C	27: B	37: B	47: C	
8: A	18: B	28: A	38: D	48: A	
9: C	19: A	29: E	39: E	49: C	
10: D	20: B	30: A	40: C	50: E	

Light and Optics

1: A	11: D	21: B	31: C	41: B	51: D
2: A	12: A	22: E	32: B	42: E	52: A
3: B	13: D	23: C	33: B	43: C	53: B
4: D	14: E	24: E	34: A	44: B	54: E
5: E	15: A	25: C	35: E	45: E	55: C
6: B	16: E	26: C	36: B	46: D	
7: C	17: C	27: A	37: C	47: A	
8: D	18: B	28: E	38: A	48: C	
9: A	19: D	29: A	39: E	49: E	
10: D	20: B	30: E	40: B	50: D	

Notes for active learning

Kinematics and Dynamics – Detailed Explanations

1. **D is correct.**

    $t = (v_f - v_i) / a$

    $t = (60 \text{ mi/h} - 0 \text{ mi/h}) / (13.1 \text{ mi/h} \cdot \text{s})$

    $t = 4.6 \text{ s}$

Acceleration is in mi/h·s, so miles and hours cancel, and the answer is in units of seconds.

2. **B is correct.**

At the top of the parabolic trajectory, the vertical velocity $v_{yf} = 0$

The initial upward velocity is the vertical component of the initial velocity:

   $v_{yi} = v \sin \theta$

   $v_{yi} = (20 \text{ m/s}) \sin 30°$

   $v_{yi} = (20 \text{ m/s}) \cdot (0.5)$

   $v_{yi} = 10 \text{ m/s}$

   $t = (v_{yf} - v_{yi}) / a$

   $t = (0 - 10 \text{ m/s}) / (-10 \text{ m/s}^2)$

   $t = (-10 \text{ m/s}) / (-10 \text{ m/s}^2)$

   $t = 1 \text{ s}$

3. **B is correct.**

   $\Delta d = 31.5 \text{ km} = 31,500 \text{ m}$

   $1.25 \text{ hr} \times 60 \text{ min/hr} = 75 \text{ min}$

   $\Delta t = 75 \text{ min} \times 60 \text{ s/min} = 4,500 \text{ s}$

   $v_{avg} = \Delta d / \Delta t$

   $v_{avg} = 31,500 \text{ m} / 4,500 \text{ s}$

   $v_{avg} = 7 \text{ m/s}$

4. **A is correct.**

Instantaneous speed is the scalar magnitude of the velocity.

Instantaneous speed can only be positive or zero (because magnitudes cannot be negative).

5. C is correct.

$$d = (v_f^2 - v_i^2) / 2a$$

$$d = [(21 \text{ m/s})^2 - (5 \text{ m/s})^2] / [2(3 \text{ m/s}^2)]$$

$$d = (441 \text{ m}^2/\text{s}^2 - 25 \text{ m}^2/\text{s}^2) / 6 \text{ m/s}^2$$

$$d = (416 \text{ m}^2/\text{s}^2) / 6 \text{ m/s}^2$$

$$d = 69 \text{ m}$$

6. E is correct.

$$a = (v_f - v_i) / t$$

$$a = [0 - (-30 \text{ m/s})] / 0.15 \text{ s}$$

$$a = (30 \text{ m/s}) / 0.15 \text{ s}$$

$$a = 200 \text{ m/s}^2$$

To represent the acceleration in terms of g, divide a by 9.8 m/s²:

$$\text{\# of } g = (200 \text{ m/s}^2) / 9.8 \text{ m/s}^2$$

$$\text{\# of } g = 20 \text{ } g$$

The initial velocity (v_i) is negative due to the acceleration of the car being a positive value.

Since the car is decelerating, its acceleration is opposite of its initial velocity.

7. D is correct.

When a bullet is fired, it is in projectile motion.

The only force in projectile motion (if air resistance is ignored) is the force of gravity.

8. B is correct.

When a car is slowing down, it is decelerating, equivalent to acceleration in the opposite direction.

9. A is correct

Uniform acceleration:

$$a = \text{change in velocity} / \text{change in time}$$

$$a = \Delta v / \Delta t$$

$$\Delta v = a \Delta t$$

$$\Delta v = (20 \text{ m/s}^2) \cdot (1 \text{ s})$$

$$\Delta v = 20 \text{ m/s}$$

10. B is correct.

Uniform acceleration:

a = change in velocity / change in time

$a = \Delta v / \Delta t$

$a = (40 \text{ m/s} - 15 \text{ m/s}) / 10 \text{ s}$

$a = (25 \text{ m/s}) / 10 \text{ s}$

$a = 2.5 \text{ m/s}^2$

11. C is correct.

$t = d / v$

$t = (540 \text{ mi}) / (65 \text{ mi/h})$

$t = 8.3 \text{ h}$

The time she can stop is the difference between her total allowed time and the time t it takes to make the trip:

$t_{\text{stop}} = 9.8 \text{ h} - 8.3 \text{ h}$

$t_{\text{stop}} = 1.5 \text{ h}$

12. B is correct.

Average velocity is the change in position with respect to time:

$v = \Delta x / \Delta t$

After one lap, the racecar's final position is the same as its initial position.

Thus,

$x = 0$, which implies the average velocity of 0 m/s.

13. A is correct.

$d = v_i \Delta t + \tfrac{1}{2} a \Delta t^2$

$d = (0.2 \text{ m/s}) \cdot (5 \text{ s}) + \tfrac{1}{2}(-0.05 \text{ m/s}^2) \cdot (5 \text{ s})^2$

$d = 1 \text{ m} + \tfrac{1}{2}(-0.05 \text{ m/s}^2) \cdot (25 \text{ s}^2)$

$d = 1 \text{ m} + (-0.625 \text{ m})$

$d = 0.375 \text{ m} \approx 0.38 \text{ m}$

Decelerating is set to negative.

The net displacement is the difference between the final and initial positions after 5 s.

14. C is correct.

Uniform acceleration:

a = change in velocity / change in time

$a = \Delta v / \Delta t$

15. B is correct.

Convert the final speed from km/h to m/s:

$v_f = (210 \text{ km/h}) \times [(1{,}000 \text{ m}/1 \text{ km})] \times [(1 \text{ h}/3{,}600 \text{ s})]$

$v_f = 58.33$ m/s

Calculate the acceleration necessary to reach this speed:

$a = (v_f^2 - v_i^2) / 2d$

$a = [(58.33 \text{ m/s})^2 - (0 \text{ m/s})^2] / 2(1{,}800 \text{ m})$

$a = (3{,}402.39 \text{ m}^2/\text{s}^2) / (3{,}600 \text{ m})$

$a = 0.95 \text{ m/s}^2$

16. D is correct.

The distance the rocket travels during its acceleration upward is calculated by:

$d_1 = \tfrac{1}{2} a t^2$

$d_1 = \tfrac{1}{2}(22 \text{ m/s}^2) \cdot (4 \text{ s})^2$

$d_1 = 176$ m

The distance from when the motor shuts off to when the rocket reaches maximum height can be calculated using the conservation of energy:

$mgd_2 = \tfrac{1}{2} m v^2$

cancel m from both sides of the expression

$gd_2 = \tfrac{1}{2} v^2$

where $v = at$

$gd_2 = \tfrac{1}{2}(at)^2$

$d_2 = \tfrac{1}{2}(at)^2 / g$

$d_2 = \tfrac{1}{2}[(22 \text{ m/s}^2) \cdot (4 \text{ s})]^2 / (10 \text{ m/s}^2)$

Magnitudes are not vectors but scalars, so no direction is needed

$d_2 = 387$ m

continued…

For the maximum elevation, add the two distances:

$h = d_1 + d_2$

$h = 176 \text{ m} + 387 \text{ m}$

$h = 563 \text{ m}$

17. E is correct.

Speed is a scalar (i.e., one-dimensional physical property), while velocity is a vector (i.e., has both magnitude and direction).

18. B is correct.

Acceleration due to gravity is constant and independent of mass.

19. D is correct.

As an object falls, its acceleration is constant due to gravity.

However, the magnitude of the velocity increases due to the acceleration of gravity, and the displacement increases because the object is going further from its starting point.

20. C is correct.

The man moves at constant velocity (no acceleration), so the net force is zero.

The only objects interacting with the man directly are Earth and the floor of the elevator.

The cable does not touch the man; it pulls the elevator car up, and the elevator floor pushes on the man.

21. E is correct.

Horizontal velocity (v_x):

$v_x = d_x / t$

$v_x = (44 \text{ m}) / (2.9 \text{ s})$

$v_x = 15.2 \text{ m/s}$

The x component of a vector is calculated by:

$v_x = v \cos \theta$

Rearrange the equation to determine the initial velocity of the ball:

$v = v_x / \cos \theta$

$v = (15.2 \text{ m/s}) / (\cos 45°)$

$v = (15.2 \text{ m/s}) / 0.7$

$v = 21.4 \text{ m/s}$

22. A is correct.

Conservation of energy:

$$mgh = \tfrac{1}{2}mv_f^2$$

cancel m from both sides of the expression

$$gh = \tfrac{1}{2}v_f^2$$

$$(10 \text{ m/s}^2)h = \tfrac{1}{2}(14 \text{ m/s})^2$$

$$(10 \text{ m/s}^2)h = \tfrac{1}{2}(196 \text{ m}^2/\text{s}^2)$$

$$h = (98 \text{ m}^2/\text{s}^2) / (10 \text{ m/s}^2)$$

$$h = 9.8 \text{ m} \approx 10 \text{ m}$$

23. B is correct.

$$d = v_i t + \tfrac{1}{2}at^2$$

$$d = (20 \text{ m/s}) \cdot (7 \text{ s}) + \tfrac{1}{2}(1.4 \text{ m/s}^2) \cdot (7 \text{ s})^2$$

$$d = (140 \text{ m}) + \tfrac{1}{2}(1.4 \text{ m/s}^2) \cdot (49 \text{ s}^2)$$

$$d = 174.3 \text{ m} \approx 174 \text{ m}$$

24. D is correct.

Force is not a scalar because it has a magnitude and direction.

25. B is correct.

$$d = \tfrac{1}{2}at^2$$

$$d_A = \tfrac{1}{2}at^2$$

$$d_B = \tfrac{1}{2}a(2t)^2$$

$$d_B = \tfrac{1}{2}a(4t^2)$$

$$d_B = 4 \times \tfrac{1}{2}at^2$$

$$d_B = 4d_A$$

26. E is correct.

$$d = v_{\text{average}} \times \Delta t$$

$$d = \tfrac{1}{2}(v_i + v_f)\Delta t$$

$$d = \tfrac{1}{2}(5 \text{ m/s} + 30 \text{ m/s}) \cdot (10 \text{ s})$$

$$d = 175 \text{ m}$$

27. C is correct.

If there is no acceleration, then velocity is constant.

28. D is correct.

The gravitational force between two objects in space, each having masses of m_1 and m_2, is:

$F_G = Gm_1m_2 / r^2$

where G is the gravitational constant and r is the distance between the two objects

Doubling the distance between the two objects:

$F_{G2} = Gm_1m_2 / (2r)^2$

$F_{G2} = Gm_1m_2 / (4r^2)$

$F_{G2} = ¼Gm_1m_2 / r^2$

$F_{G2} = ¼Gm_1m_2 / r^2$

$F_{G2} = ¼F_G$

Therefore, when the distance between the objects is doubled, the force (F_G) is one-fourth as much.

29. D is correct.

I: If the velocity is constant, the instantaneous velocity equals the average velocity.

II and III: If the velocity increases, the average value of velocity over an interval must lie between the initial velocity and the final velocity.

In going from its initial value to its final value, the instantaneous velocity must cross the average value at one point, regardless of whether the velocity is changing at a constant rate or changing irregularly.

30. E is correct.

velocity = acceleration × time

$v = at$

$v = (10 \text{ m/s}^2) \cdot (10 \text{ s})$

$v = 100 \text{ m/s}$

31. B is correct.

velocity = distance / time

$v = d / t$

d is constant, while t decreases by a factor of 3

32. C is correct.

The equation for distance, given a constant acceleration and the initial and final velocity, is:

$$d = (v_i^2 + v_f^2) / 2a$$

Since the car is coming to rest,

$$v_f = 0$$

$$d = v_i^2 / 2a$$

If the initial velocity is doubled while acceleration and final velocity remain unchanged, the new distance traveled is:

$$d_2 = (2v_i)^2 / 2a$$

$$d_2 = 4(v_i^2 / 2a)$$

$$d_2 = 4d_1$$

Another method to solve this problem:

$$d_1 = (29 \text{ mi/h})^2 / 2a$$

$$d_2 = (59 \text{ mi/h})^2 / 2a$$

$$d_2 / d_1 = [(59 \text{ mi/h})^2 / 2a] / [(29 \text{ mi/h})^2 / 2a]$$

$$d_2 / d_1 = (59 \text{ mi/h})^2 / (29 \text{ mi/h})^2$$

$$d_2 / d_1 = (3{,}481 \text{ mi/h}) / (841 \text{ mi/h})$$

$$d_2 / d_1 = 4$$

33. D is correct.

$$\text{speed}_{average} = \text{total distance} / \text{time}$$

$$\text{speed} = (400 \text{ m}) / (20 \text{ s}) = 20 \text{ m/s}$$

If this were velocity, it would be 0.

34. E is correct.

$$\Delta v = a\Delta t$$

$$(v_f - v_i) = a\Delta t$$

where $v_f = 0$ m/s (when the car stops)

$a = -0.1$ m/s² (negative because deceleration), $\Delta t = 5$ s

$$v_i = v_f - a\Delta t$$

$$v_i = [(0 \text{ m/s}) - (-0.1 \text{ m/s}^2)] \cdot (5 \text{ s})$$

$$v_i = (0.1 \text{ m/s}^2) \cdot (5 \text{ s}) = 0.5 \text{ m/s}$$

35. C is correct.

If acceleration is constant, then the velocity *vs.* time graph is linear, and the average velocity is the average of the final and initial velocity.

$$v_{average} = v_f - v_i / \Delta t$$

If acceleration is not constant, then the velocity *vs.* time graph is nonlinear.

$$v_{average} \neq v_f - v_i / \Delta t$$

36. E is correct.

Find velocity of thrown rock:

$$v_{f1}^2 - v_i^2 = 2ad$$

$$v_{f1}^2 = v_i^2 + 2ad$$

$$v_{f1}^2 = (10 \text{ m/s})^2 + [2(9.8 \text{ m/s}^2) \cdot (300 \text{ m})]$$

$$v_{f1}^2 = 100 \text{ m}^2/\text{s}^2 + 5,880 \text{ m}^2/\text{s}^2$$

$$v_{f1}^2 = 5,980 \text{ m}^2/\text{s}^2$$

$$v_{f1} = 77.33 \text{ m/s}$$

$$t_1 = (v_f - v_i) / a$$

$$t_1 = (77.33 \text{ m/s} - 10 \text{ m/s}) / 9.8 \text{ m/s}^2$$

$$t_1 = (67.33 \text{ m/s}) / (9.8 \text{ m/s}^2)$$

$$t_1 = 6.87 \text{ s}$$

Find velocity of dropped rock:

$$v_{f2} = \sqrt{2ad}$$

$$v_{f2} = \sqrt{[(2) \cdot (9.8 \text{ m/s}^2) \cdot (300 \text{ m})]}$$

$$v_{f2} = 76.7 \text{ m/s}$$

$$t_2 = (76.7 \text{ m/s}) / (9.8 \text{ m/s}^2)$$

$$t_2 = 7.82 \text{ s}$$

$$\Delta t = (7.82 \text{ s} - 6.87 \text{ s})$$

$$\Delta t = 0.95 \text{ s}$$

37. D is correct.

$$F = ma$$

Force and acceleration are directly proportional, so doubling force doubles acceleration.

38. B is correct.

Velocity is defined as having speed and direction.

If either, or both, of these changes, then the object is experiencing acceleration.

39. E is correct.

The acceleration is negative because it acts to slow the car down against the $+y$ direction.

It is unclear if the acceleration decreases in magnitude from the data provided.

40. E is correct.

Total distance is represented by the area under the velocity-time curve with respect to the x-axis.

This graph can be broken up into sections; calculate the area under the curve.

$d_{total} = d_A + d_B + d_C + d_D$

$d_A = ½(4 \text{ m/s})·(2 \text{ s}) = 4 \text{ m}$

$d_B = ½(4 \text{ m/s} + 2 \text{ m/s})·(2 \text{ s}) = 6 \text{ m}$

$d_C = (2 \text{ m/s})·(4 \text{ s}) = 8 \text{ m}$

Since the total distance traveled needs to be calculated, the area under the curve when the velocity is negative is calculated as a positive value.

Distance is a scalar quantity and therefore has no direction.

$d_D = ½(2 \text{ m/s})·(1 \text{ s}) + ½(2 \text{ m/s})·(1 \text{ s}) = 2 \text{ m}$

$d_{total} = 4 \text{ m} + 6 \text{ m} + 8 \text{ m} + 2 \text{ m}$

$d_{total} = 20 \text{ m}$

If the question asked to find the displacement, the area under the curve would be calculated as negative, and the answer would be 18 m.

41. C is correct.

The two bullets have different velocities when hitting the water but only experience the force due to gravity.

Thus, the acceleration due to gravity is the same for each bullet.

42. A is correct.

$v_f = v_i + at$

$v_f = 0 + (2.5 \text{ m/s}^2)·(9 \text{ s})$

$v_f = 22.5 \text{ m/s}$

43. E is correct.

The equation for impulse is used for contact between two objects over a specified time period:

$F\Delta t = m\Delta v$

$ma\Delta t = m(v_f - v_i)$

cancel m from both sides of the expression

$a\Delta t = (v_f - v_i)$

$a = (v_f - v_i) / \Delta t$

$a = (-2v - v) / (0.45 \text{ s})$

$a = (-3v) / (0.45 \text{ s})$

$a = (-6.7 \text{ s}^{-1})v$

Ratio $a : v = -6.7 \text{ s}^{-1} : 1$

44. B is correct.

The time for the round trip is 4 s.

The weight reaches the top of its path in ½ time:

½(4 s) = 2 s

where $v = 0$

$a = \Delta v / t$ for the first half of the trip

$a = (v_f - v_i) / t$

$a = (0 - 3.2 \text{ m/s}) / 2 \text{ s}$

$a = -1.6 \text{ m/s}^2$

$|a| = 1.6 \text{ m/s}^2$

Acceleration is a vector, and the negative direction only indicates direction.

45. A is correct.

$\Delta v = a\Delta t$

$\Delta v = (0.3 \text{ m/s}^2) \cdot (3 \text{ s})$

$\Delta v = 0.9 \text{ m/s}$

46. D is correct.

Velocity, displacement, acceleration, and force are vectors with magnitude and direction.

47. E is correct.

$$d = d_0 + (v_i^2 + v_f^2) / 2a$$

$$d = 64 \text{ m} + (0 \text{ m/s} + 60 \text{ m/s})^2 / 2(9.8 \text{ m/s}^2)$$

$$d = 64 \text{ m} + (3{,}600 \text{ m}^2/\text{s}^2) / (19.6 \text{ m/s}^2)$$

$$d = 64 \text{ m} + 184 \text{ m}$$

$$d = 248 \text{ m}$$

48. C is correct.

$$a = (v_f^2 + v_i^2) / 2d$$

$$a = [(60 \text{ m/s})^2 + (0 \text{ m/s})^2] / [2(64 \text{ m})]$$

$$a = (3{,}600 \text{ m}^2/\text{s}^2) / 128 \text{ m}$$

$$a = 28 \text{ m/s}^2$$

49. D is correct.

Expression for the time interval during constant acceleration upward:

$$d = \tfrac{1}{2}at^2$$

Solving for acceleration:

$$a = (v_f^2 + v_i^2) / 2d$$

$$a = [(60 \text{ m/s})^2 + (0 \text{ m/s})^2] / [2(64 \text{ m})]$$

$$a = (3{,}600 \text{ m}^2/\text{s}^2) / (128 \text{ m})$$

$$a = 28.1 \text{ m/s}^2$$

Solving for time:

$$t^2 = 2d/a$$

$$t^2 = 2(64 \text{ m}) / 28.1 \text{ m/s}^2$$

$$t^2 = 4.5 \text{ s}^2$$

$$t = 2.1 \text{ s}$$

50. E is correct.

$$d = (v_i^2 + v_f^2) / 2a$$

where $v_i = 0$

$$d = v_f^2 / 2a$$

For half the final velocity:

$$d_2 = (v_f / 2)^2 / 2a$$
$$d_2 = \tfrac{1}{4} v_f^2 / 2a$$
$$d_2 = \tfrac{1}{4} d$$

51. A is correct.

$v_{average} = \Delta d / \Delta t$

52. A is correct.

Use an equation that relates v, d and t:

$$d = vt$$
$$v = d / t$$

If v increases by a factor of 3, then t decreases by a factor of 3.

Another method to solve this problem:

$d = vt$, t = original time and t_N = new time

$$d = 3vt_N$$
$$vt = d = 3vt_N$$
$$vt = 3vt_N$$
$$t = 3t_N$$
$$t / 3 = t_N$$

Thus, if v increases by a factor of 3, then the original time decreases by a factor of 3.

53. E is correct.

$$v_f = v_i + at$$

$$t = (v_f - v_i) / a$$

Since the ball is thrown straight up, its initial speed upward equals its final speed downward (just before hitting the ground): $v_f = -v_i$

$$t = [39 \text{ m/s} - (-39 \text{ m/s})] / 9.8 \text{ m/s}^2$$

$$t = (78 \text{ m/s}) / 9.8 \text{ m/s}^2$$

$$t = 8 \text{ s}$$

54. D is correct.

Since the speed is changing, the velocity is changing, and therefore there *is* an acceleration.

Since the speed *decreases,* the acceleration must be *in the reverse direction* (i.e., opposite to the direction of travel).

Since the particle is moving to the right, the acceleration vector points to the left.

If the speed *increases*, the acceleration is in the *same* direction as the direction of travel, and the acceleration vector points to the right.

55. A is correct.

The only force that Larry applies to the package is the normal force due to his hand (there is no horizontal force as the package moves with constant velocity.)

The normal force due to his hand points upward.

The displacement of the package is horizontal:

$$W = Fd \cos \theta$$

where θ is the angle between the force and the displacement.

$$\theta = 90°$$

Since $\cos 90° = 0$

$$W = 0 \text{ J}$$

Notes for active learning

Notes for active learning

Force, Motion, Gravitation – Detailed Explanations

1. B is correct.

The tension of the string keeps the weight traveling in a circular path; otherwise, it would move linearly on a tangent path to the circle. Without the string, there are no horizontal forces on the weight and no horizontal acceleration. The horizontal motion of the weight is in a straight line at a constant speed.

2. E is correct.

The vertical force on the garment bag from the left side of the clothesline is:

$T_{y,\text{left}} = T \cos \theta$

Similarly, for the right side:

$T_{y,\text{right}} = T \cos \theta$

where $T = 10$ N (tension) and $\theta = 60°$.

Since the garment bag is at rest, its acceleration is zero. Therefore, according to Newton's second law:

$T_{y,\text{left}} + T_{y,\text{right}} - mg = 0 = 2T (\cos \theta) - mg$

Or: $2T (\cos \theta) = mg$

$m = 2T (\cos \theta) / g$

$m = 2(10 \text{ N}) \cdot (\cos 60°) / (10.0 \text{ m/s}^2)$

$m = 2(10 \text{ N}) \cdot (0.5) / (10.0 \text{ m/s}^2)$

$m = 1$ kg

3. A is correct.

An object's inertia is its resistance to change in motion, and the milk carton has enough inertia to overcome the force of static friction.

4. C is correct.

$(F_{\text{net}})_y = (F_N)_y - (F_g)_y$

The car is not moving up or down, so $a_y = 0$:

$(F_{\text{net}})_y = 0$

$0 = (F_N)_y - (F_g)_y$

$F_N = (F_g)_y$

$F_N = F_g \cos \theta$

$F_N = mg \cos \theta$

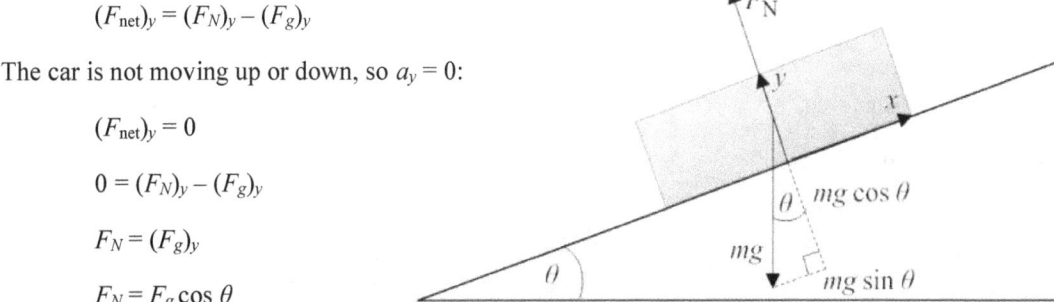

The normal force is perpendicular to the plane of contact (the slope).

5. E is correct.

$F = ma$

$F = (27 \text{ kg}) \cdot (1.7 \text{ m/s}^2)$

$F = 46 \text{ N}$

6. D is correct.

The mass on the table causes a tension force in the string that acts against the force of gravity.

7. A is correct.

Although the net force acting on the object decreases with time and the magnitude of the object's acceleration decreases, there is positive acceleration.

Therefore, the object's speed continues to increase.

8. E is correct.

An object moving at constant velocity experiences zero net force.

9. A is correct.

The sine of an angle is equal to the opposite side over the hypotenuse:

$\sin \theta = \text{opposite} / \text{hypotenuse}$

$\sin \theta = h / L$

$h = L \sin \theta$

10. C is correct.

A car accelerating horizontally does not rely on the force of gravity to move it.

Since mass does not depend on gravity, a car on Earth and a car on the Moon that experiences the same horizontal acceleration experience the same force.

11. A is correct.

$a = (v_f - v_i) / t$

$a = (3.5 \text{ m/s} - 1.5 \text{ m/s}) / (3 \text{ s})$

$a = (2 \text{ m/s}) / (3 \text{ s})$

$a = 0.67 \text{ m/s}^2$

12. E is correct.

An object with uniform circular motion (i.e., constant angular velocity) only experiences centripetal acceleration directed toward the center of the circle.

13. B is correct.

$F = ma$

So, zero force means zero acceleration in any direction.

14. C is correct.

$F = ma$

$a = F / m$

$a = 9 \text{ N} / 9 \text{ kg}$

$a = 1 \text{ m/s}^2$

15. E is correct.

The only force acting on a projectile in motion is the force due to gravity.

Since that force always acts downward, there is only a downward acceleration.

16. A is correct.

$F_{net} = ma$

If an object moves with constant v, its $a = 0$, so:

$F_{net} = 0$

Since gravity pulls down on the can with a force of mg:

$F_g = mg$

$F_g = (10 \text{ kg}) \cdot (10 \text{ m/s}^2)$

$F_g = 100 \text{ N}$

The rope pulls *up* on the can with the same magnitude of force, so the tension is 100 N for a net force = 0.

17. D is correct.

$F = ma$

$F = (1{,}000 \text{ kg}) \cdot (2 \text{ m/s}^2)$

$F = 2{,}000 \text{ N}$

18. C is correct.

$a_{cent} = v^2 / r$

$a_{cent} = (4 \text{ m/s})^2 / (4 \text{ m})$

$a_{cent} = (16 \text{ m}^2/\text{s}^2) / (4 \text{ m})$

$a_{cent} = 4 \text{ m/s}^2$

19. E is correct.

Solve for m_1:

$F_{net} = 0$

$m_2 g = F_T$

$m_1 g \sin\theta + F_f = F_T$

$m_1 g \sin\theta + \mu_s m_1 g \cos\theta = m_2 g$

cancel g from each side

$m_1(\sin\theta + \mu_s \cos\theta) = m_2$

$m_1 = m_2 / (\sin\theta + \mu_s \cos\theta)$

$m_1 = 2 \text{ kg} / [\sin 20° + (0.55) \cos 20°]$

$m_1 = 2 \text{ kg} / 0.86$

$m_1 = 2.3 \text{ kg}$

Kinetic friction is only used when the mass is in motion.

20. B is correct.

Since the masses are identical, the force of gravity on each is the same.

The force of gravity on one of the masses produces the tension force in the string, which in turn pulls on the other mass.

Since this tension force equals the force of gravity, there is no net force, and the objects remain at rest.

21. B is correct.

Newton's Third Law states that for every action, there is an equal and opposite reaction.

22. A is correct.

Newton's Third Law states that for every action, there is an equal and opposite reaction.

23. C is correct.

If w denotes the magnitude of the box's weight, then the component of this force parallel to the inclined plane is $w \sin\theta$, where θ is the incline angle.

If θ is less than 90°, then $\sin\theta$ is less than 1.

The component of w parallel to the inclined plane is less than w.

24. E is correct.

The package experiences projectile motion upon leaving the truck, so it experiences no horizontal forces, and its initial velocity of 30 m/s remains unchanged.

25. D is correct.

f = revolutions / unit of time

The time (period) for one complete revolution is:

$T = 1 / f$

Each revolution represents a length of $2\pi r$.

Velocity is the distance traveled in one revolution over duration of one revolution (circumference over period):

$v = 2\pi r / t$

$v = 2\pi r f$

If f doubles, then v doubles.

26. A is correct.

$F = ma$

$m = F / a$

$m = 4{,}500 \text{ N} / 5 \text{ m/s}^2$

$m = 900 \text{ kg}$

27. E is correct.

Newton's First Law states that every object will remain at rest or in uniform motion unless acted upon by an outside force.

In this case, Steve and the bus are in uniform constant motion until the bus stops due to sudden deceleration (the ground exerts no frictional force on Steve).

There is no force acting upon Steve.

However, his inertia carries him forward because he is still in uniform motion while the bus comes to a stop.

28. D is correct.

The ball is in a state of rest, so $F_{net} = 0$

$F_{down} = F_{up}$

$F_{external} + F_w = F_{buoyant}$

$F_{external} = F_{buoyant} - F_w$

$F_{external} = 8.4 \text{ N} - 4.4 \text{ N}$

$F_{external} = 4 \text{ N}$, in the same direction as the weight

29. A is correct.

The luggage and the train move at the same speed, so when the luggage moves forward with respect to the train, it means the train has slowed down while the luggage continues to move at its original speed.

30. D is correct.

The mass does not change by changing the object's location.

Since the object is outside of Earth's atmosphere, the object's weight is represented by:

$$F_g = GmM_{Earth} / R^2$$

If the altitude is $2R_{Earth}$, then the distance from the center of the Earth is $3R_{Earth}$.

The gravitational acceleration decreases by a factor of $3^2 = 9$ ($g = GmM / R^2$).

Weight decreases by a factor of 9.

$$\text{New weight} = 360 \text{ N} / 9 = 40 \text{ N}$$

31. C is correct.

The velocity of the rock just after its release is the same as the truck.

Once in free fall, there are no horizontal forces on the rock.

The rock's velocity remains unchanged and equal to that of the truck.

32. E is correct.

The acceleration of Jason due to thrust is:

$$F_{net} = ma_1$$

$$ma_1 = F_{ski} - \mu_k mg$$

$$a_1 = (F_{ski} - \mu_k mg) / m$$

$$a_1 = [200 \text{ N} - (0.1)\cdot(75 \text{ kg})\cdot(9.8 \text{ m/s}^2)] / 75 \text{ kg}$$

$$a_1 = (126.5 \text{ N}) / 75 \text{ kg}$$

$$a_1 = 1.69 \text{ m/s}^2$$

The distance traveled during the acceleration stage is:

$$d_1 = \tfrac{1}{2} a_1 t^2$$

$$d_1 = \tfrac{1}{2}(1.69 \text{ m/s}^2)\cdot(67 \text{ s})^2$$

$$d_1 = 3{,}793 \text{ m}$$

The distance traveled after the skis run out of fuel is:

$$d_2 = (v_f^2 - v_i^2) / 2a_2$$

continued…

a_2 is Jason's acceleration after the fuel runs out:

$$F_{net} = ma_2$$

$$ma_2 = -\mu_k mg$$

cancel m from each side of the expression

$$a_2 = -\mu_k g$$

$$a_2 = -(0.1) \cdot (9.8 \text{ m/s}^2)$$

$$a_2 = -0.98 \text{ m/s}^2$$

The acceleration is negative since the frictional force opposes the direction of motion.

v_i is the velocity at the moment when the fuel runs out:

$$v_i = a_1 t$$

$$v_i = (1.69 \text{ m/s}^2) \cdot (67 \text{ s})$$

$$v_i = 113.2 \text{ m/s}$$

Substitute a_2 and v_i into the equation for d_2:

$$d_2 = [(0 \text{ m/s})^2 - (113.2 \text{ m/s})^2] / 2(-0.98 \text{ m/s}^2)$$

$$d_2 = (-12{,}814.2 \text{ m}^2/\text{s}^2) / -1.96 \text{ m/s}^2$$

$$d_2 = 6{,}538 \text{ m}$$

The total distance Jason traveled is:

$$d_{total} = d_1 + d_2$$

$$d_{total} = 3{,}793 \text{ m} + 6{,}538 \text{ m}$$

$$d_{total} = 10{,}331 \text{ m}$$

33. D is correct.

Using force analysis:

$$F_{net} = F_g + F_{fk}$$

$$F_g = mg \sin \theta$$

$$F_g = (0.2 \text{ kg}) \cdot (-9.8 \text{ m/s}^2) \sin 30°$$

$$F_g = (0.2 \text{ kg}) \cdot (-9.8 \text{ m/s}^2) \cdot (1/2)$$

$$F_g = -1 \text{ N}$$

$$F_{fk} = \mu_k F_N$$

$$F_{fk} = \mu_k mg \cos \theta$$

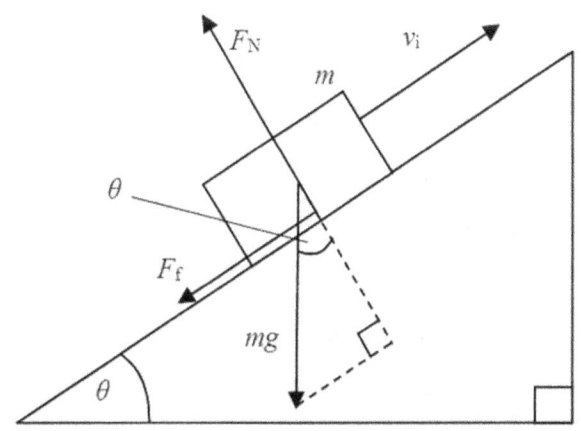

continued…

$F_{fk} = (0.3) \cdot (0.2 \text{ kg}) \cdot (-9.8 \text{ m/s}^2) \cos 30°$

$F_{fk} = (0.3) \cdot (0.2 \text{ kg}) \cdot (-9.8 \text{ m/s}^2) \cdot (0.866)$

$F_{fk} = -0.5 \text{ N}$

$F_{net} = -1 \text{ N} + (-0.5 \text{ N})$

$F_{net} = -1.5 \text{ N}$

$a = F_{net} / m$

$a = -1.5 \text{ N} / 0.2 \text{ kg}$

$a = -7.5 \text{ m/s}^2$

The distance it travels until it reaches a velocity of 0 at its maximum height:

$d = (v_f^2 - v_i^2) / 2a$

$d = [(0 \text{ m/s})^2 - (63 \text{ m/s})^2] / 2(-7.5 \text{ m/s}^2)$

$d = (-4{,}000 \text{ m}^2/\text{s}^2) / (-15 \text{ m/s}^2)$

$d = 267 \text{ m}$

The *vertical height* is:

$h = d \sin \theta$

$h = (267 \text{ m}) \sin 30°$

$h = (267 \text{ m}) \cdot (0.5)$

$h = 130 \text{ m}$

Using energy to solve the problem:

$KE = PE + W_f$

$\frac{1}{2}mv^2 = mgd \sin \theta + \mu_k mgd \cos \theta$

cancel *m* from the expression

$\frac{1}{2}v^2 = gd \sin \theta + \mu_k gd \cos \theta$

$\frac{1}{2}v^2 = d(g \sin \theta + \mu_k g \cos \theta)$

$d = v^2 / [2g(\sin \theta + \mu_k \cos \theta)]$

$d = (63 \text{ m/s})^2 / [(2) \cdot (9.8 \text{ m/s}^2) \cdot (\sin 30° + 0.3 \times \cos 30°)]$

$d = 267 \text{ m}$

$h = d \sin \theta$

$h = (267 \text{ m}) \sin 30°$

$h = 130 \text{ m}$

34. A is correct.

$F = ma$

$a = F / m$

$a_1 = F / 4 \text{ kg}$

$a_2 = F / 10 \text{ kg}$

$4a_1 = 10a_2$

$a_1 = 2.5a_2$

35. D is correct.

Mass is independent of gravity; however, weight is not; as a person moves farther from any stars or planets, the gravitational pull decreases and, therefore, her weight decreases.

36. A is correct.

Newton's Third Law describes that when one object pushes on another, the second object pushes back with the same force.

Mathematically:

$F_{\text{AonB}} = -F_{\text{BonA}}$

In this scenario, the force that the truck exerts on the car is in the opposite direction to the force that the car exerts on the truck (since they push on each other), and crucially, the *magnitudes* of the two forces are the same.

This may seem counterintuitive since it is known that the car will get far more damaged than the truck.

To understand this apparent contradiction, consider Newton's Second Law, which states that the car will *accelerate* at a much higher rate (since it is less massive than the truck). It is this extreme acceleration that causes the car to be destroyed.

Therefore, to understand this situation thoroughly, two Newton's laws must be applied:

The Third Law, which states that each vehicle experiences a force of the same magnitude, and The Second Law, which describes why the car *responds* to that force more violently due to its smaller mass.

37. D is correct.

$m = F / a_{\text{Earth}}$

$m = 20 \text{ N} / 3 \text{ m/s}^2$

$m = 6.67 \text{ kg}$

$F_{\text{Moon}} = mg_{\text{Moon}}$

$F_{\text{Moon}} = (6.67 \text{ kg}) \cdot (1.62 \text{ m/s}^2)$

$F_{\text{Moon}} \approx 11 \text{ N}$

38. B is correct.

weight = mass × gravity

$w = (0.4 \text{ kg}) \cdot (9.8 \text{ m/s}^2)$

$w \approx 4 \text{ N}$

39. B is correct.

Need an expression that connects time and mass.

Given information for F, v_1, and d:

$a = F / m$

$d = v_1 t + \tfrac{1}{2} a t^2$

Combine the expressions and set $v_i = 0$ m/s because initial velocity is zero:

$d = \tfrac{1}{2} a t^2$

$a = F / m$

$d = \tfrac{1}{2}(F / m) t^2$

$t^2 = 2dm / F$

$t = \sqrt{2dm / F}$

If m increases by a factor of 4, t increases by a factor of $\sqrt{4} = 2$

40. C is correct.

$a = (v_f^2 - v_i^2) / 2d$

$a = [(0 \text{ m/s})^2 - (27 \text{ m/s})^2] / 2(578 \text{ m})$

$a = (-729 \text{ m}^2/\text{s}^2) / 1{,}056 \text{ m}$

$a = -0.63 \text{ m/s}^2$

$F = ma$

$F = (1{,}100 \text{ kg}) \cdot (-0.63 \text{ m/s}^2)$

$F = -690 \text{ N}$

The car is decelerating, so the acceleration (and therefore the force) is negative.

41. A is correct.

Constant speed upward means no net force.

Tension = weight (equals Mg)

42. C is correct.

Weight = mg

75 N = mg

m = 75 N / 9.8 m/s^2

m = 7.65 kg

$F_{net} = F_{right} - F_{left}$

F_{net} = 50 N – 30 N

F_{net} = 20 N

$F_{net} = ma$

$a = F_{net} / m$

a = 20 N / 7.65 kg

a = 2.6 m/s^2

43. B is correct.

The string was traveling at the same velocity as the plane with respect to the ground outside.

When the plane began accelerating backward (decelerating), the string continued to move forward at its original velocity and appeared to go towards the front of the plane.

Since the string is attached to the ceiling at one end, only the bottom of the string moved.

44. C is correct.

If the object slides down the ramp with a constant speed, velocity is constant.

Acceleration and the net force = 0

$F_{net} = F_{grav\ down\ ramp} - F_{friction}$

$F_{net} = mg \sin \theta - \mu_k mg \cos \theta$

$F_{net} = 0$

$mg \sin \theta - \mu_k mg \cos \theta = 0$

$mg \sin \theta = \mu_k mg \cos \theta$

$\mu_k = \sin \theta / \cos \theta$

45. A is correct.

The net force on an object in free fall equals its weight.

46. C is correct.

$$a = \Delta v / \Delta t$$

$$a = (v_f - v_i) / t$$

$$a = (20 \text{ m/s} - 0 \text{ m/s}) / (10 \text{ s})$$

$$a = (20 \text{ m/s}) / (10 \text{ s})$$

$$a = 2 \text{ m/s}^2$$

47. E is correct.

Since the object does not move, it is in equilibrium, so forces act on it equal and oppose the force F that Yana applies to the object.

48. E is correct.

Newton's Third Law describes that when one object pushes on another, the second object pushes back with the same force.

Mathematically, it can be expressed as:

$$F_{AonB} = -F_{BonA}$$

If one pushes on an object with force F, the object must push back on them equally strongly (magnitude is F) and in the opposite direction (hence the negative sign); therefore, the force vector of the object is just $-F$.

49. C is correct.

$$F = mg$$

$$m = F / g$$

$$m = 685 \text{ N} / 9.8 \text{ m/s}^2$$

$$m = 69.9 \text{ kg} \approx 70 \text{ kg}$$

50. A is correct.

$$m_{Bob} = 4m_{Sarah}$$

Conservation of momentum, since the system (Bob and Sarah combined) initially had a total momentum of 0, in the final state, Sarah's momentum and Bob's momentum must add to 0 (i.e., they will be the same magnitude, but opposite directions):

$$m_{Bob}v_{Bob} = m_{Sarah}v_{Sarah}$$

$$4m_{Sarah} v_{Bob} = m_{Sarah}v_{Sarah}$$

$$4v_{Bob} = v_{Sarah}$$

51. C is correct.

For most surfaces, the coefficient of static friction is higher than the coefficient of kinetic friction.

Thus, the force needed to overcome static friction and start the object's motion is higher than the amount of force needed to overcome kinetic friction and keep the object moving at a constant velocity.

52. B is correct.

Weight on Jupiter:

$$W = mg$$
$$W = m(3g)$$
$$W = (100 \text{ kg}) \cdot (3 \times 10 \text{ m/s}^2)$$
$$W = 3,000 \text{ N}$$

53. D is correct.

Neither Joe nor Bill is moving, so the net force is zero:

$$F_{net} = F_{Joe} - F_T$$
$$0 = F_{Joe} - F_T$$
$$F_{Joe} = F_T$$
$$F_T = 200 \text{ N}$$

54. E is correct.

Tension in the rope is equal to F_T.

The net force on block A to the right is:

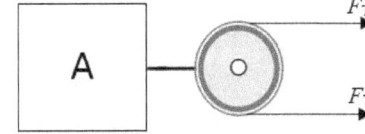

$$F_{right} = m_A a_A = 2F_T$$

The net force of block B downward is:

$$F_{down} = m_B a_B = m_B g - F_T$$

Since block A is connected to both the pulley at the end of the table and the wall, it uses twice the rope length to travel the same distance as block B.

Therefore, the distance block A moves is half that of block B, the velocity of block A is half the velocity of block B, and the acceleration of block A is half the acceleration of block B:

$$a_A = a_B / 2$$
$$F_{right} = m_A (a_B / 2)$$
$$m_A (a_B / 2) = 2F_T$$
$$m_A a_B = 4F_T$$

continued…

$F_T = ¼m_A a_B$

$m_B a_B = m_B g - ¼m_A a_B$

$m_B a_B + ¼m_A a_B = m_B g$

$a_B [m_B + ¼m_A] = m_B g$

$a_B = m_B g / [m_B + ¼m_A]$

$a_B = (5 \text{ kg}) \cdot (9.8 \text{ m/s}^2) / [5 \text{ kg} + ¼(4 \text{ kg})]$

$a_B = 49 \text{ N} / 6 \text{ kg}$

$a_B = 8.2 \text{ m/s}^2$

$a_A = a_B / 2$

$a_A = (8.2 \text{ m/s}^2) / 2$

$a_A = 4.1 \text{ m/s}^2$

55. B is correct.

Force exerted by one surface on another has a perpendicular component (i.e., normal force) and a parallel component (i.e., friction force).

The force of kinetic friction on an object acts opposite to the direction of its velocity relative to the surface.

Notes for active learning

Notes for active learning

Equilibrium and Momentum – Detailed Explanations

1. A is correct.

The rate of *change of angular momentum* of a system equals the net external torque:

$\tau_{net} = \Delta L / \Delta t$

If the angular momentum is constant, then the net external torque must be zero.

2. D is correct.

If the velocity is 7 m/s down the mountain, the *horizontal component* v_x is:

$v_x = v \cos \theta$

1.8 m/s = (7 m/s) cos θ

cos θ = 0.26

$\theta \approx 75°$

3. E is correct.

The hill exerts a normal force on the sled. However, this force is *perpendicular* to the surface of the hill.

There is no parallel force that the hill exerts because it is frictionless.

4. C is correct.

Assume water flows tangent to the wheel; it is perpendicular to the radius vector at the point of contact.

The torque around the center of the wheel is:

$\tau = rF$

τ = (10 m)·(300 N)

τ = 3,000 N·m

5. E is correct.

    1 revolution = 360°, 1 min = 60 s

    33 rpm = 33 revs/min

    (33 revs/min)·(360°/rev) = 11,880°/min

    (11,880°/min)·(1 min/60 s) = 198°/s

Degrees per second is a *rate*:

    rate × time = total degrees

    (198°/s)·(0.32 s) ≈ 63°

6. B is correct.

> momentum = mass × velocity
>
> $p = mv$

Since momentum is directly proportional to mass, doubling the mass doubles the momentum.

7. D is correct.

The total momentum before the collision is:

> $p_{total} = m_I v_I + m_{II} v_{II} + m_{III} v_{III}$
>
> $p_{before} = (1 \text{ kg}) \cdot (0.5 \text{ m/s}) + (1.5 \text{ kg}) \cdot (-0.3 \text{ m/s}) + (3.5 \text{ kg}) \cdot (-0.5 \text{ m/s})$
>
> $p_{before} = (0.5 \text{ kg·m/s}) + (-0.45 \text{ kg·m/s}) + (-1.75 \text{ kg·m/s})$
>
> $p_{before} = -1.7 \text{ kg·m/s}$

8. A is correct.

The collision of I and II does not affect the momentum of the system:

> $p_{before} = p_{after}$
>
> $p_{I \& II} = (1 \text{ kg}) \cdot (0.5 \text{ m/s}) + (1.5 \text{ kg}) \cdot (-0.3 \text{ m/s})$
>
> $p_{I \& II} = (0.5 \text{ kg·m/s}) - (0.45 \text{ kg·m/s})$
>
> $p_{I \& II} = 0.05 \text{ kg·m/s}$
>
> $p_{III} = (3.5 \text{ kg}) \cdot (-0.5 \text{ m/s})$
>
> $p_{III} = -1.75 \text{ kg·m/s}$
>
> $p_{net} = p_{I \text{ and } II} + p_{III}$
>
> $p_{net} = (0.05 \text{ kg·m/s}) + (-1.75 \text{ kg·m/s})$
>
> $p_{net} = -1.7 \text{ kg·m/s}$

Momentum is always conserved.

9. B is correct.

Set the initial momentum equal to the final momentum after the collisions have occurred.

> $p_{before} = p_{after}$
>
> $p_{before} = (m_I + m_{II} + m_{III}) v_f$
>
> $-1.7 \text{ kg·m/s} = (1 \text{ kg} + 1.5 \text{ kg} + 3.5 \text{ kg}) v_f$
>
> $v_f = (-1.7 \text{ kg·m/s}) / (6 \text{ kg})$
>
> $v_f = -0.28 \text{ m/s}$

10. C is correct.

Momentum is conserved in this system.

The momentum of each car is given by *mv*, and the sum of the momenta before the collision must equal the sum of the momenta after the collision:

$p_{before} = p_{after}$

Solve for the velocity of the first car after the collision.

Each car travels in the same direction before and after the collision, so each velocity value has the same sign.

$m_1 v_{i1} + m_2 v_{i2} = m_1 v_{f1} + m_2 v_{f2}$

(480 kg)·(14.4 m/s) + (570 kg)·(13.3 m/s) = (480 kg)·(v_{f2}) + (570 kg)·(17.9 m/s)

(480 kg)·(v_{f2}) = (480 kg)·(14.4 m/s) + (570 kg)·(13.3 m/s) − (570 kg)·(17.9 m/s)

v_{f2} = [(480 kg)·(14.4 m/s) + (570 kg)·(13.3 m/s) − (570 kg)·(17.9 m/s)] / (480 kg)

v_{f2} = 8.9 m/s ≈ 9 m/s

11. E is correct.

Impulse is a force acting over some time:

$J = F \Delta t$

An impulse changes a system's momentum, so:

$F \Delta t = \Delta p_{system}$

The moving block with the lodged bullet stops when it compresses the spring, losing all momentum.

The initial velocity of the block and bullet separately can be determined by the conservation of energy.

The two values of interest are the KE of the block and bullet and the PE of the spring.

$(KE + PE)_{before} = (KE + PE)_{after}$

½mv^2 + 0 = 0 + ½kx^2

x = distance of compression of the spring

k = spring constant

½(4 kg + 0.008 kg)v^2 = ½(1,400 N/m)·(0.089 m)^2

v^2 = (1,400 N/m)·(0.089 m)^2 / (4.008 kg)

v^2 = 2.76 m^2/s^2

v = 1.66 m/s

Thus, the block with the lodged bullet hits the spring with an initial velocity of 1.66 m/s.

Since there is no friction, the block is sent in the opposite direction with the same speed of 1.66 m/s when the spring decompresses.

continued…

Calculate the momentum, with initial momentum toward the spring and final momentum away from the spring.

$\Delta p = p_{final} - p_{initial}$

$\Delta p = (4.008 \text{ kg}) \cdot (-1.66 \text{ m/s}) - (4.008 \text{ kg}) \cdot (1.66 \text{ m/s})$

$\Delta p = (-6.65 \text{ kg·m/s}) - (6.65 \text{ kg·m/s})$

$\Delta p \approx -13 \text{ kg·m/s}$

$\Delta p \approx -13 \text{ N·s}$

Since $F\Delta t = \Delta p$, the impulse is $-13 \text{ kg·m/s} = -13 \text{ N·s}$

The negative sign signifies the coordinate system chosen in this calculation: toward the spring is the positive direction, and away from the spring is the negative direction.

12. C is correct.

For a rotating body, kinetic energy is:

$K = \frac{1}{2} I \omega^2$

Angular momentum is:

$L = I \omega$

Therefore:

$I = L / \omega$

Replacing this for I in the expression for kinetic energy:

$K = L^2 / 2I$

Taking the ice to be frictionless, there is no external torque on the skater.

Thus, angular momentum is conserved and does not change as she brings in her arms.

The moment of inertia of a body of a given mass is smaller if its mass is more concentrated toward the rotation axis (e.g., when she draws her arms in close).

Therefore, the moment of inertia of the skater decreases. Consequently, the skater's kinetic energy increases.

13. D is correct.

The centripetal force is the net force required to maintain an object in a uniform circular motion.

$F_{centripetal} = mv^2/r$

where r is the radius of the circular path

Since m is constant and r remains unchanged, the centripetal force is proportional to v^2.

$2^2 = 4$

Thus, if v is doubled, then $F_{centripetal}$ is quadrupled.

14. B is correct.

$1 \text{ J} = \text{kg} \cdot \text{m}^2/\text{s}^2$

$p = mv = \text{kg} \cdot \text{m/s}$

$\text{J} \cdot \text{s/m} = (\text{kg} \cdot \text{m}^2/\text{s}^2) \cdot (\text{s/m})$

$\text{J} \cdot \text{s/m} = \text{kg} \cdot \text{m/s}$

$\text{kg} \cdot \text{m/s} = p$

$\text{J} \cdot \text{s/m} = p$

15. E is correct.

Impulse is a change in momentum.

$J = \Delta p$

$J = m\Delta v$

Impulse is also the product of average force and time.

$J = F\Delta t$

$F\Delta t = m\Delta v$

$ma\Delta t = m\Delta v$

cancel m from each side of the expression

$a\Delta t = \Delta v$

Because acceleration g is constant, impulse depends only upon time and velocity.

The speed of the apple affects the impulse, as this is included in the Δv term.

Bouncing results in a change in direction, a greater change in velocity (the Δv term), so the impulse is greater.

The time of impulse changes the impulse as it is included in the Δt term.

16. D is correct.

$F\Delta t = m\Delta v$

$F = m\Delta v / \Delta t$

Choosing toward the wall as positive direction, the initial velocity is 25 m/s, and the final velocity is –25 m/s:

$F = m(v_\text{f} - v_\text{i}) / \Delta t$

$F = (0.8 \text{ kg}) \cdot (-25 \text{ m/s} - 25 \text{ m/s}) / (0.05 \text{ s})$

$F = -800 \text{ N}$

Thus, the wall exerts an average force of 800 N on the ball in the negative direction.

From Newton's Third Law, the ball exerts a force of 800 N on the wall in the opposite direction.

17. B is correct.

Momentum:

$$p = mv$$

Sum momentum:

$$p_{total} = m_1v_1 + m_2v_2 + m_3v_3$$

All objects moving to the left have negative velocity.

$$p_{total} = (7 \text{ kg}) \cdot (6 \text{ m/s}) + (12 \text{ kg}) \cdot (3 \text{ m/s}) + (4 \text{ kg}) \cdot (-2 \text{ m/s})$$

$$p_{total} = (42 \text{ kg·m/s}) + (36 \text{ kg·m/s}) + (-8 \text{ kg·m/s})$$

$$p_{total} = 70 \text{ kg·m/s}$$

18. E is correct.

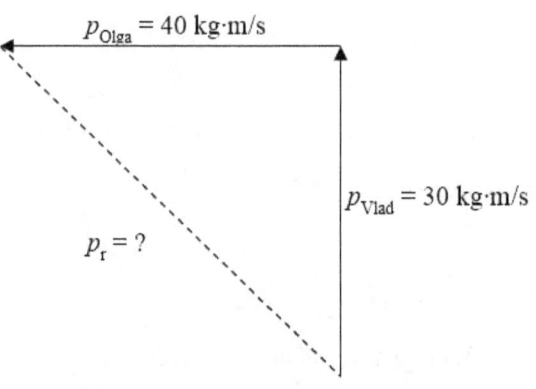

Use conservation of momentum to determine the momentum after the collision. Since they stick, treat it as a perfectly inelastic collision.

Before the collision, Vladimir's momentum is: $(60 \text{ kg}) \cdot (0.5 \text{ m/s}) = 30 \text{ kg·m/s}$ pointing North

Before the collision, Olga's momentum is:

$$(40 \text{ kg}) \cdot (1 \text{ m/s}) = 40 \text{ kg·m/s} \text{ pointing West}$$

Write two expressions: one for the conservation of momentum on the y-axis (North-South) and one for the conservation of momentum on the x-axis (East-West). They do not interact since they are perpendicular.

Since Olga and Vladimir stick, the final mass is the sum of their masses.

Use the Pythagorean Theorem:

$$a^2 + b^2 = c^2$$

$$(30 \text{ kg·m/s})^2 + (40 \text{ kg·m/s})^2 = p^2$$

$$900 \text{ (kg·m/s)}^2 + 1{,}600 \text{ (kg·m/s)}^2 = p^2$$

$$2{,}500 \text{(kg·m/s)}^2 = p^2$$

$$p = 50 \text{ kg·m/s}$$

Solve for velocity:

$$p = mv$$

$$50 \text{ kg·m/s} = (100 \text{ kg})v$$

$$v = 50 \text{ kg·m/s} / 100 \text{ kg}$$

$$v = 0.5 \text{ m/s}$$

continued…

This problem can be solved *algebraically*:

Conservation of meomentum:

$$p_{before} = p_{after}$$

$$p = mv$$

On the *y*-coordinate:

$$(60 \text{ kg}) \cdot (0.5 \text{ m/s}) = (60 \text{ kg} + 40 \text{ kg})v_y$$

$$v_y = (30 \text{ kg} \cdot \text{m/s}) / (100 \text{ kg})$$

$$v_y = 0.3 \text{ m/s}$$

On the *x*-coordinate:

$$(40 \text{ kg}) \cdot (1 \text{ m/s}) = (60 \text{ kg} + 40 \text{ kg})v_x$$

$$v_x = (40 \text{ kg} \cdot \text{m/s}) / (100 \text{ kg})$$

$$v_x = 0.4 \text{ m/s}$$

Combine these final velocity components using the Pythagorean Theorem since they are perpendicular.

$$v^2 = v_x^2 + v_y^2$$

$$v^2 = (0.4 \text{ m/s})^2 + (0.3 \text{ m/s})^2$$

$$v = 0.5 \text{ m/s}$$

19. B is correct.

Use conservation of momentum to determine the momentum after the collision.

Since they stick, treat it as a perfectly inelastic collision.

Before collision, Vladimir's momentum is $(60 \text{ kg}) \cdot (0.5 \text{ m/s}) = 30 \text{ kg} \cdot \text{m/s}$ pointing North

Before collision, Olga's momentum is $(40 \text{ kg}) \cdot (1 \text{ m/s}) = 40 \text{ kg} \cdot \text{m/s}$ pointing West

Write two expressions: one for the conservation of momentum on the *y* coordinate (North-South) and one for the conservation of momentum on the *x* coordinate (East-West).

They do not interact since they are perpendicular to each other. Since they stick, the final mass is the sum of their masses.

Simply use the Pythagorean Theorem:

$$a^2 + b^2 = c^2$$

$$(30 \text{ kg} \cdot \text{m/s})^2 + (40 \text{ kg} \cdot \text{m/s})^2 = p^2$$

$$900(\text{kg} \cdot \text{m/s})^2 + 1{,}600(\text{kg} \cdot \text{m/s})^2 = p^2$$

$$2{,}500(\text{kg} \cdot \text{m/s})^2 = p^2$$

$$p = 50 \text{ kg} \cdot \text{m/s}$$

continued...

This problem can be solved *algebraically*:

$p_{before} = p_{after}$

$p = mv$

On the *y*-coordinate:

(60 kg)·(0.5 m/s) = (60 kg + 40 kg)v_y

v_y = (30 kg·m/s) / (100 kg)

v_y = 0.3 m/s

On the *x*-coordinate:

(40 kg)·(1 m/s) = (60 kg + 40 kg)v_x

v_x = (40 kg·m/s) / (100 kg)

v_x = 0.4 m/s

Combine these final velocity components using the Pythagorean Theorem since they are perpendicular:

$v^2 = v_x^2 + v_y^2$

$v^2 = (0.4 \text{ m/s})^2 + (0.3 \text{ m/s})^2$

v = 0.5 m/s

Use the final weight and final velocity to find the final momentum directly after the collision:

$p = mv$

p = (60 kg + 40 kg)·(0.5 m/s)

p = 50 kg·m/s

20. C is correct.

$p_0 = mv$

If *m* and *v* are doubled:

$p = (2m) \cdot (2v)$

$p = 4mv$

$p = 4p_0$

The momentum increases by a factor of 4.

21. D is correct.

Balance forces on box Q to solve for tension on box P cable:

$$m_Q a = F - T_P$$

$$T_P = F - m_Q a$$

$$0 < T_P < F$$

Thus, the tension on the cable connected to box P is less than F because it equals the difference between F and $m_Q a$ but is not equal because the boxes are accelerating.

22. B is correct.

At all points on a rotating body, the *angular velocity* is equal.

The speed at different points along a rotating body is directly proportional to the radius.

$$v = \omega r$$

where v = speed, ω = angular velocity and r = radius

Thus, Melissa and her friend have different speeds due to their different radial locations.

23. A is correct.

Impulse is directly proportional to force and change in time:

$$J = F\Delta t$$

Increasing the change in time lowers the impact force while decreasing the change in time increases the force.

24. E is correct.

Angular momentum is conserved unless a system experiences a net torque greater than zero; the rotational equivalent of Newton's First Law of motion.

25. D is correct.

$$F\Delta t = m\Delta v$$

$$F = (m\Delta v) / (\Delta t)$$

$$F = (6.8 \text{ kg}) \cdot (-3.2 \text{ m/s} - 5.4 \text{ m/s}) / (2 \text{ s})$$

$$F = (-58.48 \text{ kg·m/s}) / (2 \text{ s})$$

$$F = -29.2 \text{ N}$$

$$|F| = 29.2 \text{ N}$$

26. A is correct.

Before the collision, the total momentum of the system = 0 kg·m/s.

Momentum is conserved in the explosion.

The momentum of the moving rifle and bullet are in opposite directions:

Therefore,

$p = 0$

The total momentum after the explosion = 0 kg·m/s.

27. E is correct.

$p = mv$

Conservation of momentum:

$p_{initial} = p_{final}$

0 kg·m/s = (0.01 kg)·(300 m/s) + (4 kg)v_{recoil}

0 kg·m/s = 3 kg·m/s + (4 kg)v_{recoil}

−3 kg·m/s = (4 kg)v_{recoil}

(−3 kg·m/s) / (4 kg) = v_{recoil}

v_{recoil} = −0.75 m/s

Velocity is negative since the gun recoils in the opposite direction of the bullet.

28. C is correct.

Since the initial velocity only has a horizontal component, the *y*-component of the initial velocity = 0.

Use 24 m to calculate the time the ball is in the air:

$d_y = \frac{1}{2}at^2$

$t^2 = 2d_y / a$

$t^2 = 2(24 \text{ m}) / (9.8 \text{ m/s}^2)$

$t^2 = 4.89 \text{ s}^2$

$t = 2.21$ s

Use the time in the air and the horizontal distance to calculate the horizontal speed of the ball:

$v_x = d_x / t$

$v_x = (18 \text{ m}) / (2.21 \text{ s})$

$v_x = 8.1$ m/s

29. A is correct.

An object moving in a circle at constant speed is undergoing uniform circular motion.

In a uniform circular motion, the acceleration is due to centripetal acceleration and points inward towards the center of a circle.

30. B is correct.

Impulse:

$$J = F\Delta t$$
$$J = \Delta p$$

where p is momentum

31. C is correct.

Conservation of energy:

$$KE_i + PE_i = KE_f + PE_f$$
$$KE_i + PE_i = KE_f + 0$$
$$KE_f = \tfrac{1}{2}mv_i^2 + mgh_i$$
$$KE_f = \tfrac{1}{2}(4 \text{ kg})\cdot(20 \text{ m/s})^2 + (4 \text{ kg})\cdot(10 \text{ m/s}^2)\cdot(10 \text{ m})$$
$$KE_f = 800 \text{ J} + 400 \text{ J}$$
$$KE_f = 1{,}200 \text{ J}$$

32. E is correct.

The *force* needed to stop a car can be related to KE and work:

$$KE = W$$
$$\tfrac{1}{2}mv^2 = Fd$$
$$F = \tfrac{1}{2}mv^2 / d$$

Momentum is included in the KE term.

$$p = mv$$
$$F = \tfrac{1}{2}(mv)v / d$$
$$F = \tfrac{1}{2}(p)v / d$$

If there is less stopping distance, the force increases as they are inversely proportional.

If the momentum or mass increases, the force increases as they are directly proportional.

33. C is correct.

Impulse:

$$J = F\Delta t$$

Based on Newton's Third Law, the force experienced by these two objects is equal and opposite.

Therefore, the magnitudes of impulse are the same.

34. B is correct.

Balance the counterclockwise (CCW) torque with the clockwise (CW) torque.

Let the axis of rotation be at the point where the rope attaches to the bar.

This placement causes the torque from the rope to be zero since the lever arm is zero.

$$\Sigma \tau : \tau_1 - \tau_2 = 0$$

$$\tau_1 = \tau_2$$

The CCW torque due to the weight of the 6 kg mass:

$$\tau = r_1 F_1$$

$$r_1 F_1 = (x) \cdot (6 \text{ kg}) \cdot (9.8 \text{ m/s}^2)$$

The CW torque due to the weight of the 30 kg mass:

$$r_2 F_2 = (5 \text{ m} - x) \cdot (30 \text{ kg}) \cdot (9.8 \text{ m/s}^2)$$

Set the two expressions equal

$$(9.8 \text{ m/s}^2) \cdot (x) \cdot (6 \text{ kg}) = (5 \text{ m} - x) \cdot (30 \text{ kg}) \cdot (9.8 \text{ m/s}^2)$$

cancel g and kg from each side of the equation:

$$6x = 30(5 \text{ m} - x)$$

$$6x = 150 \text{ m} - 30x$$

$$36x = 150 \text{ m}$$

$$x = 4.2 \text{ m}$$

35. E is correct.

If the block is at rest, then the force of static friction equals the force of gravity at angle θ.

$$F_f = mg \sin \theta$$

36. C is correct.

$F_{net} = 0$ is necessary to maintain a constant velocity.

If 45 N must be exerted on the block to maintain a constant velocity, the force due to kinetic friction against the block equals 45 N.

For a horizontal surface and no other vertical forces acting, the normal force on the block equals its weight.

$N = mg$

$F_{friction} = \mu_k N$

$F_{friction} = \mu_k mg$

$\mu_k = (F_{friction}) / mg$

$\mu_k = (45 \text{ N}) / [(30 \text{ kg}) \cdot (10 \text{ m/s}^2)]$

$\mu_k = 0.15$

37. B is correct.

Newton's Second Law:

$F = ma$

The impulse-momentum relationship can be derived by multiplying Δt on both sides:

$F\Delta t = ma\Delta t$

$F\Delta t = m\Delta v$

$J = m\Delta v$

Thus, the impulse equals the change in momentum.

38. C is correct.

Force X acts perpendicular to the short arm of the rectangle; this is the *lever arm*.

$\tau = rF$

$\tau = (0.5 \text{ m}) \cdot (15 \text{ N})$

$\tau = 7.5 \text{ N·m}$

Since the torque causes the plate to rotate clockwise, its sign is negative.

$\tau = -7.5 \text{ N·m}$

39. E is correct.

$\tau = rF$

Force Z acts directly at the pivot, so the lever arm equals zero.

$\tau = (0 \text{ m}) \cdot (30 \text{ N})$

$\tau = 0 \text{ N·m}$

40. A is correct.

$\tau = rF$

Force Y acts perpendicular to the long arm of the rectangle; this is the lever arm.

$\tau = (0.6 \text{ m}) \cdot (25 \text{ N})$

$\tau = 15 \text{ N·m}$

The torque is clockwise, so its sign is negative.

$\tau = -15 \text{ N·m}$

41. B is correct.

The tension in the string provides the centripetal force.

$T = mv^2 / r$

$m = 50 \text{ g} = 0.05 \text{ kg}$

$T = [(0.05 \text{ kg}) \cdot (20 \text{ m/s})^2] / (2 \text{ m})$

$T = [(0.05 \text{ kg}) \cdot (400 \text{ m}^2/\text{s}^2)] / (2 \text{ m})$

$T = (20 \text{ kg·m}^2/\text{s}^2) / (2 \text{ m})$

$T = 10 \text{ N}$

42. A is correct.

Newton's Third Law states that each force is paired with an equal and opposite reaction force.

Therefore, the small car and the truck each receive the same force.

43. C is correct.

Choose the axis of rotation at the point where the bar attaches to the wall.

Since the lever arm of the force that the wall exerts is zero, the torque at that point is zero and can be ignored.

The two other torques present arise from the weight of the bar exerting a force downward and the cable exerting force upward.

The weight of the bar acts at the center of mass, so its lever arm is 1 m.

continued...

The lever arm for the cable is 2 m since it acts the full 2 m from the wall at the end of the bar.

Torque is the product of the length of the lever arm and the component of force perpendicular to the arm.

The torque applied by the wire is:

$F_T l \sin \theta$

The sum of torques = 0 since the bar is in rotational equilibrium.

Let the torque of the cable be positive and the torque of the weight be negative.

$(F_T \sin 30°) \cdot (2 \text{ m}) - (10 \text{ kg}) \cdot (10 \text{ m/s}^2) \cdot (1 \text{ m}) = 0$

$F_T = [(10 \text{ kg}) \cdot (10 \text{ m/s}^2) \cdot (1 \text{ m})] / [(2 \text{ m}) \cdot (\sin 30°)]$

$F_T = [(10 \text{ kg}) \cdot (10 \text{ m/s}^2) \cdot (1 \text{ m})] / [(2 \text{ m}) \cdot (0.5)]$

$F_T = 100 \text{ N}$

44. B is correct.

Momentum is defined as:

$p = mv$

$m_A = 2m_B$

$p_A = 2m_B v$

$p_B = m_B v$

$p_A = 2p_B$

If both objects reach the ground at the same time, they have equal velocities.

However, because A is twice the mass, it has twice the momentum as object B.

45. D is correct.

Use conservation of momentum to make equations for momenta along the *x*-axis and the *y*-axis.

Since the mass ratio is 1 : 4, one car has a mass of *m*, and the other has a 4*m*.

The entangled cars after the collision have a combined mass of 5*m*.

Let the car of mass *m* be traveling in the positive *x*-direction, and the car of mass 4*m* be traveling in the positive *y*-direction.

The choice of directions here is arbitrary, but the angle of impact is important.

$p_{initial} = p_{final}$ for the *x*- and *y*-axes

$p = mv$

continued...

For the *x*-axis:

$$m_i v_i = m_f v_{fx}$$

$$m(12 \text{ m/s}) = 5mv_x$$

cancel *m* from both sides of the expression

$$12 \text{ m/s} = 5v_x$$

$$v_x = 2.4 \text{ m/s}$$

For the *y*-axis:

$$m_i v_i = m_f v_{fy}$$

$$4m(12 \text{ m/s}) = 5mv_y$$

cancel *m* from both sides of the expression

$$4(12 \text{ m/s}) = 5v_y$$

$$v_y = 9.6 \text{ m/s}$$

The question asks for the magnitude of the final velocity, so combine the *x* and *y* components of the final velocity using the Pythagorean Theorem.

$$v^2 = (2.4 \text{ m/s})^2 + (9.6 \text{ m/s})^2$$

$$v^2 = 5.76 \text{ m}^2/\text{s}^2 + 92.16 \text{ m}^2/\text{s}^2$$

$$v = 9.9 \text{ m/s}$$

46. C is correct.

Use conservation of momentum on the horizontal plane.

Before the throw, the total momentum of the skater-ball system is zero.

Thus, after the throw, the total horizontal momentum must sum to zero: the horizontal component of the ball's momentum equals the momentum of the skater moving the opposite way.

Use m_s for the skater's mass and $m_s/3$ for the ball's mass.

$$p = mv$$

$$p_{\text{skater}} = p_{\text{ball}}$$

$$m_s v_s = m_b v_b$$

$$m_s(2.9 \text{ m/s}) = (1/3)m_s v \cos 5°$$

cancel *m* from both sides of the expression

$$v = (2.9 \text{ m/s}) \cdot (3) / (\cos 5°)$$

$$v = (2.9 \text{ m/s}) \cdot (3) / (0.996)$$

$$v = 8.73 \text{ m/s}$$

47. B is correct.

weight = mass × gravity

W = mg

m = W / g

m = (98 N) / (9.8 m/s^2)

m = 10 kg

Newton's Second Law:

F = ma

F = (10 kg)·(10 m/s^2)

F = 100 N

48. A is correct.

KE is constant because speed is constant.

PE increases because the cart is at a higher height at point B.

The cart as a system is not isolated since the winch does work on it, so its energy is not conserved.

Conservation of energy: PE increase of the cart = work done by the winch

49. D is correct.

The vertical component of the initial velocity:

v_{iy} = (140 m/s) sin 35°

v_{iy} = (140 m/s)·(0.57)

v_{iy} = 79.8 m/s

The initial velocity upward, the time elapsed, and acceleration due to gravity is known.

Determine the final velocity after 4 s.

v_y = v_{iy} + at

v_y = 79.8 m/s + (−9.8 m/s^2)·(4 s)

v_y = 41 m/s

50. C is correct.

impulse = force × time

J = FΔt

51. D is correct.

Conservation of momentum: the momentum of the fired bullet is equal and opposite to that of the rifle.

$p = mv$

$p_{before} = p_{after}$

$0 = p_{rifle} + p_{bullet}$

$-p_{rifle} = p_{bullet}$

$-(2 \text{ kg})v = (0.01 \text{ kg}) \cdot (220 \text{ m/s})$

$v = (0.01 \text{ kg}) \cdot (220 \text{ m/s}) / (-2 \text{ kg})$

$v = -1.1 \text{ m/s}$

Thus, the velocity of the rifle is 1.1 m/s in the opposite direction as the bullet.

52. D is correct.

Airbags reduce force by increasing the time of contact between the passenger and surface.

In a collision, an impulse is experienced by a passenger:

$J = F\Delta t$

$F = J / \Delta t$

The impulse is a constant, but the force experienced by the passenger is inversely related to the time of contact. Airbags increase the time of impact and thus reduce the forces experienced by the person.

53. A is correct.

Since Force I is perpendicular to the beam, the force produces torque without any horizontal force component.

$\tau = rF$

$\tau = (0.5 \text{ m}) \cdot (10 \text{ N})$

$\tau = 5 \text{ N·m}$

Because the force causes the beam to rotate clockwise against the positive counterclockwise direction, the torque sign should be negative:

$\tau = -5 \text{ N·m}$

54. E is correct.

To calculate torque, use the 35° angle.

For *torque*:

$\tau = rF \sin \theta$

$\tau = (1 \text{ m}) \cdot (5 \text{ N}) \sin 35°$

$\tau = 2.9 \text{ N·m}$

The torque is counterclockwise, so the sign is positive.

55. B is correct.

Force III acts purely in tension with the beam and has no component acting vertically against the beam.

Torque can only be calculated using a force with some component perpendicular to the length vector.

Because Force III has no perpendicular component to the length vector, torque is zero.

$\tau = rF$

$\tau = (1 \text{ m}) \cdot (0 \text{ N})$

$\tau = 0 \text{ N·m}$

Notes for active learning

Rotational Motion – Detailed Explanations

1. D is correct.

An object is rolling down an incline experiences three forces, and hence three *torques*.

The forces are 1) the force of gravity acting on the center of mass of the object, 2) the normal force between the incline and the object, and 3) the force of friction between the incline and the object.

If the origin is taken to be the center of the object, the force of gravity provides zero torque.

This can be seen by noting that the distance between the origin and the point of application of the force is zero.

$\tau_{gravity} = F_{gravity} r = mg(0) = 0$

Similarly, the normal force contributes zero torque because the direction of the force is directed through the origin (pivot point).

$\tau_{normal} = F_{normal}\, r \sin \theta = F_{normal}(R) \cdot (\sin 180°) = F_{normal}(R) \cdot (0) = 0$

Use a coordinate system in which the *x*-axis is parallel to the incline, and the *y*-axis is perpendicular.

The object is rolling in the positive *x*-direction.

The dynamical equation for linear motion along the *x*-direction:

$F_{net} = ma$

$(mg \sin \theta - f) = ma$

Note that the normal force is only in the *y*-direction and thus does not directly contribute to the acceleration in the *x*-direction.

The dynamical equation for rotational motion:

$\tau_{net} = I\alpha$

$fR = I\alpha$ (Note that the frictional force is perpendicular to the *r* vector, and sin 90° = 1)

where R is the radius of the object, f is the force of friction, and I is the moment of inertia

A relation coupling these two dynamical equations is needed.

This is the equation of constraint imposed by the restriction that the object rolls without slipping:

$\alpha = a / R$

To find the linear acceleration, use the equation of constraint to eliminate α from the rotational equation by replacing it with a / R:

$fR = I(a / R)$

The force of friction is of no interest, so rearrange this last expression:

$f = Ia / R^2$

continued…

Substitute this into the linear dynamic equation from above in place of f:

$$(mg \sin \theta - Ia / R^2) = ma$$

Solving for a:

$$a = mg \sin \theta / [m + (I / R^2)]$$

$$a = g \sin \theta / [1 + (I / mR^2)]$$

The moment of inertia of a circular object:

$$NmR^2$$

where N is some real number different for different shapes

For example, for a sphere,

$$I = (2/5)mR^2$$

so for a sphere $N = 2/5$

So, for a rolling object, the linear acceleration is:

$$a = g \sin \theta / (1 + N)$$

which depends on neither the radius nor the mass of the object.

Only the shape of the object is important.

2. B is correct.

Use the conservation of energy. The initial and final states are the sphere at the top and bottom of the ramp.

Take the zero of gravitational potential energy to be the sphere's configuration at the bottom.

The potential energy at the bottom is zero.

The sphere starts from rest, so the kinetic energy at the top is zero. So:

$$mgh = K_{linear} + K_{rotation}$$

$$mgh = \tfrac{1}{2}mv^2 + \tfrac{1}{2}I\omega^2$$

For a sphere,

$$I = (2/5)mr^2$$

Because the sphere rolls without slipping,

$$v = r\omega$$

Substituting these into the conservation of energy equation:

$$mgh = \tfrac{1}{2}mr^2\omega^2 + \tfrac{1}{2}(2/5)mr^2\omega^2$$

$$mgh = \tfrac{1}{2}mr^2\omega^2 + (2/10)mr^2\omega^2$$

$$mgh = (7/10)mr^2\omega^2$$

continued…

Isolating ω:

$$\omega = \sqrt{(10gh / 7r^2)}$$

$$\omega = \sqrt{[10 \cdot (9.8 \text{ m/s}^2) \cdot (5.3 \text{ m})] / [7 \cdot (1.7 \text{ m})^2]}$$

$$\omega = 5.1 \text{ rad/s}$$

3. A is correct.

There is no torque on the ball during the fall; therefore, its rotational speed does not change.

The *rotational kinetic energy* just before the ball hits the floor is the same as rolling on the horizontal surface.

The rotational kinetic energy when it was rolling on the surface can be calculated directly.

Recall that the *moment of inertia of a solid sphere* is:

$$I = 2/5 mR^2$$

$$K_{rot} = \tfrac{1}{2} I \omega^2$$

$$K_{rot} = \tfrac{1}{2}(2/5) mR^2 \omega^2$$

Since the ball is rolling without slipping,

$$\omega = v / R$$

$$K_{rot} = \tfrac{1}{2}(2/5) mR^2 (v/R)^2$$

$$K_{rot} = (2/10) mv^2$$

$$K_{rot} = (2/10) mv^2$$

$$K_{rot} = (2/10) \cdot (0.125 \text{ kg}) \cdot (4.5 \text{ m/s})^2$$

$$K_{rot} = 0.51 \text{ J}$$

4. D is correct.

The angular momentum of an object in a circular motion is:

$$L = I\omega$$

where I is the moment of inertia with respect to the center of motion and ω is the angular speed

The moment of inertia of a point mass is:

$$I = mr^2$$

The angular momentum is:

$$L = mr^2 \omega$$

Angular speed is in rev/s. Express in rad/s:

$$1.2 \text{ rev/s} \cdot (2\pi \text{ rad/rev}) = 7.540 \text{ rad/s}$$

continued...

Finally:

$$L = (0.38 \text{ kg}) \cdot (1.3 \text{ m})^2 \cdot (7.540 \text{ rad/s})$$

$$L = 4.8 \text{ kg m}^2/\text{s}$$

5. C is correct.

Conservation of angular momentum requires:

$$L_f = L_i$$

$$I_f \omega_f = I_i \omega_i$$

The final angular speed is:

$$\omega_f = \omega_i (I_i / I_f)$$

$$\omega_f = (3.0 \text{ rev/s}) \cdot (5.0 \text{ kg} \cdot \text{m}^2) / (2.0 \text{ kg m}^2)$$

$$\omega_f = 7.5 \text{ rev/s}$$

6. C is correct.

An external torque changes the angular velocity of a system ($\alpha = \sum \tau / I$), and hence its angular momentum.

To maintain a constant angular momentum, the sum of external torques must be zero.

7. E is correct.

For a *rotating circular object*:

$$\omega = v / r$$

$$\omega = v / (d / 2)$$

$$\omega = (4.0 \text{ m/s}) / [(0.60 \text{ m}) / 2]$$

$$\omega = 13.3 \text{ rad/s}$$

8. A is correct.

$$K = \tfrac{1}{2} I \omega^2$$

The moment of inertia of a rod with respect to its "short axis" is $I = (1/12)ml^2$

$$K = (1/24) m l^2 \omega^2$$

$$K = (1/24) \cdot (0.4500 \text{ kg}) \cdot (1.20 \text{ m})^2 \cdot (3.60 \text{ rad/s})^2$$

$$K = 0.350 \text{ J}$$

9. C is correct.

The *moment of inertia* can be found from the dynamic relation:

$\tau = I\alpha$

$I = \tau / \alpha$

where τ is the torque applied to the pulley, and α is the pulley's angular acceleration

The *torque* is defined as:

$FR \sin \theta$

In this case, force F is the tension force from the rope, R is the radius of the wheel, and $\theta = 90°$.

Thus, the torque is just the product of the tension of the rope and the radius of the pulley:

$\tau = TR$

The *angular acceleration* is related to the acceleration of a point on the circumference of the pulley:

$\alpha = a / R$

where a is the linear acceleration at the circumference, and R is the pulley's radius

Combining these two results, the *moment of inertia* is:

$I = TR^2 / a$

If the rope does not slip on the pulley, then the rope, and hence the hanging mass, has an acceleration a.

Find the acceleration and the tension.

The tension is found by applying Newton's Second Law to the hanging mass.

Two forces on the hanging mass: 1) force of gravity pointing down and 2) tension of the rope pointing up.

From Newton's Second Law (with down as the positive direction):

$(mg - T) = ma$

$T = m(g - a)$

With that, the moment of inertia becomes:

$I = mR^2 [(g - a) / a]$

The acceleration can be found from the kinematic information given about the movement of the hanging mass.

The relation needed is:

$\Delta y = \frac{1}{2} a (\Delta t)^2 + v_0 (\Delta t)$

$a = 2\Delta y / (\Delta t)^2$

$a = 2 \cdot (10 \text{ m}) / (2 \text{ s})^2$

$a = 5.000 \text{ m/s}^2$

continued...

Calculate the *moment of inertia*:

$I = (14 \text{ kg}) \cdot (2.0 \text{ m})^2 \cdot [(9.8 \text{ m/s}^2 - 5.000 \text{ m/s}^2) / (5.000 \text{ m/s}^2)]$

$I = 53.76 \text{ kg·m}^2$

$I = 53.8 \text{ kg·m}^2$

10. B is correct.

The *final speed of the string* can be found if the acceleration is known:

$v_f^2 = v_i^2 + 2ad = 0 + 2ad$

$v_f = \sqrt{2ad}$

where d is the distance over which the acceleration occurs

The acceleration of the string is related to the *acceleration of the pulley*:

$a = r\alpha$

The *angular acceleration* follows from the dynamical equation for the rotational motion:

$\tau = I\alpha$

The *torque* is the force applied times the radius of the pulley:

$Fr = I\alpha$

Combining these equations gives:

$a = r^2 F / I$

The final velocity of the string is:

$v_f = \sqrt{2r^2 F d / I}$

$v_f = \sqrt{[2 \cdot (0.125 \text{ m})^2 \cdot (5.00 \text{ N}) \cdot (1.25 \text{ m}) / (0.0352 \text{ kg·m}^2)]}$

$v_f = 2.36 \text{ m/s}$

11. A is correct.

The angle of every point remains fixed relative to all other points.

The tangential acceleration increases moving *away* from the center ($a_t = \alpha r$).

The radial (or centripetal) acceleration depends on the distance r from the center ($a_c = \omega^2 r$).

The only choice that does not depend on r (i.e., is the same for all points in the object) is I.

12. E is correct.

Linear velocity is related to angular velocity by $v = r\omega$.

13. C is correct.

For a rotating object:

$K = \frac{1}{2}I\omega^2$

The moment of inertia of a cylinder is:

$I = \frac{1}{2}mr^2$

Combining these:

$K = \frac{1}{4}mr^2\omega^2$

Solving for the angular speed:

$\omega = \sqrt{4K / mr^2}$

$\omega = \sqrt{4 \cdot (3.2 \times 10^7 \text{ J}) / [(400.0 \text{ kg})(0.60 \text{ m})^2]}$

$\omega = 940$ rad/s

14. B is correct.

For a rotating object subject to a constant torque which has undergone a total angular displacement of $\Delta\theta$.

The *work done on the wheel* is:

$W = \tau\Delta\theta$

Since work is the change in energy of the wheel from external forces, and since the wheel started with $E = 0$ ("from rest"), the final kinetic energy can be written as:

$K = \tau\Delta\theta$

By *rotational kinematics*:

$\Delta\theta = \frac{1}{2}\alpha t^2$

The equation of *rotational dynamics* is:

$\tau = I\alpha$, or $\alpha = \tau / I$

So:

$\Delta\theta = t^2\tau / 2I$

and

$K = t^2\tau^2 / 2I$

(Note that this expression can be developed by finding the final angular velocity and using the definition of rotational kinetic energy.)

$K = [(8.0 \text{ s})^2 \cdot (3.0 \text{ N·m})^2] / [2 \cdot (5.0 \text{ kg·m}^2)]$

$K = 58$ J

15. D is correct.

Tangential speed depends on the distance of the point from the fixed axis, so points at different radii have different tangential speeds.

Angular speed and *acceleration* of a rigid object do not depend on radius and are the same for all points (since each point on the object must rotate through the same angle in the same time interval, or it would not be rigid).

16. D is correct.

The *angular momentum of a rotating object* is:

$$L = I\omega$$

For a cylinder,

$$I = \tfrac{1}{2}mr^2$$

giving:

$$L = \tfrac{1}{2}mr^2\omega$$

$$L = (0.5)\cdot(15.0 \text{ kg})\cdot(1.4 \text{ m})^2\cdot(2.4 \text{ rad/s})$$

$$L = 35 \text{ kg m}^2/\text{s}$$

17. E is correct.

The speed of an accelerating object is related to the distance covered by the kinematic relation:

$$v_f^2 - v_i^2 = 2ad$$

In this case, the initial speed is zero.

so:

$$v_f = \sqrt{2ad}$$

Thus, find the linear acceleration of the disk.

An object rolling down an incline experiences three forces, and hence three torques.

The forces are the force of gravity acting on the center of mass of the object, the normal force between the incline and the object, and the force of friction between the incline and the object.

If the origin is taken to be the center of the object, the force of gravity provides zero torque.

This can be seen by noting that the distance between the origin and the point of application of the force is zero.

$$\tau_{gravity} = F_{gravity} r = mg(0) = 0$$

Similarly, the normal force contributes zero torque because the direction of the force is directed through the origin (pivot point).

$$\tau_{normal} = F_{normal}\, r \sin\theta = F_{normal}(R)\cdot(\sin 180°) = F_{normal}(R)\cdot(0) = 0$$

continued...

Use a coordinate system where the x-axis is parallel to the incline and the y-axis is perpendicular.

The object is rolling in the positive x-direction.

The dynamical equation for linear motion along the x-direction is:

$F_{net} = ma$

$(mg \sin \theta - f) = ma$

The normal force is only in the y-direction and does not directly contribute to the acceleration in the x-direction.

The dynamical equation for rotational motion is:

$\tau_{net} = I\alpha$

$fR = I\alpha$ (Note that the frictional force is perpendicular to the r vector, and $\sin 90° = 1$)

where R is the radius of the object, f is the force of friction, and I is the moment of inertia

A relation coupling these two dynamical equations is needed.

This is the equation of constraint imposed by the restriction that the object rolls without slipping:

$\alpha = a / R$

To find the linear acceleration, use the equation of constraint to eliminate α from the rotational equation by replacing it with a / R:

$fR = I(a / R)$

The force of friction is of no interest, so rearrange this last expression:

$f = Ia / R^2$

Substitute this into the linear dynamic equation from above in place of f:

$(mg \sin \theta - Ia / R^2) = ma$

Solving this for a:

$a = mg \sin \theta / [m + (I / R^2)]$

$a = g \sin \theta / [1 + (I / mR^2)]$

For a disk,

$I = \frac{1}{2}mR^2$

So:

$a = g \sin \theta / (1 + \frac{1}{2}) = (2/3)g \sin \theta$

continued...

Using this in the kinematic equation above:

$v_f = \sqrt{(4gd \sin \theta / 3)}$

$v_f = \sqrt{[(4/3) \cdot (9.8 \text{ m/s}^2) \cdot (3.0 \text{ m}) \cdot \sin(25°)]}$

$v_f = \sqrt{[(4/3) \cdot (9.8 \text{ m/s}^2) \cdot (3.0 \text{ m}) \cdot (0.4226)]}$

$v_f = 4.1$ m/s

18. A is correct.

The center of the tire is moving at velocity v, but the bottom is in contact with the ground without slipping, so the speed at the bottom is 0 m/s.

Thus, with respect to the ground, the tire is *instantaneously* rotating about the point of contact with the ground, and all points in the tire have the same instantaneous angular speed.

The top of the tire is twice the distance from the ground as the center.

For the center of the tire:

$v = r\omega$

At the top:

$v_{top} = (2r)\omega = 2(r\omega) = 2v$

19. D is correct.

The string does not slip. This means that the speed of the string is the same as the speed of a point on the circumference of the pulley.

The angular speed of the pulley (radius R) and the speed of a point on its circumference are related:

$\omega = v / R$

$\omega = (5.0 \text{ m/s}) / (0.050 \text{ m})$

$\omega = 100$ rad/s

20. D is correct.

One way of expressing the magnitude of angular momentum is:

$L = rp \sin \theta$

where r is the magnitude of the object's absolute position vector, p is the magnitude of the object's linear momentum, and θ is the angle between the position vector and the momentum vector

The magnitude of the position vector is:

$r = \sqrt{(r_x^2 + r_y^2)}$

$r = \sqrt{[(2.00 \text{ m})^2 + (3.10 \text{ m})^2]}$

$r = 3.689$ m

continued...

Its angle with respect to the positive x axis is:

$\theta_r = \text{atan}(r_y / r_x)$

$\theta_r = \text{atan}(3.10 / 2.00)$

$\theta_r = 0.99783 \text{ rad} = 57.17°$

The magnitude of the momentum vector is:

$p = mv$

$p = (1.4 \text{ kg})(4.62 \text{ m/s})$

$p = 6.468 \text{ kg·m/s}$

The angle of the momentum vector is given in the problem:

$\theta_p = 45°$

Thus, the angle between the two vectors is:

$\theta = \theta_r - \theta_p$

$\theta = 57.17° - 45°$

$\theta = 12.17°$

The angular momentum is then:

$L = (3.689 \text{ m}) \cdot (6.468 \text{ kg m/s}) \cdot \sin(12.17°)$

$L = 5.0 \text{ kg·m}^2/\text{s}$

21. B is correct.

Average angular acceleration is defined by:

$\alpha_{avg} = \Delta\omega / \Delta t$

$\alpha_{avg} = |(6.3 \text{ rad/s} - 10.0 \text{ rad/s}) / (5.0 \text{ s})|$

$\alpha_{avg} = 0.74 \text{ rad/s}$

22. D is correct.

The kinematic equation for *angular velocity* is:

$\omega_f = \omega_i + \alpha\Delta t$

Note that the sign of angular velocity is opposite from the sign of angular acceleration. That means that the wheel is slowing down.

The wheel must slow and continue beyond zero speed to gain speed in the opposite direction for the final kinetic energy to be larger than the initial kinetic energy.

That is, the final angular velocity must be negative.

continued...

The *kinetic energy of a rotating wheel* is:

$$K = \tfrac{1}{2} I \omega^2$$

The kinetic energy scales as the square of the angular speed.

To double the kinetic energy, the angular speed must increase by a factor of $\sqrt{2}$:

$$\omega_f = -\omega_i \sqrt{2}$$

Substitution into the kinematic equation:

$$-\omega_i \sqrt{2} = \omega_i + \alpha \Delta t$$

Solving for Δt:

$$\Delta t = -(1 + \sqrt{2})\omega_i / \alpha)$$

$$\Delta t = -(1 + \sqrt{2}) \cdot (26.0 \text{ rad/s}) / (-0.43 \text{ rad/s}^2)$$

$$\Delta t = 146 \text{ s}$$

23. B is correct.

Apply conservation of energy.

Take zero of gravitational potential energy to be the configuration when the disk is at the bottom of the ramp.

Conservation of energy demands:

$$E_{\text{top}} = E_{\text{bottom}}$$

At the top of the ramp, the disk is at rest, so KE = 0.

The total energy at the top is:

$$E_{\text{top}} = K_{\text{top}} + U_{\text{top}}$$

$$E_{\text{top}} = 0 + mgh$$

At the bottom of the ramp, the gravitational potential energy is zero, and the kinetic energy is the sum of the linear and rotational kinetic energies:

$$E_{\text{bottom}} = K_{\text{bottom}} + U_{\text{bottom}}$$

$$E_{\text{bottom}} = K_{\text{linear}} + K_{\text{rotational}} + 0$$

$$E_{\text{bottom}} = \tfrac{1}{2} mv^2 + \tfrac{1}{2} I\omega^2$$

Final linear velocity is not given, but rather the final angular velocity is given.

Eliminate the linear velocity in favor of the angular velocity by applying the constraint to a circular object rolling without slipping.

$$v = \omega r$$

$$E_{\text{bottom}} = \tfrac{1}{2} m\omega^2 r^2 + \tfrac{1}{2} I\omega^2$$

continued...

The moment of inertia of a disk is:

$$I = \tfrac{1}{2}mr^2$$

Thus:

$$E_{bottom} = \tfrac{1}{2}m\omega^2 r^2 + \tfrac{1}{2}(\tfrac{1}{2}mr^2)\omega^2$$

$$E_{bottom} = \tfrac{1}{2}m\omega^2 r^2 + \tfrac{1}{4}mr^2\omega^2$$

$$E_{bottom} = \tfrac{3}{4}m\omega^2 r^2$$

Combining results into the expression for conservation of energy:

$$mgh = \tfrac{3}{4}m\omega^2 r^2$$

or

$$h = 3\omega^2 r^2 / 4g$$

Note that the mass has canceled, a common occurrence in mechanics problems.

The chance of error is reduced by proceeding algebraically (rather than plugging in numbers in the beginning), by which mass can be canceled.

Note that diameter is given, not the radius.

$$r = d/2 = 1.6 \text{ m}$$

$$h = [3(4.27 \text{ rad/s})^2 \cdot (1.60 \text{ m})^2] / [4(9.8 \text{ m/s}^2)]$$

$$h = 3.57 \text{ m}$$

24. A is correct.

The direction of angular velocity is taken by convention to be given by applying the right-hand rule to the rotation.

In the case of a wheel of a forward-moving bicycle, that direction is to the left of the rider.

25. A is correct.

Angular acceleration can be found if linear acceleration is calculated from:

$$\alpha = a/R$$

The linear acceleration follows from the kinematic relationship:

$$v_f^2 - v_i^2 = 2ad$$

In this case, v_f is zero.

$$-v_i^2 = 2ad$$

Only the absolute value of the acceleration is needed; drop the minus sign and solve for a:

$$a = v_i^2 / 2d$$

continued...

The *angular acceleration* is:

$$\alpha = v_i^2 / 2dR$$

$$\alpha = (8.4 \text{ m/s})^2 / [2(115.0 \text{ m})\cdot(0.34 \text{ m})]$$

$$\alpha = 0.90 \text{ rad/s}^2$$

26. A is correct.

For a *rotating object*:

$$K = \tfrac{1}{2}I\omega^2$$

The *moment of inertia of a cylinder* is:

$$I = \tfrac{1}{2}mr^2$$

Combining these:

$$K = \tfrac{1}{4}mr^2\omega^2$$

The angular speed is given in rpm but needs to be in rad/s:

$$33.4 \text{ rpm}\cdot(1 \text{ min} / 60 \text{ s})\cdot(2\pi \text{ rad/rev}) = 3.498 \text{ rad/s}$$

Thus:

$$K = (0.25)\cdot(3.0 \text{ kg})\cdot(0.10 \text{ m})^2\cdot(3.489 \text{ rad/s})^2$$

$$K = 0.091 \text{ J}$$

27. A is correct.

The angular momentum of a spinning object can be written as:

$$L = I\omega$$

The moment of inertia of a long thin uniform object of length l about an axis through the center perpendicular to the long axis is:

$$I = (1/12)ml^2$$

Giving:

$$L = (1/12)ml^2\omega$$

$$L = (1/12)\cdot(0.1350 \text{ kg})\cdot(1.000 \text{ m})^2\cdot(3.5 \text{ rad/s})$$

$$L = 0.0394 \text{ kg}\cdot\text{m}^2/\text{s}$$

28. C is correct.

The *period of a rotating object* can be expressed as:

$$T = 2\pi / \omega$$

ω can be extracted from the definition of centripetal force:

$$F = mv^2 / r$$

Combining this with the relationship $v = \omega r$ gives:

$$F = mr\omega^2$$

Thus:

$$\omega = \sqrt{(F / mr)}$$

Then:

$$T = 2\pi\sqrt{(mr / F)}$$

$$T = 2\pi\sqrt{[(23.0 \text{ kg}) \cdot (1.3 \text{ m}) / (51.0 \text{ N})]}$$

$$T = 4.8 \text{ s}$$

29. C is correct.

Angular acceleration is:

$$\alpha = \Delta\omega / \Delta t$$

$$\alpha = (38.0 \text{ rad/s} - 0.00 \text{ rad/s}) / (10.0 \text{ s})$$

$$\alpha = 3.80 \text{ rad/s}$$

The other information given in the question is not needed.

30. C is correct.

The magnitude of *torque* can be expressed as:

$$\tau = rF \sin \theta$$

where r is the distance from the origin (taken here to be the pivot point) to the point of application of the force F, and θ is the angle between the position vector of the point of application and the force vector

$$\tau = (0.63 \text{ m}) \cdot (17.0 \text{ N}) \cdot \sin (45°)$$

$$\tau = 7.6 \text{ N m}$$

31. D is correct.

Use the conservation of energy.

Take the zero of potential energy to be the configuration in which the sphere is at the bottom.

The potential energy at the bottom is zero.

The disk starts from rest, so the kinetic energy at the top is zero.

So:

$$mgh = K_{linear} + K_{rotation}$$

$$mgh = \tfrac{1}{2}mv^2 + \tfrac{1}{2}I\omega^2$$

For a disk,

$$I = \tfrac{1}{2}mr^2$$

Because the disk rolls without slipping,

$$v = r\omega$$

Substituting these into the conservation of energy equation:

$$mgh = \tfrac{1}{2}m(r^2\omega^2) + \tfrac{1}{2}(\tfrac{1}{2}mr^2)\omega^2$$

$$mgh = \tfrac{1}{2}mr^2\omega^2 + \tfrac{1}{4}mr^2\omega^2$$

$$mgh = \tfrac{3}{4}mr^2\omega^2$$

Isolating h:

$$h = 3r^2\omega^2 / 4g$$

$$h = [3 \cdot (1.60 \text{ m})^2 \cdot (4.27 \text{ rad/s})^2] / [4 \cdot (9.8 \text{ m/s}^2)]$$

$$h = 3.57 \text{ m}$$

32. D is correct.

The *moment of inertia* is proportional to the square of the distance of an object from the center of rotation.

As Paul moves toward the center, the moment of inertia decreases.

Angular momentum,

$$L = I\omega \text{ is conserved}$$

As I decreases, ω, the angular speed increases to compensate.

33. B is correct.

Converting units:

$$210.0 \text{ rpm} \cdot (1 \text{ min} / 60 \text{ s}) \cdot (2\pi \text{ rad/revolution}) = 22.0 \text{ rad/s}$$

34. C is correct.

Use the conservation of energy:

$$E_f = E_i$$

$$KE_f + PE_f = KE_i + PE_i$$

Take the zero potential energy at the initial height so that PE_i is zero.

The kinetic energy of a *rolling object* is the sum of the KE of translation plus the kinetic energy of rotation:

$$KE = \tfrac{1}{2}mv^2 + \tfrac{1}{2}I\omega^2$$

If the object rolls without slipping, as is the example,

$$v = r\omega$$

where R is the radius of the sphere

The *moment of inertia of a sphere* is:

$$I = (2/5)m\omega^2$$

Write the kinetic energy of a sphere that rolls without slipping:

$$KE = \tfrac{1}{2}mv^2 + \tfrac{1}{2}(2/5)mr^2(v/r)^2$$

$$KE = \tfrac{1}{2}mv^2 + (1/5)mv^2$$

$$KE = (7/10)mv^2$$

Conservation of energy becomes:

$$(7/10)mv_f^2 + mgh = (7/10)mv_i^2$$

Canceling the mass and solving for the final speed:

$$v_f = \sqrt{[v_i^2 - (10/7)gh]}$$

The height, h, is related to the distance traveled and the angle of incline:

$$h = d \sin\theta$$

So:

$$v_f = \sqrt{[v_i^2 - (10/7)gd \sin\theta]}$$

$$v_f = \sqrt{[(5.5 \text{ m/s})^2 - (10/7)\cdot(9.8 \text{ m/s}^2)\cdot(3.0 \text{ m})\cdot\sin(25°)]}$$

$$v_f = \sqrt{[(5.5 \text{ m/s})^2 - (10/7)\cdot(9.8 \text{ m/s}^2)\cdot(3.0 \text{ m})\cdot(0.4226)]}$$

$$v_f = 3.5 \text{ m/s}$$

35. D is correct.

The dynamical relation for rotational motion is:

$\tau = I\alpha$

$I = \tau / \alpha$

where τ is the torque applied to the wheel, and α is the wheel's angular acceleration

The torque is the product of the force and the radius of the pulley:

$\tau = FR$ (note that sin $\theta = 1$, since $\theta = 90°$ [the problem mentions the force is applied tangentially]).

So:

$I = \tau / \alpha$

$I = FR / \alpha$

$I = (16.88 \text{ N}) \cdot (0.340 \text{ m}) / (1.20 \text{ rad/s}^2)$

$I = 4.78 \text{ kg} \cdot \text{m}^2$

36. B is correct.

The energy required to bring a rotating object to rest is:

$E = \frac{1}{2}I\omega_0^2$

The moment of inertia of a cylinder is:

$I = \frac{1}{2}mr^2$

So, the energy needed to stop the object is:

$E = \frac{1}{4}mr^2\omega_0^2$

Solving for the mass:

$m = 4E / r^2\omega_0^2$

We are given the angular speed in rpm, but should be in rad/s:

500.0 rpm·(1 min / 60 sec)·(2π rad / 1 rev) = 52.36 rad/s

The mass of the object is:

$m = 4(3900 \text{ J}) / (1.2 \text{ m})^2 \cdot (52.36 \text{ rad/s})^2$

$m = 4.0 \text{ kg}$

37. C is correct.

The torque at P$_2$ due to the weight of the billboard is:

$\tau = r_{perp}F$

$\tau = r_{perp}mg$

where r_{perp} is the perpendicular distance between P$_2$ and the line of application of the force

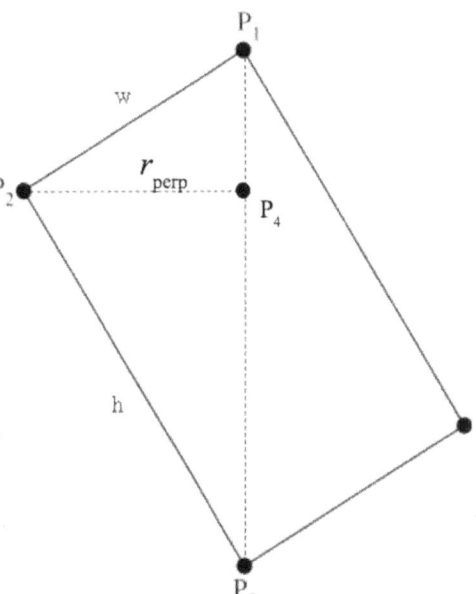

The weight of the billboard acts as if the mass is concentrated at the center of mass, which is the center of the billboard.

Thus, the line of application of the force of gravity is P$_1$P$_3$, the vertical line that passes through P$_1$ and P$_3$.

The length r_{perp} is the length of the horizontal line segment between P$_2$ and the line of application of the force (P$_1$P$_3$).

This line segment is perpendicular to P$_1$P$_3$ and intersects P$_1$P$_3$ at point P$_4$.

To find r_{perp} note that the triangle P$_1$P$_2$P$_3$ is similar to the triangle formed by P$_1$P$_4$P$_2$.

The ratios of the lengths of corresponding sides of similar triangles are equal.

Forming the ratio of the length of the long side to the length of the hypotenuse of these two triangles:

$h / \sqrt{(h^2 + w^2)} = r_{perp} / w$

$r_{perp} = hw / \sqrt{(h^2 + w^2)}$

Giving an expression for torque:

$\tau = mg[hw / \sqrt{(h^2 + w^2)}]$

$\tau = (5.0 \text{ kg}) \cdot (9.8 \text{ m/s}^2) \cdot (0.20 \text{ m}) \cdot (0.11 \text{ m}) / \sqrt{[(0.20 \text{ m})^2 + (0.11 \text{ m})^2]}$

$\tau = 4.7 \text{ N m}$

38. B is correct.

An object is rolling down an incline experiences three forces, and hence three torques.

The forces are 1) the force of gravity acting on the center of mass of the object, 2) the normal force between the incline and the object, and 3) the force of friction between the incline and the object.

If the origin is taken to be the center of the object, the force of gravity provides zero torque.

This can be seen by noting that the distance between the origin and the point of application of the force is zero.

$\tau_{gravity} = F_{gravity}r = mg(0) = 0$

continued...

Similarly, the normal force contributes zero torque because the direction of the force is directed through the origin (pivot point).

$$\tau_{normal} = F_{normal}\, r \sin\theta = F_{normal}(R)\cdot(\sin 180°) = F_{normal}(R)\cdot(0) = 0$$

Use a coordinate system in which the x-axis is parallel to the incline, and the y-axis is perpendicular.

The object is rolling in the positive x-direction.

The dynamical equation for linear motion along the x-direction is:

$$F_{net} = ma$$

$$(mg \sin\theta - f) = ma$$

The normal force is only in the y-direction and thus does not contribute to the acceleration in the x-direction.

The dynamical equation for rotational motion is:

$$\tau_{net} = I\alpha$$

$$fR = I\alpha \quad \text{(Note: the frictional force is perpendicular to the } r \text{ vector, and } \sin 90° = 1\text{)}$$

where R is the radius of the object, f is the force of friction, and I is the moment of inertia

A relation coupling these two dynamical equations is necessary.

The equation of constraint imposed by the restriction that the object rolls without slipping:

$$\alpha = a / R$$

To find the linear acceleration, use the equation of constraint to eliminate α from the rotational equation by replacing it with a / R:

$$fR = I(a / R)$$

The force of friction is of no interest, so rearrange this last expression:

$$f = Ia / R^2$$

Substitute this into the linear dynamic equation from above in place of f:

$$(mg \sin\theta - Ia / R^2) = ma$$

Solving this for a:

$$a = mg \sin\theta / [m + (I / R^2)]$$

$$a = g \sin\theta / [1 + (I / mR^2)]$$

For a sphere,

$$I = (2/5)mR^2$$

continued…

So:

$$a_{sphere} = g \sin \theta / [1 + (2/5)]$$

$$a_{sphere} = (5/7)g \sin \theta$$

$$a_{sphere} = (0.714)g \sin \theta$$

For the disk, $I = \frac{1}{2} mR^2$, so:

$$a_{disk} = g \sin \theta / (1 + \frac{1}{2})$$

$$a_{disk} = (2/3)g \sin \theta$$

$$a_{disk} = (0.667)g \sin \theta$$

For the hoop, $I = mR^2$, so:

$$a_{hoop} = g \sin \theta / (1 + 1)$$

$$a_{hoop} = (1/2)g \sin \theta$$

$$a_{hoop} = (0.500)g \sin \theta$$

The object with the largest acceleration reaches the bottom first.

The order is the sphere, disk, hoop.

39. B is correct.

The *kinetic energy of a rotating object* is:

$$K = \frac{1}{2}I\omega^2$$

The moment of inertia is given in SI units, but the *angular speed is in rpm*, which is not an SI unit.

Convert:

$$96.0 \text{ rpm} \cdot (1 \text{ m}/ 60 \text{ s}) \cdot (2\pi \text{ rad/rev}) = 10.05 \text{ rad/s}$$

$$K = (0.5) \cdot (6.0 \times 10^{-3} \text{ kg·m}^2) \cdot (10.05 \text{ rad/s})^2$$

$$K = 0.30 \text{ J}$$

40. D is correct.

If a wheel of radius r rolls without slipping on the pavement, the relationship between angular speed ω and translational speed is:

$$v = r\omega$$

$$\omega = v / r$$

$$\omega = (6.00 \text{ m/s}) / (0.120 \text{ m})$$

$$\omega = 50.00 \text{ rad/s}$$

continued…

Converting to rpm:

$$\omega = (50.00 \text{ rad/s}) \cdot (1 \text{ rev} / 2\pi \text{ rad}) \cdot (60 \text{ s} / 1 \text{ min})$$

$$\omega = 477.5 \text{ rpm}$$

$$\omega = 478 \text{ rpm}$$

41. C is correct.

Her *moment of inertia* does not remain constant because the radial position of her hands is changing.

Angular momentum does remain constant because there is no torque on her, assuming the ice is frictionless.

Her *kinetic energy* changes. To note this, notice that her hands initially execute uniform circular motion, and hence there is no tangential force. If this condition is maintained, her kinetic energy will be constant.

However, *as she pulls her hands in,* they are no longer in a uniform circular motion.

During this time, there is a tangential component of force, and work will be done.

Another way to think about it is to recognize that because L is conserved,

$$L_0 = L_f$$

$$I_0 \omega_0 = I_f \omega_f$$

I will decrease by some amount, and ω will multiply by that same amount, e.g., I halves and ω doubles.

When examining $KE = \frac{1}{2} I \omega^2$, I has gone down by half, but ω has doubled.

Since KE depends on ω squared, the change to ω has a greater impact on KE.

KE increases.

42. D is correct.

There are two perpendicular components to Tanya's acceleration.

Centripetal acceleration:

$$a_c = r\omega^2$$

and *tangential acceleration*:

$$a_t = r\alpha$$

Since these two acceleration components are perpendiculars, find the magnitude of the total linear acceleration:

$$a = \sqrt{(a_c^2 + a_t^2)}$$

$$a = \sqrt{[(r\omega^2)^2 + (r\alpha)^2]}$$

$$a = r\sqrt{(\omega^4 + \alpha^2)}$$

$$a = (4.65 \text{ m}) \cdot \sqrt{[(1.25 \text{ rad/s})^4 + (0.745 \text{ rad/s}^2)^2]}$$

$$a = 8.05 \text{ m/s}^2$$

43. C is correct.

For a *rotating object*:

$$\Delta\theta = \omega \Delta t$$

The *angular speed* is given in rpm, but needs to be in deg/s:

33.0 rpm·(1 min / 60 s)·(360 deg / rev) = 198.0 deg/s

$\Delta\theta = (198.0 \text{ deg/s})\cdot(0.32 \text{ s})$

$\Delta\theta = 63°$

44. C is correct.

The question asks to find the *magnitude* of the force of the floor on the bottom of the ladder.

Two forces act on the bottom of the ladder: the normal force of the floor, N, which acts in the vertical direction, and the force of friction, f, which acts horizontally.

These two forces are directed in mutually perpendicular directions, so the magnitude of the net force of the floor on the bottom of the ladder is:

$F_{\text{net, bottom}} = \sqrt{(N^2 + f^2)}$

The values of N and f can be found by applying the laws of rotational and linear static equilibrium.

Assume the ladder is static (not moving), but this is not a valid assumption until it is known whether the force of friction exceeds the limit imposed by static friction. This is verified at the end.

If the ladder is in equilibrium, the sum of the forces in the *x*- and *y*-directions is zero, and the sum of the torques is zero:

$\Sigma F_x = 0; \Sigma F_y = 0; \Sigma \tau = 0$

The force equations are:

$\Sigma F_x = F_W - f = 0$

$\Sigma F_y = N - Mg - mg = 0$

where F_W is the normal force from the wall on the top of the ladder, M is the mass of the hanging block, and m is the mass of the ladder

Since the problem statement states the wall is "smooth," there is no frictional force from the wall on the ladder.

The ΣF_y equation has only one unknown quantity and can be solved for N:

$N = (M + m)g$

$N = (80\text{kg} + 50\text{kg})\cdot(9.8\text{m/s}^2) = 1{,}274$ N

For the torque equation, choose the origin where the top of the ladder touches the wall.

continued...

With that choice, the torque due to the normal force of the wall is zero, and the torque due to the hanging block is zero (since these forces are exerted right at the origin).

There are three non-zero torques.

One is the torque due to gravity on the ladder. This acts as if all the mass of the ladder is concentrated at the center of mass.

The next torque is due to the normal force of the floor on the bottom of the ladder.

The third is the force of friction acting on the bottom of the ladder.

Using the general formula for torque:

$$\tau = Fr \sin \phi$$

where F is the force, r is the distance from the origin to the point where the force is applied, and ϕ is the angle between the force vector and the r vector (which points from the origin to the point where the force is applied):

$$\sum \tau = 0 + 0 - mg(L/2) \sin \beta - fL \sin \alpha + NL \sin \beta = 0$$

where m is the mass of the ladder and α and β are the angles in the figure shown

Take counterclockwise torques to be positive and clockwise torques to be negative.

Use trigonometry to find α and β:

$$\alpha = \sin^{-1}(h/L)$$

$$\alpha = \sin^{-1}(3.7/5)$$

$$\alpha = 47.73°$$

$$\beta = 90° - \alpha = 42.27°$$

There is only one unknown quantity in the torque equation, so, dividing by L and rearranging, solve for f:

$$f \sin \alpha = N \sin \beta - \tfrac{1}{2}mg \sin \beta$$

$$f = [(N - \tfrac{1}{2}mg) \cdot \sin \beta] / \sin \alpha$$

$$f = \frac{[1274 \text{ N} - \tfrac{1}{2}(50 \text{ kg}) \cdot (9.8 \text{ m/s}^2) \cdot (\sin 42.27°)]}{[\sin 47.73°]}$$

$$f = 935.3 \text{ N}$$

Calculate the magnitude of the force on the bottom of the latter due to the floor:

$$F_{\text{net, bottom}} = \sqrt{(N^2 + f^2)}$$

$$F_{\text{net, bottom}} = \sqrt{[(1274 \text{ N})^2 + (935.3 \text{ N})^2]}$$

$$F_{\text{net, bottom}} = 1580 \text{ N}$$

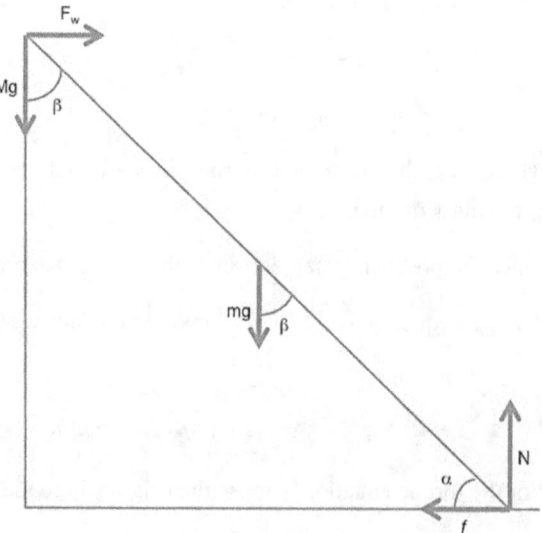

continued...

Check if the force of friction is lower than the limit imposed by static friction.

$f_{max} = \mu_s N$

$f_{max} = (0.750) \cdot (1274 \text{ N})$

$f_{max} = 955 \text{ N}$

The calculated value for static friction force is 935.3 N, lower than the limit imposed by static friction, so the assumption that the system is in static equilibrium is valid.

If this number had been lower than the value of f, the static frictional force would *not* be able to hold the ladder stationary.

45. B is correct.

The maximum displacement occurs when the acceleration stops forward motion, and the wheel reverses direction.

$\theta_0 = 0$

$\Delta\theta = \theta - 0 = \theta$

The kinematic relation is:

$\omega_f^2 - \omega_i^2 = 2\alpha\theta$

The final angular speed is zero at the instant that the wheel changes direction.

$-\omega_i^2 = 2\alpha\theta$

Thus, the angular displacement at that instant is:

$\theta = -(\omega_i^2 / 2\alpha)$

$\theta = -(29.0 \text{ rad/s})^2 / [2 \cdot (-0.52 \text{ rad/s}^2)]$

$\theta = 809 \text{ rad}$

46. C is correct.

The simplest way to determine the direction of the angular momentum for a rotating object is to use the *right-hand rule*.

Take the fingers of your right hand and curl them in the direction that the object is rotating and stick your thumb out perpendicular to your fingers (as in a "thumbs-up" or "thumbs down" signal).

The direction your thumb points in the direction of the angular momentum – in this case, down.

For a mathematical approach, use the definition:

$L = r \times p$ is applied to a point on the outer edge of the object

Taking a point on the disk along the positive x-axis, the position vector r is positive x-direction.

The momentum vector p is in the positive z-direction.

The cross-product of these two vectors points down in the negative y-direction.

Notes for active learning

Work and Energy – Detailed Explanations

1. D is correct.

The final velocity in projectile motion is related to the maximum height of the projectile through conservation of energy:

$KE = PE$

$½mv^2 = mgh$

When the stone thrown straight up passes its starting point on its way back down, its downward speed equals its initial upward velocity (2D motion).

The stone thrown straight downward contains the same magnitude of initial velocity as the stone is thrown upward, and thus the stone thrown upward and the stone thrown downward to have the same final speed.

A stone is thrown horizontally (or, for example, a stone was thrown at 45°) does not achieve the same height h as a stone thrown straight up, so it has a smaller final vertical velocity.

2. B is correct.

Work = force × displacement × cos θ

$W = Fd \cos \theta$

where θ is the angle between the vectors F and d

$W = (5 \text{ N}) \cdot (10 \text{ m}) \cos 45°$

$W = (50 \text{ J}) \cdot (0.7)$

$W = 35 \text{ J}$

3. A is correct.

$KE = ½mv^2$

KE is influenced by mass and velocity. However, since velocity is squared, its influence on KE is greater than the influence of mass.

4. B is correct.

Work = force × displacement × cos θ

$W = Fd \cos \theta$

$\cos 90° = 0$

$W = 0$

Since the force of gravity acts perpendicular to the distance traveled by the ball, the force due to gravity does no work in moving the ball.

5. E is correct.

$$KE = \tfrac{1}{2}mv^2$$

$$KE = \tfrac{1}{2}(5 \text{ kg}) \cdot (2 \text{ m/s})^2$$

$$KE = 10 \text{ J}$$

6. A is correct.

$$W = Fd \cos \theta$$

$$\cos \theta = 1$$

$$F = W / d$$

$$F = (360 \text{ J}) / (8 \text{ m})$$

$$F = 45 \text{ N}$$

$$F = ma$$

$$m = F / a$$

$$m = (45 \text{ N}) / (10 \text{ m/s}^2)$$

$$m = 4.5 \text{ kg}$$

7. D is correct.

On a displacement (x) vs. force (F) graph, the displacement is the y-axis, and the force is the x-axis.

The slope is x / F (in units of m/N), which is the reciprocal of the spring constant k, measured in N/m.

8. C is correct.

Work done by a spring equation:

$$W = \tfrac{1}{2}kx^2$$

$$W = \tfrac{1}{2}(22 \text{ N/m}) \cdot (3 \text{ m})^2$$

$$W = 99 \text{ J}$$

9. A is correct.

The force of gravity always points down.

When the ball is moving upwards, the direction of its displacement is opposite of that of the force of gravity, and therefore the work done by gravity is negative.

On the way down, the direction of displacement is the same as that of the force of gravity, and therefore the work done by gravity is positive.

10. B is correct.

Work done by gravity is an object's change in gravitational PE.

$$W = -PE$$

$$A_1 = 400 \text{ J}$$

By the work-energy theorem,

$$W = KE$$

$$B_1 = 400 \text{ J}$$

11. D is correct.

Work is calculated as the product of force and displacement parallel to the direction of the applied force:

$$W = Fd \cos \theta$$

where some component of d is in the direction of the force

12. B is correct.

Work only depends on force and distance:

$$W = Fd \cos \theta$$

Power = W / t is the amount of work done in a unit of time.

13. A is correct.

The area under the curve on a graph is the product of the values of $y \times x$.

Here, the y value is force, and the x value is distance:

$$Fd = W$$

14. C is correct.

This is the conservation of energy. The only force acting on the cat is gravity.

$$KE = PE_g$$

$$KE = mgh$$

$$KE = (3 \text{ kg}) \cdot (10 \text{ m/s}^2) \cdot (4 \text{ m})$$

$$KE = 120 \text{ J}$$

15. B is correct.

Although the book is stationary with respect to the plank, the plank is applying a force to the book, causing it to accelerate in the direction of the force. Since the displacement of the point of application of the force is in the same direction as the force, the work done is positive.

Choices D and E are not correct because work is a scalar and has no direction.

16. D is correct.

$$W = Fd$$
$$d = W / F$$
$$d = (350 \text{ J}) / (900 \text{ N})$$
$$d = 0.39 \text{ m}$$

17. E is correct.

Conservation of energy between kinetic energy and potential energy:

$$KE = PE$$
$$KE = \tfrac{1}{2}mv^2 \text{ and } PE = mgh$$

Set the equations equal:

$$\tfrac{1}{2}mv^2 = mgh$$

cancel m from each side

$$\tfrac{1}{2}v^2 = gh$$

h is only dependent on the initial v, equal between both objects, so the two objects rise to the same height.

18. A is correct.

$$\text{Work} = \text{Power} \times \text{time}$$
$$P_1 = W / t$$
$$P_2 = (3 \text{ W}) / (1/3 \text{ } t)$$
$$P_2 = 3(3/1) \cdot (W / t)$$
$$P_2 = 9(W / t)$$
$$P_2 = 9(P_1)$$

19. D is correct.

Conservation of energy:

$$KE = PE$$
$$KE = mgh$$
$$W = mg$$
$$KE = Wh$$
$$KE = (450 \text{ N}) \cdot (9 \text{ m})$$
$$KE = 4{,}050 \text{ J}$$

20. A is correct.

$$F_1 = -kx_1$$

Solve for the spring constant k:

$$k = F / x_1$$
$$k = (160 \text{ N}) / (0.23 \text{ m})$$
$$k = 696 \text{ N/m}$$
$$F_2 = -kx_2$$
$$F_2 = (696 \text{ N/m}) \cdot (0.34 \text{ m})$$
$$F_2 = 237 \text{ N}$$

21. B is correct.

There is a frictional force since the net force = 0

The mule pulls in the same direction as the direction of travel, so $\cos \theta = 1$

$$W = Fd \cos \theta$$
$$d = v \Delta t$$
$$W = Fv \Delta t$$

22. D is correct.

$$W = Fd \cos \theta$$
$$F_T = W / (d \times \cos \theta)$$
$$F_T = (540 \text{ J}) / (18 \text{ m} \times \cos 32°)$$
$$F_T = (540 \text{ J}) / (18 \text{ m} \times 0.848)$$
$$F_T = 35 \text{ N}$$

23. B is correct.

The spring force balances the gravitational force on the mass.

Therefore:

$$Fg = -kx$$
$$mg = -kx$$

By adding an extra 120 grams, the mass is doubled:

$$(2m)g = -kx$$

Since the weight mg and the spring constant k are constant, only x changes.

continued…

Thus, after the addition of 120 g, x doubles:

$$PE_1 = \tfrac{1}{2}kx^2$$

$$PE_2 = \tfrac{1}{2}k(2x)^2$$

$$PE_2 = \tfrac{1}{2}k(4x^2)$$

$$PE_2 = 4(\tfrac{1}{2}kx^2)$$

The potential energy increases by a factor of 4.

24. C is correct.

In each case, the car's energy is reduced to zero by the work done by the frictional force, or:

$$KE + (-W) = 0$$

$$KE = W$$

Each car starts with kinetic energy $KE = (1/2)mv^2$.

The initial speed is the same for each car, so the Ferrari has the most KE due to the differences in mass.

Thus, reducing the Ferrari's energy to zero requires the most work.

25. E is correct.

The hammer does work on the nail as it drives it into the wood.

The amount of work done equals the amount of kinetic energy lost by the hammer:

$$\Delta KE = \Delta W$$

26. A is correct.

The only force doing work is the road's friction, so the work done by the road's friction is the total work.

This work equals the change in KE.

$$W = \Delta KE$$

$$W = KE_f - KE_i$$

$$W = \tfrac{1}{2}mv_2^2 - \tfrac{1}{2}mv_1^2$$

$$W = 0 - [\tfrac{1}{2}(1{,}500 \text{ kg}) \cdot (25 \text{ m/s})^2]$$

$$W = -4.7 \times 10^5 \text{ J}$$

27. D is correct.

$$KE = \tfrac{1}{2}mv^2$$

$$KE_{car} = \tfrac{1}{2}(1,000 \text{ kg}) \cdot (4.72 \text{ m/s})^2$$

$$KE_{car} = 11,139 \text{ J}$$

Calculate the KE of the 2,000 kg truck with 20 times the KE:

$$KE_{truck} = KE_{car} \times 20$$

$$KE_{truck} = (11,139 \text{ J}) \times 20$$

$$KE_{truck} = 222.7 \text{ kJ}$$

Calculate the speed of the 2,000 kg truck:

$$KE = \tfrac{1}{2}mv^2$$

$$v^2 = 2KE / m$$

$$v^2 = 2(222.7 \text{ kJ}) / (2,000 \text{ kg})$$

$$v_{truck} = \sqrt{[2(222.7 \text{ kJ}) / (2,000 \text{ kg})]}$$

$$v_{truck} = 14.9 \text{ m/s}$$

28. E is correct.

Gravity and the normal force are balanced, vertical forces.

Since the car is slowing (i.e., accelerating backward), there is a net force backward due to friction (i.e., braking).

Newton's First Law of Motion states that the car would keep moving forward in the absence of any forces.

29. B is correct.

Energy is always conserved, so the work needed to lift the piano is 0.15 m equals the work to pull the rope 1 m:

$$W_1 = W_2$$

$$F_1 d_1 = F_2 d_2$$

$$F_1 d_1 / d_2 = F_2$$

$$F_2 = (6,000 \text{ N}) \cdot (0.15 \text{ m}) / 1 \text{ m}$$

$$F_2 = 900 \text{ N}$$

30. C is correct.

The area under the curve on a graph is the product of the values of $y \times x$.

Here, the y value is force, and the x value is distance:

$$Fd = W$$

31. B is correct.

Most of the Earth's energy comes from the sun, which produces radiation that penetrates the Earth's atmosphere. Likewise, radiation is emitted from the Earth's atmosphere.

32. C is correct.

$$W = Fd$$

$$W = \Delta KE$$

$$F \times d = \tfrac{1}{2}mv^2$$

If v is doubled:

$$F \times d_2 = \tfrac{1}{2}m(2v)^2$$

$$F \times d_2 = \tfrac{1}{2}m(4v^2)$$

$$F \times d_2 = 4(\tfrac{1}{2}mv^2)$$

For equations to remain equal, d_2 must be 4 times d.

33. D is correct.

$$\text{Work} = \text{Power} \times \text{time}$$

$$P = W / t$$

$$W = Fd$$

$$P = (Fd) / t$$

$$P = [(2{,}000 \text{ N}) \cdot (320 \text{ m})] / (60 \text{ s})$$

$$P = 10{,}667 \text{ W} = 10.7 \text{ kW}$$

34. E is correct.

Solution using the principle of conservation of energy.

Assuming the system consists of the barbell alone, the force of gravity and the force of the hands raising the barbell are external forces. Since the system contains only a single object, potential energy is not defined.

The net power expended is:

$$P_{net} = W_{ext} / \Delta t$$

Conservation of energy requires:

$$W_{ext} = \Delta KE$$

$$W_{ext} = \tfrac{1}{2}m(v_f^2 - v_i^2)$$

For constant acceleration situations:

$$(v_f + v_i) / 2 = v_{average} = \Delta y / \Delta t$$

continued…

$(v_f + 0.0 \text{ m/s}) / 2 = 3.0 \text{ m} / 3.0 \text{ s}$

$v_f = 2.0 \text{ m/s}$

Therefore:

$W_{ext} = \frac{1}{2}(25 \text{ kg}) \cdot (2.0 \text{ m/s})^2$

$W_{ext} = 50.0 \text{ J}$

The net power expended is:

$P_{net} = 50.0 \text{ J} / 3.0 \text{ s}$

$P_{net} = 17 \text{ W}$

Solution using work.

The power expended in raising the barbell is:

$P_{net} = W_{net} / \Delta t$

The net work is defined as:

$W_{net} = F_{net}\Delta y$

By Newton's Second law:

$F_{net} = ma$

Find the acceleration:

$\Delta y = \frac{1}{2}a\Delta t^2$

$a = (2) \cdot (3.0 \text{ m}) / (3.0 \text{ s})^2$

$a = 0.67 \text{ m/s}^2$

The net force on the barbell is:

$F_{net} = (25 \text{ kg}) \cdot (0.67 \text{ m/s}^2)$

$F_{net} = (50 / 3) \text{ N}$

The net work is:

$W_{net} = F_{net}\Delta y$

$W_{net} = [(50 / 3) \text{ N}] \cdot (3.0 \text{ m})$

$W_{net} = 50.0 \text{ J}$

The net power expended:

$P_{net} = 50.0 \text{ J} / 3.0 \text{ s}$

$P_{net} = 17 \text{ W}$

35. B is correct.

The bag was never lifted off the ground and moved horizontally at a constant velocity.

$F = 0$

$W = Fd$

$W = 0$ J

Because there is no acceleration, the force is zero, and thus the work is zero.

36. B is correct.

Using energy conservation to solve the problem:

$W = |\Delta KE|$

$Fd = |\frac{1}{2}m(v_f^2 - v_0^2)|$

$d = |m(v_f^2 - v_0^2) / 2F|$

$d = |(1{,}000 \text{ kg}) \cdot [(22 \text{ m/s})^2 - (30 \text{ m/s})^2] / (2) \cdot (9{,}600 \text{ N})|$

$d = |(1{,}000 \text{ kg}) \cdot (484 \text{ m}^2/\text{s}^2 - 900 \text{ m}^2/\text{s}^2) / 19{,}200 \text{ N}|$

$d = 22$ m

Kinematic approach to solve the problem:

$F = ma$

$a = F / m$

$a = (9{,}600 \text{ N}) / (1{,}000 \text{ kg})$

$a = 9.6$ m/s²

$v_f^2 = v_0^2 + 2a\Delta d$

$(v_f^2 - v_0^2) / 2a = \Delta d$

Note that acceleration is negative due to it acting opposite the velocity.

$\Delta d = [(22 \text{ m/s})^2 - (30 \text{ m/s})^2] / 2(-9.6 \text{ m/s}^2)$

$\Delta d = (484 \text{ m}^2/\text{s}^2 - 900 \text{ m}^2/\text{s}^2) / (-19.2 \text{ m/s}^2)$

$\Delta d = (-416 \text{ m}^2/\text{s}^2) / (-19.2 \text{ m/s}^2)$

$\Delta d = 21.7$ m ≈ 22 m

37. C is correct.

$W = 100$ J

Work = Power × time

$P = W / t$

$P = 100$ J $/ 50$ s

$P = 2$ W

38. D is correct.

The original potential energy (with respect to the bottom of the cliff) is converted into kinetic energy.

$mgh = \frac{1}{2} mv_f^2$

Therefore:

$v_f = \sqrt{2gh}$

$v_f = \sqrt{(2) \cdot (10 \text{ m/s}^2) \cdot (58 \text{ m})}$

$v_f = 34$ m/s

Kinematic approach:

$v_f^2 = v_0^2 + 2a\Delta x$

$v_f^2 = 0 + 2a\Delta x$

$v_f = \sqrt{2a\Delta x}$

$v_f = \sqrt{[2(10 \text{ m/s}^2) \cdot (58 \text{ m})]}$

$v_f = \sqrt{(1{,}160 \text{ m}^2/\text{s}^2)}$

$v_f = 34$ m/s

39. E is correct.

$PE = mgh$

If height and gravity are constant, then potential energy is directly proportional to mass.

As such, if the second stone has four times the mass of the first, then it must have four times the potential energy of the first stone.

$m_2 = 4m_1$

$PE_2 = 4PE_1$

Therefore, the second stone has four times the potential energy.

40. B is correct.

$W = Fd$

$W = mgh$, work done by gravity

$W = (1.3 \text{ kg}) \cdot (10 \text{ m/s}^2) \cdot (6 \text{ m})$

$W = 78 \text{ J}$

41. A is correct.

Potential energy (PE) is the energy associated with the relative positions of pairs of objects, regardless of their state of motion.

Kinetic energy (KE) is the energy associated with the motion of single particles, regardless of their location.

42. A is correct.

$F_{spring} = F_{centripetal}$

$F_{spring} = kx$

$kx = 15 \text{ N}$

$x = (15 \text{ N}) / (65 \text{ N/m})$

$x = 0.23 \text{ m}$

$PE_{spring} = \frac{1}{2}kx^2$

$PE_{spring} = \frac{1}{2}(65 \text{ N/m}) \cdot (0.23 \text{ m})^2$

$PE_{spring} = 1.7 \text{ J}$

43. C is correct.

total time = (3.5 h/day)·(7 days)·(5 weeks)

total time = 122.5 h

cost = (8.16 cents/kW·h)·(122.5 h)·(0.12 kW)

cost = 120 cents = $1.20

44. E is correct.

$x = 5.1 \text{ m} \times (\cos 32°)$

$x = 4.33 \text{ m}$

$h = 5.1 \text{ m} - 4.33 \text{ m}$

$h = 0.775 \text{ m}$

$W = Fd$

$W = mg \times h$

$m = W / gh$

$m = (120 \text{ J}) / (9.8 \text{ m/s}^2) \cdot (0.775 \text{ m})$

$m = 15.8 \text{ kg}$

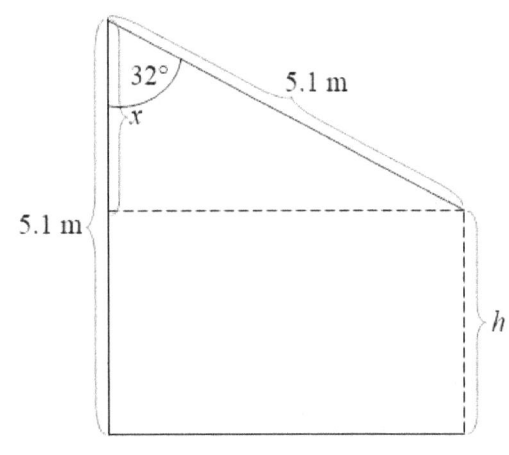

45. D is correct.

Potential energy of spring:

$PE_i + W = PE_f$

$½ k x_i^2 + 111 \text{ J} = ½ k x_f^2$

$111 \text{ J} = ½ k (x_f^2 - x_i^2)$

$111 \text{ J} = ½ k [(2.9\text{m})^2 - (1.4\text{m})^2]$

$111 \text{ J} = ½ k [(8.41\text{m}^2) - (1.96\text{m}^2)]$

$111 \text{ J} = ½ k (6.45\text{m}^2)$

$k = 2(111 \text{ J}) / (6.45 \text{ m}^2)$

$k = 34 \text{ N/m}$

Unit check:

$J = \text{kg} \cdot \text{m}^2/\text{s}^2$

$J/\text{m}^2 = (\text{kg} \cdot \text{m}^2/\text{s}^2) \cdot (1/\text{m}^2)$

$J/\text{m}^2 = (\text{kg}/\text{s}^2)$

$N/\text{m} = (\text{kg} \cdot \text{m}/\text{s}^2) \cdot (1/\text{m})$

$N/\text{m} = (\text{kg}/\text{s}^2)$

46. E is correct.

Potential energy, kinetic energy and work are all measured in joules:

$$J = kg \cdot m^2/s^2$$

$$KE = \tfrac{1}{2}mv^2 = kg(m/s)^2 = J$$

$$PE = mgh$$

$$PE = kg(m/s^2) \cdot (m) = J$$

$$W = Fd = J$$

47. A is correct.

Potential energy of spring:

$$PE = \tfrac{1}{2}kx^2$$

Kinetic energy of mass:

$$KE = \tfrac{1}{2}mv^2$$

Set equal and rearrange:

$$\tfrac{1}{2}kx^2 = \tfrac{1}{2}mv^2$$

cancel ½ from each side of the expression

$$kx^2 = mv^2$$

$$x^2 = (mv^2) / k$$

$$x^2 = (m/k)v^2$$

Since m/k is provided:

$$x^2 = (0.038 \text{ kg} \cdot \text{m/N}) \cdot (18 \text{ m/s})^2$$

$$x^2 = 12.3 \text{ m}^2$$

$$x = \sqrt{12.3} \text{ m}$$

$$x = 3.5 \text{ m}$$

48. A is correct.

$m_t = 2m_c$

$v_t = 2v_c$

KE of the truck:

$KE_t = \frac{1}{2}m_t v_t^2$

Replace the mass and velocity of the truck with the equivalent mass and velocity of the car:

$KE_t = \frac{1}{2}(2m_c)\cdot(2v_c)^2$

$KE_t = \frac{1}{2}(2m_c)\cdot(4v_c^2)$

$KE_t = \frac{1}{2}(8m_c v_c^2)$

The truck has 8 times the kinetic energy of the car.

49. E is correct.

When a car stops, the KE equals the work done by the force of friction from the brakes.

Through friction, the KE is transformed into heat.

50. B is correct.

When the block comes to rest at the end of the spring, the upward force of the spring balances the downward force of gravity.

$F = kx$

$mg = kx$

$x = mg / k$

$x = (30 \text{ kg})\cdot(10 \text{ m/s}^2) / 900 \text{ N/m}$

$x = 0.33 \text{ m}$

51. D is correct.

$KE = \frac{1}{2}mv^2$

$KE = \frac{1}{2}(0.33 \text{ kg})\cdot(40 \text{ m/s})^2$

$KE = 264 \text{ J}$

52. C is correct.

Work is the area under a force *vs.* position graph.

$$\text{area} = Fd = W$$

The area of the triangle as the object moves from 0 to 4 m:

$$A = \tfrac{1}{2}bh$$

$$A = \tfrac{1}{2}(4 \text{ m})(10 \text{ N})$$

$$A = 20 \text{ J}$$

$$W = 20 \text{ J}$$

53. C is correct.

$$KE = PE$$

$$\tfrac{1}{2}mv^2 = mgh$$

$$v^2 / 2g = h$$

If v is doubled:

$$h_B = v_B^2 / 2g$$

$$v_J = 2v_B$$

$$(2v_B)^2 / 2g = h_J$$

$$4(v_B^2 / 2g) = h_J$$

$$4h_B = h_J$$

James's ball travels 4 times higher than Bob's ball.

54. B is correct.

Hooke's Law is given as:

$$F = -kx$$

The negative is only by convention to demonstrate that the spring force is a restoring force.

Graph B is correct because the force is linearly increasing with increasing distance.

All other graphs are constant or exponential.

55. C is correct.

A decrease in the KE for the rocket causes a gain in its gravitational PE, or the transfer of heat, or a combination.

The rocket loses some KE due to air resistance (friction).

Thus, some of the rocket's KE is converted to heat, which causes the air temperature surrounding the rocket to increase.

Therefore, the average KE of the air molecules increases.

Notes for active learning

Waves and Periodic Motion – Detailed Explanations

1. B is correct.

Frequency is the number of cycles per second a wave experiences, independent of the wave's amplitude.

2. D is correct.

Hooke's Law:

$F = kx$

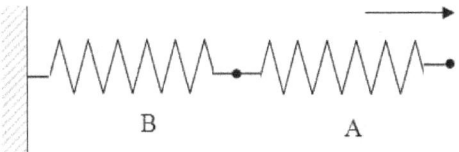

It is known that the force on each spring must be equal if they are in static equilibrium, therefore:

$F_A = F_B$

Therefore, the expression can be written as:

$k_A L_A = k_B L_B$

Solve for the spring constant of spring B:

$k_B = (k_A L_A) / L_B$

3. D is correct.

In a longitudinal wave, particles of a material are displaced parallel to the direction of the wave.

4. C is correct.

speed = wavelength × frequency

$v = \lambda f$

$v = (0.25 \text{ m}) \cdot (1{,}680 \text{ Hz})$

$v = 420 \text{ m/s}$

5. A is correct.

$E_{stored} = PE = \frac{1}{2}kA^2$

Stored energy is potential energy.

In a simple harmonic motion (e.g., a spring), the potential energy is:

$PE = \frac{1}{2}kx^2$ or $\frac{1}{2}kA^2$,

where k is a constant and A (or x) is the distance from equilibrium

A is the amplitude of a wave in simple harmonic motion (SHM).

6. E is correct.

The spring will oscillate around its new equilibrium position (3 cm below the equilibrium position with no mass hanging) with period $T = 2\pi\sqrt{m/k}$ since the mass-spring system is undergoing simple harmonic motion.

To find k, consider how much the spring stretched when the mass was hung from it.

Since the spring found a new equilibrium point 3 cm below its natural length, the upwards force from the spring (F_s) must balance the downwards gravitational force (F_g) at that displacement:

$$|F_s| = |F_g|$$

$$kd = mg$$

$$k(0.03 \text{ m}) = (11 \text{ kg}) \cdot (9.8 \text{ m/s}^2)$$

$$k = 3593 \text{ N/m}$$

Now, solve for T:

$$T = 2\pi\sqrt{m/k}$$

$$T = 2\pi\sqrt{11 \text{ kg} / 3593 \text{ N/m}}$$

$$T = 0.35 \text{ s}$$

The frequency is the reciprocal of the period:

$$f = 1/T$$

$$f = 1/(0.35 \text{ s})$$

$$f = 2.9 \text{ Hz}$$

7. C is correct.

$$T = 1/f$$

8. D is correct.

The period of a pendulum:

$$T = 2\pi\sqrt{L/g}$$

The period only depends on the pendulum's length and gravity.

In an elevator, the apparent force of gravity only changes if the elevator is accelerating in either direction.

9. A is correct.

The period is the reciprocal of the frequency:

$$T = 1/f$$

$$T = 1/100 \text{ Hz}$$

$$T = 0.01 \text{ s}$$

10. E is correct.

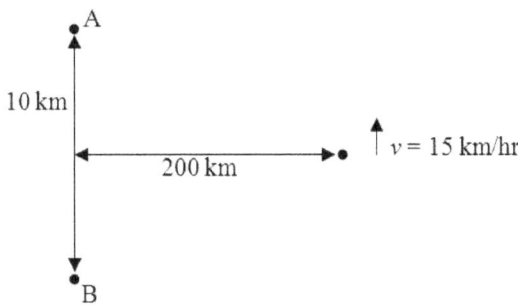

Convert v to m/s:

$v = (15 \text{ km}/1 \text{ h}) \cdot (1 \text{ h}/60 \text{ min}) \cdot (1 \text{ min}/60 \text{ s}) \cdot (10^3 \text{ m}/1 \text{ km})$

$v = 4.2$ m/s

Convert frequency to λ:

$\lambda = c / f$

$\lambda = (3 \times 10^8 \text{ m/s}) / (4.7 \times 10^6 \text{ Hz})$

$\lambda = 63.8$ m

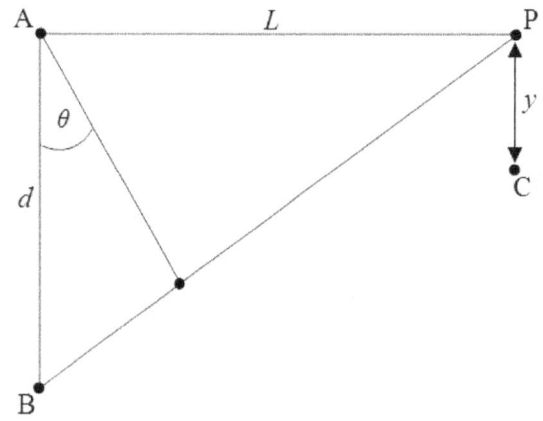

According to Young's Equation:

$\lambda = yd / mL$

where m = 0, 1, 2, 3, 4…

Solve for y by rearranging to isolate y:

$y = \lambda L m / d$

y = distance travelled by the ship:

$y = vt$

Since the first signal came at the point of maximum intensity, m = 0 at that time, at the next maximum m = 1.

Therefore:

$t = L\lambda / vd$

$t = (200{,}000 \text{ m}) \cdot (63.8 \text{ m}) / (4.2 \text{ m/s}) \cdot (10{,}000 \text{ m})$

$t = 304$ s

Convert time from seconds to minutes:

$t = (304 \text{ s}) \cdot (1 \text{ min}/60 \text{ s})$

$t = 5.06$ min ≈ 5.1 min

For m values greater than 1, the calculated times are beyond the answer choices, so 5.1 min is the answer.

11. D is correct.

The *tension* in the rope is given by:

$$T = (mv^2) / L$$

where v is the velocity of the wave, and L is the length of the rope

Substituting:

$$v = L / t$$

$$T = [m(L/t)^2] / L$$

$$T = mL / t^2$$

$$t^2 = mL / T$$

$$t = \sqrt{mL / T}$$

$$t = \sqrt{[(2.31 \text{ kg}) \cdot (10.4 \text{ m}) / 74.4 \text{ N}]}$$

$$t = \sqrt{(0.323 \text{ s}^2)} = 0.57 \text{ s}$$

12. A is correct.

$$\omega_A = 2\omega_B$$

$$\omega_B = \sqrt{g / l_B}$$

Therefore:

$$l_B = g / \omega^2_B$$

Similarly, for A:

$$l_A = g / \omega^2_A$$

$$l_A = g / (2\omega_B)^2$$

$$l_A = \tfrac{1}{4} g / \omega^2_B$$

$$l_A = \tfrac{1}{4} l_B$$

13. B is correct.

$$F = -kx$$

Since the motion is simple harmonic, the restoring force is proportional to displacement.

Therefore, if the displacement is 5 times greater, then so is the restoring force.

14. C is correct.

$$\text{Period} = (60 \text{ s}) / (10 \text{ oscillations})$$

$$T = 6 \text{ s}$$

The period is the time for one oscillation.

If 10 oscillations take 60 s, then one oscillation takes 6 s.

15. E is correct.

Conservation of Energy:

$$\text{total ME} = \Delta KE + \Delta PE = \text{constant}$$

$$\tfrac{1}{2}mv^2 + \tfrac{1}{2}kx^2 = \text{constant}$$

16. B is correct.

A displacement from the position of maximum elongation to the position of maximum compression represents *half* a cycle.

If it takes 1 s, then the time required for a complete cycle is 2 s.

$$f = 1 / T$$

$$f = 1 / 2 \text{ s}$$

$$f = 0.5 \text{ Hz}$$

17. C is correct.

Sound waves are longitudinal waves.

18. E is correct.

$$\text{speed} = \text{wavelength} \times \text{frequency}$$

$$\text{speed} = \text{wavelength} / \text{period}$$

$$v = \lambda / T$$

$$\lambda = vT$$

$$\lambda = (362 \text{ m/s}) \cdot (0.004 \text{ s})$$

$$\lambda = 1.5 \text{ m}$$

19. A is correct.

$$a = -A\omega^2 \cos(\omega t)$$

where A is the amplitude or displacement from the resting position

20. D is correct.

The *acceleration of a simple harmonic oscillation* is:

$$a = -A\omega^2 \cos(\omega t)$$

Its maximum occurs when $\cos(\omega t)$ equals 1

$$a_{max} = -\omega^2 x$$

If ω is doubled:

$$a = -(2\omega)^2 x$$

$$a = -4\omega^2 x$$

The maximum value of acceleration changes by a factor of 4.

21. B is correct.

Resonant frequency of a spring and mass system in any orientation:

$$\omega = \sqrt{k/m}$$

$$f = \omega / 2\pi$$

$$T = 1/f$$

$$T = 2\pi\sqrt{m/k}$$

Period of a spring does not depend on gravity.

The period remains constant because only mass and the spring constant affect the period.

22. E is correct.

$$v = \lambda f$$

$$\lambda = v/f$$

An increase in v and a decrease in f must increase λ.

23. B is correct.

Frequency is the measure of oscillations or vibrations per second.

frequency = 60 vibrations in 1 s

frequency = 60 Hz

speed = 30 m / 1 s

speed = 30 m/s

24. A is correct.

$T = (mv^2) / L$

$m = TL / v^2$

$m = (60 \text{ N}) \cdot (16 \text{ m}) / (40 \text{ m/s})^2$

$m = (960 \text{ N} \cdot \text{m}) / (1{,}600 \text{ m}^2/\text{s}^2)$

$m = 0.6$ kg

25. E is correct.

Amplitude is *independent* of frequency.

26. C is correct.

$f = \text{\# cycles} / \text{time}$

$f = 60 \text{ drips} / 40 \text{ s}$

$f = 1.5$ Hz

27. D is correct.

Transverse waves are characterized by their crests and valleys, which are caused by the particles of the wave traveling "up and down" with respect to the lateral movement of the wave.

The particles in longitudinal waves travel parallel to the direction of the wave.

28. B is correct.

The velocity *vs.* time graph shows that at $t = 0$, the velocity of the particle is positive, and speed is increasing.

When speed increases, velocity and acceleration point in the same direction.

Therefore, the acceleration is non-zero and positive.

Only graph B displays a positive acceleration at $t = 0$.

29. E is correct.

The speed of a wave is determined by the characteristics of the medium (and the type of wave).

Speed is independent of amplitude.

30. A is correct.

$f = 1 / \text{period}$

$f = \text{\# cycles} / \text{second}$

$f = 1 \text{ cycle} / 2 \text{ s}$

$f = \frac{1}{2}$ Hz

31. B is correct.

$$f = v / \lambda$$

$$\lambda = v / f$$

$$\lambda = (340 \text{ m/s}) / (2{,}100 \text{ Hz})$$

$$\lambda = 0.16 \text{ m}$$

32. D is correct.

$$\text{Period (T)} = 2\pi\sqrt{(L / g)}$$

The period is independent of the mass.

33. E is correct.

$$v = \omega x$$

$$\omega = v / x$$

$$\omega = (15 \text{ m/s}) / (2.5 \text{ m})$$

$$\omega = 6.0 \text{ rad/s}$$

34. D is correct.

Unpolarized light on a polarizer reduces the intensity by ½.

$$I = (½)I_0$$

The light is further reduced in intensity by the second filter.

Law of Malus:

$$I = I_0 \cos^2 \theta$$

$$(0.14\, I_0) = (0.5\, I_0) \cos^2 \theta$$

$$0.28 = \cos^2 \theta$$

$$\cos^{-1} \sqrt{(0.28)} = \theta$$

$$\theta = 58°$$

35. C is correct.

At a maximum distance from equilibrium, the energy in the system is potential energy, and the speed is zero.

Therefore, kinetic energy is zero. Since there is no kinetic energy, the mass has no velocity.

36. D is correct.

$$v = \lambda f$$

$$f = v / \lambda$$

$$f = (240 \text{ m/s}) / (0.1 \text{ m})$$

$$f = 2{,}400 \text{ Hz}$$

37. B is correct.

In a *transverse wave*, the vibrations of particles are *perpendicular* to the wave's direction of travel.

Transverse waves have *crests and troughs* that move along the wave.

In a *longitudinal wave*, the vibrations of particles are *parallel* to the wave's direction of travel.

Longitudinal waves have *compressions and rarefactions* that move along the wave.

38. C is correct.

$$v = \sqrt{(T / \mu)}$$

where μ is the linear density of the wire

$$T = v^2 \mu$$

$$\mu = \rho A$$

where A is the cross-sectional area of the wire and equals πr^2

$$\mu = (2{,}700 \text{ kg/m}^3)\pi(4.6 \times 10^{-3} \text{ m})^2$$

$$\mu = 0.18 \text{ kg/m}$$

$$T = (36 \text{ m/s})^2 \cdot (0.18 \text{ kg/m})$$

$$T = 233 \text{ N}$$

39. D is correct.

Refraction is the change in the direction of a wave caused by the change in the wave's speed.

Examples of waves include sound waves and light waves.

Refraction is often seen when a wave passes from one medium to a different medium (e.g., from air to water and vice versa).

40. C is correct.

$$f = \text{\# cycles / second}$$

$$f = 2 \text{ cycles} / 1 \text{ s}$$

$$f = 2 \text{ Hz}$$

41. A is correct.

Pitch is how the brain perceives frequency.

Pitch becomes higher as the frequency increases.

42. C is correct.

The KE is maximum when the spring is neither stretched nor compressed.

If the object is bobbing, KE is maximum at the midpoint between fully stretched and fully compressed because all the spring's energy is KE rather than a mix of KE and PE.

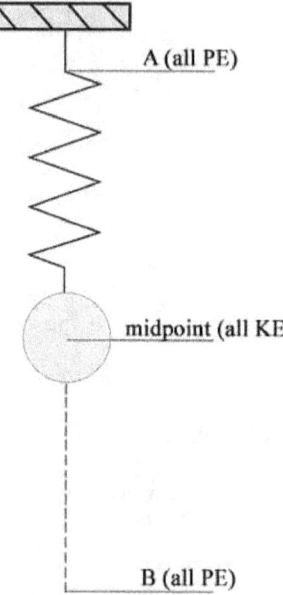

43. B is correct.

$\text{Torque} = rF \sin \theta$

$F = ma$, substitute mg for F

$\tau = rmg \sin \theta$

$\tau = (1 \text{ m}) \cdot (0.5 \text{ kg}) \cdot (10 \text{ m/s}^2) \sin 60°$

$\tau = (5 \text{ kg} \cdot \text{m}^2/\text{s}^2) \times 0.87$

$\tau = 4.4 \text{ N} \cdot \text{m}$

44. E is correct.

The Doppler effect can be observed in all types of waves.

45. A is correct.

$v = \sqrt{(T/\mu)}$

where μ is the linear density of the wire

$F_T = ma$

$F_T = (2{,}500 \text{ kg}) \cdot (10 \text{ m/s}^2)$

$F_T = 25{,}000 \text{ N}$

$v = \sqrt{(25{,}000 \text{ N} / 0.65 \text{ kg/m})}$

$v = 196 \text{ m/s}$

The weight of the wire can be assumed to be negligible compared to the cement block.

46. B is correct.

$f = \frac{1}{2\pi}[\sqrt{(g/L)}]$

Frequency is independent of mass.

47. A is correct.

$$T = 2\pi\sqrt{(L/g)}]$$
$$T = 2\pi\sqrt{(3.3 \text{ m} / 10 \text{ m/s}^2)}$$
$$T = 3.6 \text{ s}$$

48. C is correct.

$$f = (1/2\pi)\sqrt{(k/m)}$$

If k increases by a factor of 2, then f increases by a factor of $\sqrt{2}$ (or 1.41).

Increasing by a factor of 1.41 or 41%

49. D is correct.

In a simple harmonic motion (SHM), the acceleration is highest at the ends of motions (points A and D), where velocity is zero.

Velocity is highest at the nadir, where acceleration equals zero (point C).

50. A is correct.

At the lowest point, the KE is maximum, and the PE is minimum.

The loss of gravitational PE equals the gain in KE:

$$mgh = \tfrac{1}{2}mv^2$$

cancel m from each side of the expression

$$gh = \tfrac{1}{2}v^2$$
$$(10 \text{ m/s}^2)\cdot(10 \text{ m}) = \tfrac{1}{2}v^2$$
$$(100 \text{ m}^2/\text{s}^2) = \tfrac{1}{2}v^2$$
$$200 \text{ m}^2/\text{s}^2 = v^2$$
$$v = 14 \text{ m/s}$$

51. A is correct.

Pitch is a psychophysical phenomenon when the sensation of a frequency is commonly referred to as the pitch of a sound.

A perception of high-pitch sound corresponds to a high-frequency sound wave, and a low-pitch sound corresponds to a low-frequency sound wave.

Amplitude plays no role, and speed is constant.

52. C is correct.

Because wind is blowing in the reference frame of the train and observer, it does not need to be considered.

$f_{observed} = [v_{sound} / (v_{sound} - v_{source})] f_{source}$

$f_{observed} = [340 \text{ m/s} / (340 \text{ m/s} - 50 \text{ m/s})] \cdot 500 \text{ Hz}$

$f_{observed} = 586 \text{ Hz}$

$\lambda = v / f$

$\lambda = 340 \text{ m/s} / 586 \text{ Hz}$

$\lambda = 0.58 \text{ m}$

53. B is correct.

$PE = \frac{1}{2} k x^2$

Doubling the amplitude x increases PE by a factor of 4.

54. E is correct.

The elastic modulus is given by:

$E =$ tensional strength / extensional strain

$E = \sigma / \varepsilon$

55. D is correct.

Resonance is when one system transfers its energy to another at its resonant frequency (natural frequency).

It is a forced vibration that produces the highest amplitude response for a given force amplitude.

Notes for active learning

Notes for active learning

Sound – Detailed Explanations

1. B is correct.

Intensity is inversely proportional to distance (in W/m², not dB).

$$I_2 / I_1 = (d_1 / d_2)^2$$

$$I_2 / I_1 = (3 \text{ m} / 30 \text{ m})^2$$

$$100 \, I_2 = I_1$$

The intensity is 100 times greater at 3 m than 30 m.

Intensity to decibel relationship:

$$I \text{ (dB)} = 10 \log_{10} (I / I_0)$$

The intensity to dB relationship is logarithmic.

Thus, if I_1 is 100 times the initial intensity, then it is two times the dB intensity because:

$$\log_{10} (100) = 2$$

Thus, the decibel level at 3 m is:

$$I \text{ (dB)} = (2) \cdot (20 \text{ dB})$$

$$I = 40 \text{ dB}$$

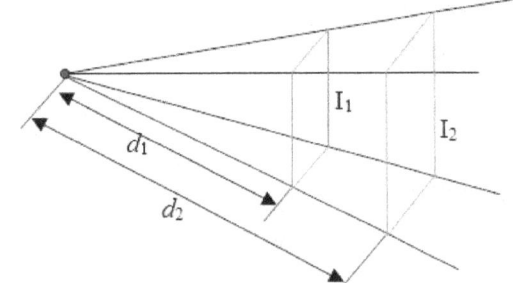

2. A is correct.

distance = velocity × time

$$d = vt$$

$$t = d / v$$

$$t = (6{,}000 \text{ m}) / (340 \text{ m/s})$$

$$t = 18 \text{ s}$$

3. B is correct.

Resonance occurs when a vibrating system is driven at its resonance frequency, resulting in a relative maximum of the vibrational energy of the system.

When the force associated with the vibration exceeds the strength of the material, the glass shatters.

4. C is correct.

The third harmonic is shown below:

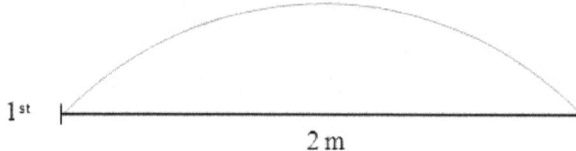
1st |————2 m————| 0.5 wavelength

2nd |——————————| 1 wavelength

3rd |——————————| 1.5 wavelength

There are $(3/2)\lambda$ in the 2 m wave in the third harmonic

$L = (n / 2)\lambda$ (for n harmonic)

$L = (3 / 2)\lambda$ (for 3rd harmonic)

$L(2 / 3) = \lambda$

$\lambda = (2 \text{ m}) \cdot (2 / 3)$

$\lambda = 4/3$ m

5. B is correct.

High-pitched sound has a high frequency.

6. E is correct.

Snell's law:

$n_1 \sin \theta_1 = n_2 \sin \theta_2$

Solve for θ_2:

$(n_1 / n_2) \sin \theta_1 = \sin \theta_2$

$\sin \theta_1 = (n_1 / n_2) \sin \theta_2$

$\theta_2 = \sin^{-1}[(n_1 / n_2) \sin \theta_1]$

Substituting the given values:

$\theta_2 = \sin^{-1}[(1 / 1.5) \sin 60°]$

$\theta_2 = \sin^{-1}(0.67 \sin 60°)$

7. D is correct.

For a standing wave, the length and wavelength are related:

$L = (n / 2)\lambda$ (for n harmonic)

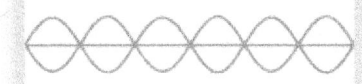

From the diagram, the wave is the 6th harmonic:

$L = (6 / 2)\lambda$

$\lambda = (2 \text{ m}) \cdot (2 / 6)$

$\lambda = 0.667$ m

$f = v / \lambda$

$f = (92 \text{ m/s}) / (0.667 \text{ m})$

$f = 138$ Hz

8. A is correct.

$v = d / t$

$v = (0.6 \text{ m}) / (0.00014 \text{ s})$

$v = 4{,}286$ m/s

$\lambda = v / f$

$\lambda = (4{,}286 \text{ m/s}) / (1.5 \times 10^6 \text{ Hz})$

$\lambda = 0.0029$ m = 2.9 mm

9. C is correct.

The wave velocity is increased by a factor of 1.3.

$v^2 = T / \rho_L$

$T = v^2 \times \rho_L$

Increasing v by a factor of 1.3:

$T = (1.3v)^2 \rho_L$

$T = 1.69 v^2 \rho_L$

T increases by 69%

10. D is correct.

$\rho_L = \rho A$

$\rho_L = \rho(\pi r^2)$

Thus, if the diameter decreases by a factor of 2, then the radius decreases by a factor of 2, and the area decreases by a factor of 4. The linear mass density decreases by a factor of 4.

11. B is correct.

The v and period (T) of wire C are equal to wire A, so the ρ_L must be equal.

$\rho_{LA} = \rho_{LC}$

$\rho_A A_A = \rho_C A_C$

$A_C = (\rho_A A_A) / \rho_C$

$(\pi / 4) \cdot (d_C)^2 = (7 \text{ g/cm}^3)(\pi / 4) \cdot (0.6 \text{ mm})^2 / (3 \text{ g/cm}^3)$

$(d_C)^2 = (7 \text{ g/cm}^3) \cdot (0.6 \text{ mm})^2 / (3 \text{ g/cm}^3)$

$d_C^2 = 0.84 \text{ mm}^2$

$d_C = \sqrt{0.84 \text{ mm}^2}$

$d_C = 0.92 \text{ mm}$

12. A is correct.

$A = \pi r^2$

If d increases by a factor of 4, r increases by a factor of 4; area increases by a factor of 16.

13. E is correct.

Since the bird is moving toward the observer, the $f_{observed}$ must be higher than f_{source}.

Doppler shift for an approaching sound source:

$f_{observed} = (v_{sound} / v_{sound} - v_{source}) f_{source}$

$f_{observed} = [340 \text{ m/s} / (340 \text{ m/s} - 10 \text{ m/s})] f_{source}$

$f_{observed} = (340 \text{ m/s} / 330 \text{ m/s}) \cdot (60 \text{ kHz})$

$f_{observed} = (1.03) \cdot (60 \text{ kHz})$

$f_{observed} = 62 \text{ kHz}$

14. C is correct.

When an approaching sound source is heard, the observed frequency is higher than the frequency from the source due to the Doppler effect.

15. D is correct.

Sound requires a medium of solid, liquid, or gas substances to be propagated through.

A vacuum is none of these.

16. E is correct.

According to the Doppler effect, frequency increases as the sound source moves towards the observer.

Higher frequency is perceived as a higher pitch.

Conversely, as the sound source moves *away* from the observer, the perceived pitch decreases.

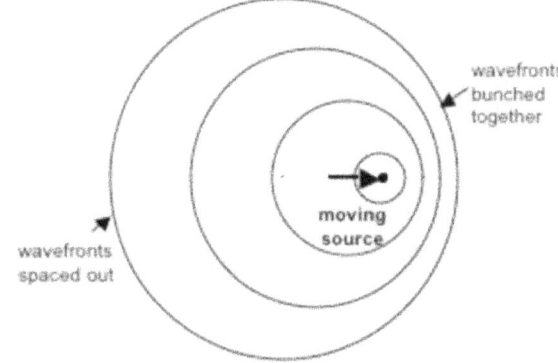

17. C is correct.

If waves are *out of phase*, the combination has its minimum amplitude of (0.6 – 0.4) Pa = 0.2 Pa.

If waves are *in-phase*, the combination has its maximum amplitude of (0.6 + 0.4) Pa = 1.0 Pa.

When the phase difference has a value between in-phase and out of phase, the amplitude is between 0.2 Pa and 1.0 Pa.

18. B is correct.

$$I = P / A$$

$$I = P / \pi d^2$$

Intensity at $2d$:

$$I_2 = P / \pi (2d)^2$$

$$I_2 = P / 4\pi d^2$$

$$I_2 = ¼ P / \pi d^2$$

The new intensity is ¼ the original.

19. A is correct.

$$\text{speed of sound} = \sqrt{[\text{resistance to compression} / \text{density}]}$$

$$v_{\text{sound}} = \sqrt{(E / \rho)}$$

Low resistance to compression and high density results in low velocity because this minimizes the term under the radical and thus minimizes velocity.

20. B is correct.

A pipe open at each end has no constraint on displacement at the ends.

Furthermore, the pressure at the ends must equal the ambient pressure.

Thus, the pressure is maximum at the ends: an antinode.

21. E is correct.

For a pipe open at both ends, the resonance frequency:

$$f_n = nf_1$$

where n = 1, 2, 3, 4…

Therefore, only a multiple of 200 Hz can be a resonant frequency.

22. D is correct.

Unlike light, sound waves require a medium to travel through, and their speed depends on the medium.

Sound is fastest in solids, then liquids, and slowest in the air.

$$v_{solid} > v_{liquid} > v_{air}$$

23. B is correct.

Currents or moving charges induce magnetic fields.

24. C is correct.

$$\lambda = v / f$$

$$\lambda = (5{,}000 \text{ m/s}) / (620 \text{ Hz})$$

$$\lambda = 8.1 \text{ m}$$

25. E is correct.

Sound intensity radiating spherically:

$$I = P / 4\pi r^2$$

If r is doubled:

$$I = P / 4\pi(2r)^2$$

$$I = \tfrac{1}{4} P / 4\pi r^2$$

The intensity is reduced by a factor of ¼.

26. D is correct.

As the sound propagates through a medium, it spreads out in an approximately spherical pattern.

Thus, the power is radiated along the surface of the sphere, and the intensity can be given by:

$I = P / (4\pi r^2)$ ← for the surface area of a sphere

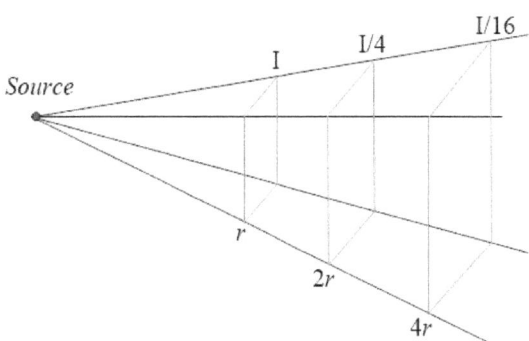

Because the surface area of a sphere contains a square component regarding radial distance, the sound intensity is inversely proportional to the square of the distance from the sound source.

27. B is correct.

The closed-end is a node, and the open end is an antinode.

$\lambda = (4 / n)L$

where n = 1, 3, 5 ...

For the fundamental n = 1:

$\lambda = (4 / 1) \cdot (1.5 \text{ m})$

$\lambda = 6 \text{ m}$

The 1.5 m tube (open at one end) is a quarter of a full wave, so the wavelength is 6 m.

28. A is correct.

The 1.5 m is ¼ a full wave, so the wavelength is 6 m, for the fundamental.

$f = v / \lambda$

$f = (960 \text{ m/s}) / 6 \text{ m}$

$f = 160 \text{ Hz}$

29. E is correct.

For a closed-ended pipe, the wavelength to the harmonic relationship is:

$\lambda = (4 / n)L$

where n = 1, 3, 5…

For the 5th harmonic n = 5

$\lambda = (4 / 5) \cdot (1.5 \text{ m})$

$\lambda_n = 1.2 \text{ m}$

Closed end tube

Harmonic # (n)	# of waves in a tube	# of nodes	# of antinodes	Wavelength to length
1	1/4	1	1	$\lambda = 4L$
3	3/4	2	2	$\lambda = 4/3 L$
5	5/4	3	3	$\lambda = 4/5 L$
7	7/4	4	4	$\lambda = 4/7 L$

30. A is correct.

$$f = v / \lambda$$

$$f = (340 \text{ m/s}) / (6 \text{ m})$$

$$f = 57 \text{ Hz}$$

31. C is correct.

Wavelength to harmonic number relationship in a standing wave on a string:

$$\lambda = (2L / n)$$

where n = 1, 2, 3, 4, 5 …

For the 3rd harmonic:

$$\lambda = (2) \cdot (0.34 \text{ m}) / 3$$

$$\lambda = 0.23 \text{ m}$$

32. D is correct.

Beat frequency equation:

$$f_{\text{beat}} = |f_2 - f_1|$$

If one of the tones increases in frequency, the beat frequency increases or decreases, but this cannot be determined unless the two tones are known.

33. A is correct.

For a closed-ended pipe, the wavelength to harmonic relationship is:

$$\lambda = (4 / n)L$$

where n = 1, 3, 5, 7…

The lowest three tones are n = 1, 3, 5

$$\lambda = (4 / 1)L;$$

$$\lambda = (4 / 3)L;$$

$$\lambda = (4 / 5)L$$

34. E is correct.

The sound was barely perceptible; the intensity at Mary's ear is $I_0 = 9.8 \times 10^{-12}$ W/m².

Since the mosquito is 1 m away, imagine a sphere 1 m in a radius around the mosquito.

If 9.8×10^{-12} W emanates from each area 1 m², the surface area is $4\pi(1\text{ m})^2$.

This is the power produced by one mosquito:

$$P = 4\pi r^2 I_0$$

$$P = 4\pi(1\text{ m})^2 \times (9.8 \times 10^{-12}\text{ W/m}^2)$$

$$P = 1.2 \times 10^{-10}\text{ W}$$

energy = power × time

$$E = Pt$$

Energy produced in 200 s:

$$Pt = (1.2 \times 10^{-10}\text{ W}) \cdot (200\text{ s})$$

$$E = 2.5 \times 10^{-8}\text{ J}$$

35. A is correct.

$$v = c / n$$

where c is the speed of light in a vacuum

$$v = \Delta x / \Delta t$$

$$\Delta x / \Delta t = c / n$$

$$\Delta t = n\Delta x / c$$

$$\Delta t = (1.33) \cdot (10^3\text{ m}) / (3 \times 10^8\text{ m/s})$$

$$\Delta t = 4.4 \times 10^{-6}\text{ s}$$

36. E is correct.

When waves interfere constructively (i.e., in-phase), the sound level is amplified.

When waves interfere destructively (i.e., out of phase), they cancel, and no sound is heard.

Acoustic engineers ensure that there are no "dead spots," and the sound waves add.

An engineer should minimize destructive interference, which can distort the sound.

37. B is correct.

The velocity of a wave on a string in tension can be calculated by:

$v = \sqrt{(TL/m)}$

Graph B gives a curve of a square root relationship, which is how velocity and tension are related.

$y = x^{1/2}$

38. D is correct.

From the diagram, the wave is a 6th harmonic standing wave.

Find wavelength:

$\lambda = (2L/n)$

$\lambda = (2)\cdot(4\text{ m})/(6)$

$\lambda = 1.3\text{ m}$

Find frequency:

$f = v/\lambda$

$f = (20\text{ m/s})/(1.3\text{ m})$

$f = 15.4\text{ Hz}$

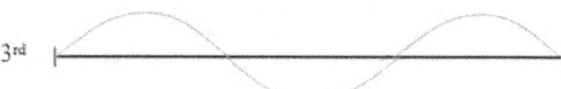

39. E is correct.

Soundwave velocity is independent of frequency and does not change.

40. C is correct.

Find the frequency of the string, then the length of the pipe excited to the second overtone using that frequency.

The speed of sound in the string is:

$v_{string} = \sqrt{T/\mu}$

where T is the tension in the string, and μ is the linear mass density.

$v_{string} = \sqrt{[(75\text{ N})/(0.00040\text{ kg})]}$

$v_{string} = 433.01\text{ m/s}$

The wavelength of a string of length L_{string} vibrating in harmonic n_{string} is:

$\lambda_{string} = 2L_{string}/n_{string}$

Therefore, the vibration frequency of the string is:

$f = v_{string}/\lambda_{string}$

$f = [(n_{string})(v_{string})]/2L_{string}$

continued…

$f = [(6)·(433.01 \text{ m/s})] / (2 \times 0.50 \text{ m})$

$f = (2{,}598.06 \text{ m/s}) / 1 \text{ m}$

$f = 2{,}598.1 \text{ Hz}$

Consider the open pipe.

The relationship of length, wavelength, and harmonic number for an open pipe is the same as a string.

Therefore:

$L_{\text{pipe}} = n_{\text{pipe}} (\lambda_{\text{pipe}} / 2)$

However, since

$\lambda_{\text{pipe}} = v_{\text{air}} / f$

$L_{\text{pipe}} = n_{\text{pipe}} (v_{\text{air}} / 2f)$

Noting that the second overtone is the third harmonic ($n_{pipe} = 3$):

$L_{pipe} = (3 \times 345 \text{ m/s}) / (2 \times 2{,}598.1 \text{ Hz})$

$L_{pipe} = 0.20 \text{ m}$

Note that it is not necessary to calculate the frequency; its value cancel.

There is less chance for error if the two steps that use frequency are skipped.

In $L_{\text{pipe}} = n_{\text{pipe}} (v_{\text{air}} / 2f)$ substitute $f = [(n_{\text{string}})·(v_{\text{string}})] / 2L_{\text{string}}$

which gives:

$L_{\text{pipe}} = L_{\text{string}} (v_{\text{air}} / v_{\text{string}})·(n_{\text{pipe}} / n_{\text{string}})$

It yields the same answer but with fewer calculations.

41. A is correct.

$v = \sqrt{(T/\mu)}$

$\mu = m/L$

$v = \sqrt{(TL/m)}$

$v_2 = \sqrt{(T(2L)/m)}$

$v_2 = \sqrt{2} \sqrt{(TL/m)}$

$v_2 = v\sqrt{2}$

42. C is correct.

For a standing wave, the resonance frequency: $f_n = nf_1$

where n is the harmonic number, n = 1, 2, 3, 4 …

Therefore, only a multiple of 500 Hz can be a resonant frequency.

43. D is correct.

The angle of incidence equals the angle of reflection.

A light beam entering a medium with a higher refractive index than the incident medium refracts *toward* the normal. Thus, the angle of refraction is less than the angles of incidence and reflection.

Snell's law:

$$n_1 \sin \theta_1 = n_2 \sin \theta_2$$

where $n_1 < n_2$

For Snell's law to be true, then:

$$\theta_1 > \theta_2$$

44. A is correct.

Speed of sound in gas:

$$v_{sound} = \sqrt{(yRT / M)}$$

where y = adiabatic constant, R = gas constant, T = temperature and M = molecular mass

The speed of sound in a gas is only dependent upon temperature and not frequency or wavelength.

45. B is correct.

Waves only transport energy and not matter.

46. E is correct.

$$v = \lambda f$$

$$\lambda = v / f$$

$$\lambda = (344 \text{ m/s}) / (700 \text{ s}^{-1})$$

$$\lambda = 0.5 \text{ m}$$

The information about the string is unnecessary, as the only contributor to the wavelength of the sound in air is the frequency and the speed.

47. C is correct.

$$v = \lambda f$$

$$f = v / \lambda$$

Distance from the sound source is not part of the equation for frequency.

48. A is correct.

Velocity of a wave in a rope:

$$v = \sqrt{T / (m/L)}$$

$$t = d/v$$

$$d = L$$

$$t = d / \sqrt{T/(m/L)}$$

$$t = (8 \text{ m}) / [40 \text{ N} / (2.5 \text{ kg} / 8 \text{ m})]^{1/2}$$

$$t = 0.71 \text{ s}$$

49. C is correct.

Intensity to decibel relationship:

$$I \text{ (dB)} = 10 \log_{10}(I_1 / I_0)$$

where I_0 = threshold of hearing

$$\text{dB} = 10\log_{10}[(10^{-5} \text{ W/m}^2) / (10^{-12} \text{ W/m}^2)]$$

$$I = 70 \text{ decibels}$$

50. E is correct.

The diagram represents the described scenario.

The wave is in the second harmonic with a wavelength of:

$$\lambda = (2/n)L$$

$$\lambda = (2/2) \cdot (1 \text{ m})$$

$$\lambda = 1 \text{ m}$$

$$f = v/\lambda$$

$$f = (3.8 \times 10^4 \text{ m/s}) / (1 \text{ m})$$

$$f = 3.8 \times 10^4 \text{ Hz}$$

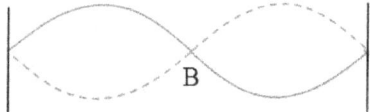

The lowest frequency corresponds to the lowest possible harmonic number.

For this problem, n = 2.

51. D is correct.

The speed of light traveling in a vacuum is c.

$$c = \lambda v$$

$$c = \lambda f$$

$$f = c / \lambda$$

Frequency and wavelength are inversely proportional, so an increase in frequency results in a decreased wavelength.

52. B is correct.

Radio waves are electromagnetic waves, while all other choices are mechanical waves.

53. D is correct.

Since the microphone is precisely equidistant from each speaker (i.e., equal path lengths), the sound waves take equal time to reach the microphone.

The speakers emit sound waves in phase (i.e., peaks are emitted simultaneously).

Since those peaks reach the microphone simultaneously (because of the equal path length), they combine constructively and add, forming a large peak or antinode.

54. C is correct.

Doppler equation for receding source of sound:

$$f_{observed} = [v_{sound} / (v_{sound} + v_{source})] f_{source}$$

$$f_{observed} = [(342 \text{ m/s}) / (342 \text{ m/s} + 30 \text{ m/s})] \cdot (1{,}200 \text{ Hz})$$

$$f_{observed} = 1{,}103 \text{ Hz}$$

The observed frequency is lower when the source is receding.

55. E is correct.

$$f_1 = 600 \text{ Hz}$$

$$f_2 = 300 \text{ Hz}$$

$$f_2 = \tfrac{1}{2} f_1$$

$$\lambda_1 = v / f_1$$

$$\lambda_2 = v / (\tfrac{1}{2} f_1)$$

$$\lambda_2 = 2 (v / f_1)$$

The wavelength of the 300 Hz frequency is twice the wavelength of the 600 Hz frequency.

Notes for active learning

Notes for active learning

Light and Optics – Detailed Explanations

1. A is correct.

Soap film that reflects a given wavelength of light exhibits constructive interference.

The expression for constructive interference of a thin film:

$$2t = (m + \tfrac{1}{2})\lambda$$

where t = thickness, m = 0, 1, 2, 3… and λ = wavelength

To find the minimum thickness set m = 0:

$$2t = (0 + \tfrac{1}{2})\lambda = \tfrac{1}{2}\lambda$$

$$t = \tfrac{1}{4}\lambda$$

2. A is correct.

By the law of reflection, the angle of incidence equals the angle of reflection.

Thus, as the angle of incidence increases, the angle of reflection increases equal to the angle of incidence.

3. B is correct.

If image is twice her height and upright, then:

$$2h_o = h_i$$

$$m = h_i / h_o$$

$$m = -d_i / d_o$$

$$m = 2h_o / h_o$$

$$m = 2$$

$$2 = -d_i / d_o$$

$$-2d_o = d_i$$

Use lens equation to solve:

$$1 / f = 1 / d_o + 1 / d_i$$

$$1 / 100 \text{ cm} = 1 / d_o + (-1 / 2 d_o)$$

$$1 / 100 \text{ cm} = 1 / 2 d_o$$

$$2d_o = 100 \text{ cm}$$

$$d_o = 50 \text{ cm}$$

4. D is correct.

If a person's eye is too long, the light entering the eye is focused in front of the retina, causing *myopia*, a condition of nearsightedness.

Hyperopia is farsightedness.

5. E is correct.

Visible light:

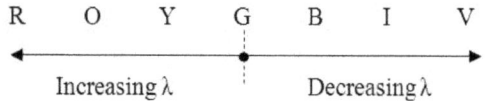

speed of light = wavelength × frequency

$c = \lambda f$

Wavelength to frequency:

$f = c / \lambda$

Frequency and wavelength are inversely proportional:

As λ increases, f decreases.

As λ decreases, f increases.

Thus, because

$E = hf$:

R   O   Y   G   B   I   V
→ Increasing energy

6. B is correct.

The lens equation:

$1 / f = 1 / d_o + 1 / d_i$

$1 / d_i = 1 / f - 1 / d_o$

$1 / d_i = -1 / 3 \text{ m} - 1 / 4 \text{ m}$

$1 / d_i = (-3 \text{ m} - 4 \text{ m}) / 12 \text{ m}$

$1 / d_i = -7 \text{ m} / 12 \text{ m}$

$d_i = -12 / 7 \text{ m}$

continued...

Magnification:

$$m = -d_i / d_o$$

$$m = -(-12/7 \text{ m}) / 4 \text{ m}$$

$$m = 3/7$$

Height of the candle image:

$$h_i = m h_o$$

$$h_i = (3/7) \cdot (18 \text{ cm})$$

$$h_i = 54/7 \text{ cm}$$

$$h_i = 7.7 \text{ cm}$$

7. **C is correct.**

$$\theta_{syrup} = \tan^{-1}(0.9 \text{ m} / 0.66 \text{ m})$$

$$\theta_s = \tan^{-1}(1.36)$$

$$\theta_s = 53.7°$$

$$\theta_{oil} = \tan^{-1}[(2 \text{ m} - 0.9 \text{ m}) / 1.58 \text{ m}]$$

$$\theta_o = \tan^{-1}(0.7)$$

$$\theta_o = 34.8°$$

$$n_o \sin \theta_o = n_{air} \sin \theta_{air}$$

$$n_o \sin 34.8° = (1) \sin 90°$$

$$n_o = 1 / (\sin 34.8°)$$

$$n_o = 1.75$$

8. **D is correct.**

$$\theta_{syrup} = \tan^{-1}(0.9 \text{ m} / 0.66 \text{ m})$$

$$\theta_s = \tan^{-1}(1.36)$$

$$\theta_s = 53.7°$$

$$\theta_{oil} = \tan^{-1}[(2 \text{ m} - 0.9 \text{ m}) / 1.58 \text{ m}]$$

$$\theta_o = \tan^{-1}(0.7)$$

$$\theta_o = 34.8°$$

$$n_o \sin \theta_o = n_{air} \sin \theta_{air}$$

$$n_o \sin 34.8° = (1) \sin 90°$$

continued…

$n_o = 1 / (\sin 34.8°)$

$n_o = 1.75$

$n_s \sin \theta_s = n_o \sin \theta_o$

$n_s = n_o \sin \theta_o / \sin \theta_s$

$n_s = (1.75) \cdot (\sin 34.8°) / (\sin 53.7°)$

$n_s = 1.24$

9. A is correct.

The wave theory of light cannot explain the photoelectric effect.

In the photoelectric effect, photons from a light source are absorbed by electrons on a metal surface and cause them to be ejected.

The energy of ejected electrons is only dependent upon photon frequency and is found by:

$KE = hf - \phi$

where h = Planck's constant, f = frequency and ϕ = stopping potential

10. D is correct.

Geometrical optics, or ray optics, describe light propagation in terms of rays and fronts to approximate the path along which light propagates in certain circumstances.

11. D is correct.

First find the critical angle:

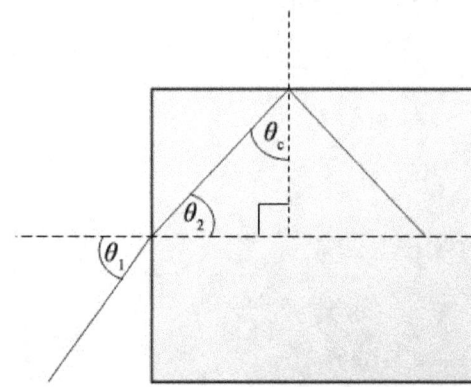

$n_{fiber} \sin \theta_c = n_{air} \sin \theta_{air}$

$(1.26) \sin \theta_c = (1) \sin 90°$

$\sin \theta_c = 1 / 1.26$

$\theta_c = \sin^{-1} (1 / 1.26)$

$\theta_c = 52.5°$

Find θ_2:

$\theta_2 + \theta_c + 90° = 180°$

$(\theta_2 + 52.5° + 90°) = 180°$

$\theta_2 = 37.5°$

Find θ_1:

$n_{air} \sin \theta_1 = n_{fiber} \sin \theta_2$

continued…

(1) sin θ_1 = (1.26) sin 37.5°

sin θ_1 = 0.77

θ_1 = sin^{-1} (0.77)

θ_1 = 50°

12. A is correct.

If the power of the lens is 10 diopters,

$1 / f$ = 10 D

where f is the focal length in m

Thin Lens Equation:

$1 / f = 1 / d_o + 1 / d_i$

10 m^{-1} = 1 / 0.5 m + 1 / d_i

1 / d_i = 10 m^{-1} – 1 / 0.5 m

1 / d_i = 8 m^{-1}

d_i = 1 / 8 m

d_i = 0.13 m

13. D is correct.

Most objects observed by humans are virtual images or objects which reflect incoming light to project an image.

14. E is correct.

An image from a convex mirror has the following characteristics, regardless of object distance:

- located behind the convex mirror
- virtual
- upright
- reduced in size from the object (image < object)

15. A is correct.

The mirror has a positive focal length, which indicates that the mirror is concave.

The object is at a distance greater than the focal length.

Therefore, it is inverted.

Use lens equation to solve image distance:

$1 / f = 1 / d_o + 1 / d_i$

continued...

$1 / 10$ m $= 1 / 20$ m $+ 1 / d_i$

$d_i = 20$ cm

The image distance is positive, so the image is real.

Use the magnification equation to determine if it is upright or inverted.

$m = -d_i / d_o$

$m = h_i / h_o$

$-(20$ m $/ 20$ m$) = h_i / h_o$

$-1 = h_i / h_o$

The object height h_o is always positive, so the image height h_i must be negative to satisfy the equation.

A negative image height indicates an inverted image.

16. E is correct.

For a *converging lens*, if an object is placed beyond $2f$ from the lens, the image is real, inverted, and reduced.

Use the lens equation to determine if the image is real (or virtual):

Assume $f = 1$ m and $d_o = 3f$ (because $d_o > 2f$)

$1 / f = 1 / d_o + 1 / d_i$

$1 / f = 1 / 3f + 1 / d_i$

$d_i = 1.5$

A positive d_i indicates a real image.

Use the *magnification equation* to determine if the image is inverted and reduced.

$m = -d_i / d_o$

$m = -(1.5$ m $/ 3$ m$)$

$m = -½$

$|m| = ½$

$|m| < 1$

A negative magnification factor with an absolute value less than 1 a reduced and inverted image.

17. C is correct.

Radio waves range from 3 kHz to 300 GHz, lower than all forms of radiation listed.

Since the energy of radiation is proportional to frequency ($E = hf$), radio waves have the lowest energy.

18. B is correct.

A medium's index of refraction is the ratio of the speed of refracted light in a vacuum to its speed in the reference medium.

$n = c / v$

$n = 2.43$

$2.43 = c / v_{diamond}$

$c = 2.43(v_{diamond})$

19. D is correct.

$1 / f = 1 / d_o + 1 / d_i$

$1 / 20 \text{ cm} = 1 / 15 \text{ cm} + 1 / d_i$

$3 / 60 \text{ cm} - 4 / 60 \text{ cm} = 1 / d_i$

$-1 / 60 \text{ cm} = 1 / d_i$

$d_i = -60 \text{ cm}$

The negative sign indicates that the image is projected back the way it came.

20. B is correct.

The red paper absorbs all colors but reflects only red light giving it the appearance of being red.

Cyan is the complementary color to red, so when the cyan light shines upon the red paper, no light is reflected, and the paper appears black.

21. B is correct.

$1 / f = 1 / d_o + 1 / d_i$

If $d_i = f$,

$1 / d_o = 0$

Thus, d_o must be large.

22. E is correct.

Since the index of refraction depends on the frequency, and the focal length depends on the refraction of the beam in the lens, dispersion causes the focal length to depend on frequency.

23. C is correct.

Use the equation for magnification:

$$m = -d_i / d_o$$

$$d_i = d_o$$

$$m = 1$$

Thus, there is no magnification, so the image is the same size as the object.

24. E is correct.

When viewed straight down (90° to the surface), an incident light ray moving from water to air is refracted 0°.

25. C is correct.

The rotating of one polarized lens 90° with respect to the other lens results in complete darkness since no light would be transmitted.

26. C is correct.

First, find the angle that the ray makes with the normal of the glass:

$$180° = x + 90° + 54°$$

$$x = 36°$$

Find θ_1:

$$\theta_1 = 90° - 36°$$

$$\theta_1 = 54°$$

Referring to the diagram, $\theta_1 = 54°$

Snell's Law:

$$n_1 \sin \theta_1 = n_2 \sin \theta_2$$

$$\sin^{-1}[(n_1 / n_2) \sin \theta_1] = \theta_2$$

$$\theta_2 = \sin^{-1}[(1.45 / 1.35) \sin 54°]$$

$$\theta_2 = 60°$$

Solve for the angle with the horizontal:

$$\theta_H = 60° - 54°$$

$$\theta_H = 6°$$

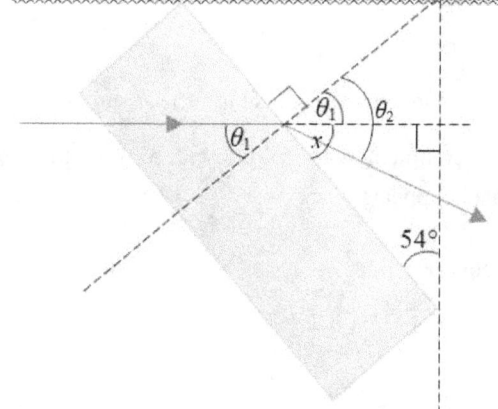

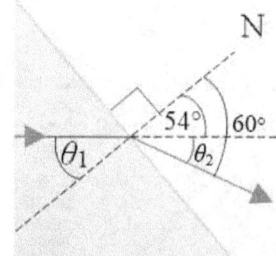

27. A is correct.

The angle at which the ray is turned is the sum of the angles if reflected off each mirror once:

$$\theta_{turned} = \theta_1 + \theta_2 + \theta_3 + \theta_4$$

By the law of reflection:

$$\theta_1 = \theta_2$$

$$\theta_3 = \theta_4$$

Note the triangle formed (sum of interior angles is 180°):

$$30° + (90° - \theta_2) + (90° - \theta_3) = 180°$$

$$\theta_2 + \theta_3 = 30°$$

Given:

$$\theta_2 + \theta_3 = \theta_1 + \theta_4$$

Thus:

$$\theta_{turned} = 30° + 30°$$

$$\theta_{turned} = 60°$$

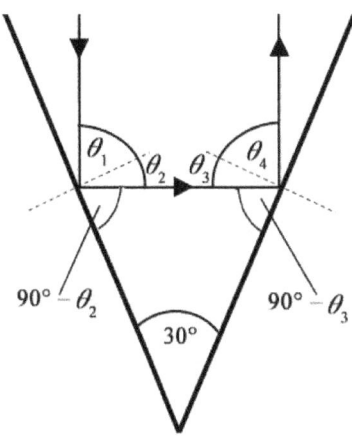

Note: figure is not to scale

In general, two plane mirrors that meet at an angle of $\theta \leq 90°$ the ray deflected off both mirrors are deflected through an angle of 2θ.

28. E is correct.

The following statement about light is true: a packet of light energy is a *photon*.

Color can be used to determine the approximate energy of visible light, and light travels through space at a speed of 3.0×10^8 m/s.

29. A is correct.

The angle of incidence is < the angle of refraction if the light travels into a *less dense* medium.

The angle of incidence is > the angle of refraction if the light travels into a *more dense* medium.

The angle of incidence is = the angle of refraction if the densities of the mediums are *equal density*.

30. E is correct.

Plane mirrors do not distort the size or the shape of an object since light is reflected at the same angle the mirror received it.

Magnification equation:

$$m = h_i / h_o$$

For a plane mirror m = 1:

$$1 = h_i / h_o$$

$$h_i = h_o$$

Therefore, the image size is the same as object size, and the image is virtual since it is located behind the mirror.

31. C is correct.

A spherical concave mirror has a focal length of:

$$f = R / 2$$

32. B is correct.

Refracted rays bend further from the normal than the original incident angle when the refracting medium is optically less dense than the incident medium.

Therefore, $n_1 > n_2$.

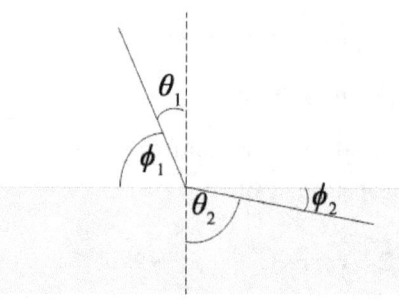

The index of refraction for a medium can never be less than 1.

33. B is correct.

If a person's eye is too short, the light entering the eye is focused behind the retina, causing farsightedness (hyperopia).

34. A is correct.

Hot air is less dense than cold air. Light traveling through both types of air experiences refractions, which appear as shimmering or "wavy" air.

35. E is correct.

Chromatic aberration occurs when a lens focuses on different wavelengths of color at different positions in the focal plane.

It occurs in the following pattern for converging lens:

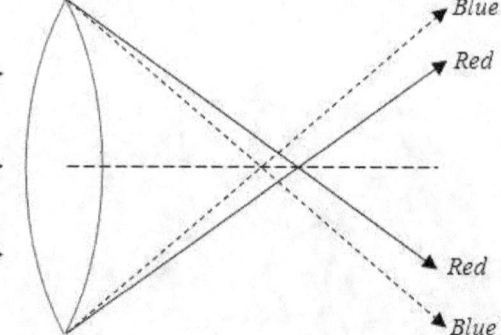

36. B is correct.

$1/f_{total} = 1/f_1 + 1/f_2$

$1/f_{total} = 1/2\text{ m} + 1/4\text{ m}$

$1/f_{total} = 3/4\text{ m}$

$f_{total} = 4/3\text{ m}$

37. C is correct.

The angle in the water respective to the normal:

$\theta = \tan^{-1}(37.5\text{ ft}/50\text{ ft})$

$\theta = \tan^{-1}(0.75)$

$\theta = 36.9°$

$n_{air}\sin(90-\theta) = n_{water}\sin\theta$

$(1)\sin(90-\theta) = (1.33)\sin 36.9°$

$\sin(90-\theta) = 0.8$

$(90-\theta) = \sin^{-1}(0.8)$

$(90-\theta) = 52.9$

$\theta = 37.1° \approx 37°$

38. A is correct.

Violet light has the highest energy and frequency, and therefore has the shortest wavelength.

39. E is correct.

Objects directly in front of plane mirrors are reflected in their likeness since plane mirrors are not curved and reflect light perpendicularly to their surface.

40. B is correct.

A virtual image is always upright and can be formed by the diverging lens and converging lens.

Diverging lens → reduced and virtual image

Converging lens → enlarged and virtual image

41. B is correct.

The neon light is emitted from neon atoms as their energized electrons cascade back down to ground level. When this occurs, energy is released in light at wavelengths known as the *emission spectrum*.

When this light is passed through a prism, a series of bright discontinuous spots or lines will be seen due to the specific wavelengths of an emission spectrum of neon.

42. E is correct.

The law of reflection states that the angle of incidence equals the angle of reflection (with respect to the normal) and is valid for all mirrors.

$$\theta_i = \theta_r$$

43. C is correct.

A concave lens always forms an image that is virtual, upright, and reduced in size.

44. B is correct.

Virtual images are upright.

There is no correlation between the size and nature – virtual or real – of an image.

Images may be larger, smaller, or the same size as the object.

45. E is correct.

Red is the light with the lowest frequency (longest wavelength) detected by your eyes from all choices listed. (ROY G BIV)

46. D is correct.

$$1/f = 1/d_o + 1/d_i$$

$$1/6\text{ m} = 1/3\text{ m} + 1/d_i$$

$$1/d_i = 1/6\text{ m} - 1/3\text{ m}$$

$$1/d_i = -1/6\text{ m}$$

$$d_i = -6\text{ m}$$

where the negative sign indicates the image is on the same side as the object

The image is upright and virtual since the rays must be extended to intersect.

47. A is correct.

A diverging lens (concave) produces a virtual, upright image and is reduced in size.

48. C is correct.

Thin lens formula:

$$1/f = 1/d_o + 1/d_i$$

d_i is negative because the image is virtual

$$1/f = 1/14\text{ cm} + 1/-5\text{ cm}$$

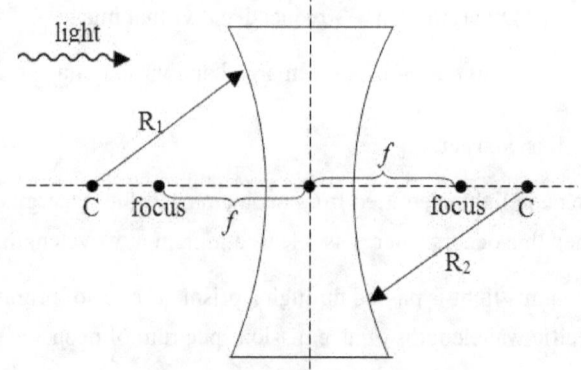

continued...

$f = -7.8$ cm

The focus is negative because the lens is diverging.

Lens maker formula:

$1/f = (n-1) \cdot (1/R_1 - 1/R_2)$

R_1 is negative by convention because the light ray passes its center of curvature before the curved surface.

$1/(-7.8 \text{ cm}) = (n-1) \cdot (1/-15 \text{ cm} - 1/15 \text{ cm})$

$(1/-7.8 \text{ cm}) \cdot (15 \text{ cm}/-2) + 1 = n$

$n = 2$

49. E is correct.

The magnification equation relates to the image and object distance:

$m = -d_i / d_o$

or

The magnification equation relates to the image and object height:

$m = h_i / h_o$

50. D is correct.

(see image)

For a concave mirror, if an object is located between the focal point and center of curvature, the image is formed beyond the center of curvature.

In this problem, Mike does not see his image because he is in front of where it forms.

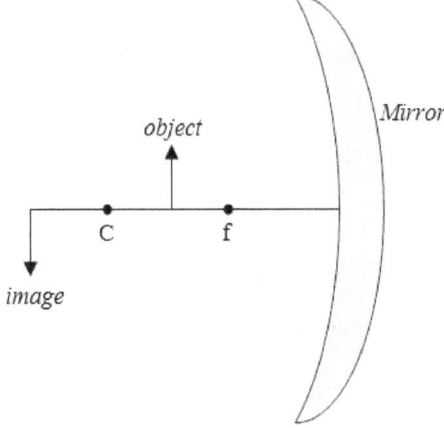

51. D is correct.

For a concave spherical mirror, the produced image characteristics depend upon the placement of the object in relation to the focal point and center of curvature.

The image can be smaller, larger, or the same size as the object.

52. A is correct.

Lens power is the reciprocal of the focal length in meters:

$P = 1/f$

If the effective focal length of the lens combination is less than the focal length of either individual lens, then the power of the combination must be greater than the power of either individual lens.

53. B is correct.

A medium's index of refraction is the ratio of the speed of refracted light in a vacuum to its speed in the reference medium.

$n = c / v$

54. E is correct.

As it is a plane mirror, the image is not distorted.

Only some light rays are reflected; the others create an image behind the mirror's surface.

For a plane mirror:

m = 1

$m = -d_i / d_o$

$1 = -d_i / d_o$

$d_o = -d_i$

The negative indicates the image is virtual and behind the mirror.

55. C is correct.

The radius length is the center of curvature,

$r = 50$ cm

Find the focal length:

$f = r / 2$

$f = 50$ cm $/ 2$

$f = 25$ cm

For a concave mirror with an object between the center of curvature and the focal length, the resulting image is real and inverted.

Notes for active learning

Notes for active learning

Answer Keys and Detailed Explanations

Diagnostic Tests

Diagnostic Test 1 – Detailed Explanations

Answer Key

1	A	Kinematics & dynamics	26	E	Electric circuits
2	E	Force, motion, gravitation	27	A	Light & geometrical optics
3	C	Heat & thermodynamics	28	C	Quantum mechanics
4	B	Work & energy	29	B	Kinematics & dynamics
5	E	Waves & periodic motion	30	D	Force, motion, gravitation
6	B	Sound	31	B	Kinematics & dynamics
7	A	Fluids & solids	32	A	Work & energy
8	A	Electrostatics & electromagnetism	33	B	Waves & periodic motion
9	D	Electric circuits	34	D	Heat & thermodynamics
10	E	Light & optics	35	B	Light & geometrical optics
11	B	Atomic & nuclear structure	36	E	Electric circuits
12	B	Equilibrium & momentum	37	C	Electrostatics & electromagnetism
13	E	Work & energy	38	A	Fluids & solids
14	A	Waves & periodic motion	39	A	Heat & thermodynamics
15	A	Sound	40	E	Waves & periodic motion
16	B	Rotational motion	41	D	Work & energy
17	C	Electrostatics & electromagnetism	42	C	Equilibrium & momentum
18	A	Electric circuits	43	C	Special relativity
19	C	Light & geometrical optics	44	A	Waves & periodic motion
20	C	Sound	45	D	Heat & thermodynamics
21	B	Heat & thermodynamics	46	C	Work & energy
22	E	Force, motion, gravitation	47	B	Sound
23	D	Sound	48	C	Electrostatics & electromagnetism
24	C	Waves & periodic motion	49	D	Fluids & solids
25	B	Electrostatics & electromagnetism	50	D	Kinematics & dynamics

51	D	Equilibrium & momentum	76	A	Fluids & solids
52	E	Light & optics	77	E	Electrostatics & electromagnetism
53	C	Sound	78	C	Electric circuits
54	E	Electric circuits	79	C	Light & optics
55	E	Rotational motion	80	A	Atomic & nuclear structure
56	E	Heat & thermodynamics	81	D	Special relativity
57	C	Kinematics & dynamics	82	C	Atomic & nuclear structure
58	D	Force, motion, gravitation	83	B	Sound
59	C	Electrostatics & electromagnetism	84	E	Fluids & solids
60	B	Atomic & nuclear structure	85	D	Electrostatics & electromagnetism
61	A	Electrostatics & electromagnetism	86	A	Heat & thermodynamics
62	B	Force, motion, gravitation	87	A	Light & optics
63	E	Heat & thermodynamics	88	B	Atomic & nuclear structure
64	C	Work & energy	89	D	Fluids & solids
65	C	Waves & periodic motion	90	D	Force, motion, gravitation
66	B	Sound	91	B	Equilibrium & momentum
67	C	Fluids & solids	92	B	Work & energy
68	D	Electrostatics & electromagnetism	93	A	Waves & periodic motion
69	C	Electric circuits	94	D	Heat & thermodynamics
70	C	Light & optics	95	D	Light & optics
71	C	Atomic & nuclear structure	96	E	Quantum mechanics
72	B	Equilibrium & momentum	97	B	Heat & thermodynamics
73	C	Work & energy	98	B	Fluids & solids
74	A	Waves & periodic motion	99	C	Sound
75	B	Sound	100	A	Waves & periodic motion

1. A is correct.

Its inertia characterizes an object's resistance to change in its state of motion.

Inertia is not a physical property but is directly related to an object's mass.

Thus, mass determines resistance to change in motion.

2. E is correct.

The three forces are in equilibrium, so the net force $F_{net} = 0$

$$F_{net} = F_1 + F_2 + F_3$$
$$0 = F_1 + F_2 + F_3$$

Since the forces F_1 and F_2 are mirror images along the x-axis, their net force in the y-direction is zero.

Therefore, F_3 is zero in the y-direction.

The net force along the x-direction must add to zero, so set the sum of the x components to zero.

The angles for F_1 and F_2 are equal and measured with respect to the x-axis, so θ_1 and θ_2 are both 20°.

Since F_3 has no y component,

$$\theta_3 = 0°$$

Note that force components to the left are set to negative in this answer, and components to the right are positive.

$$0 = F_{1x} + F_{2x} + F_{3x}$$
$$0 = F_1 \cos \theta_1 + F_2 \cos \theta_2 + F_3 \cos \theta_3$$
$$0 = (-4.6 \text{ N} \cos 20°) + (-4.6 \text{ N} \cos 20°) + (F_3 \cos 0°)$$

Since $\cos 0° = 1$:

$$0 = (-4.3 \text{ N}) + (-4.3 \text{ N}) + F_3$$
$$-F_3 = -8.6 \text{ N}$$
$$F_3 = 8.6 \text{ N, to the right}$$

3. C is correct.

Heat transfer between two materials occurs until both materials reach the same temperature, so the colder material gains the amount of heat lost by the hotter material.

4. B is correct.

$$W = Fd$$
$$W = (20 \text{ N}) \cdot (3.5 \text{ m})$$
$$W = 70 \text{ J}$$

5. E is correct.

Constructive interference occurs when two or more waves of equal frequency and phase produce a single amplitude wave that is the sum of amplitudes of the individual waves.

If there is a phase difference, the interference will not be the total amplitude of each wave.

If the phase difference is 180°, there is total destructive interference.

6. B is correct.

The expression for the Doppler shift is:

$$f = f_s[(c + v_o) / (c + v_s)]$$

where f is the frequency heard by the observer, f_s is the frequency of the source, c is the speed of sound, v_o is the velocity of the observer, v_s is the velocity of the source

The velocity of the source v_s is positive when the source moves *away* from the observer and negative when moving toward the observer

Since the train is traveling *away*, once it passes, the velocity of the source (i.e., train) is positive.

Kevin is standing still, so the velocity of the observer is zero.

$$f = f_s[(c + v_o) / (c + v_s)]$$

$$f = (420 \text{ Hz}) \cdot [(350 \text{ m/s} + 0 \text{ m/s}) / (350 \text{ m/s} + 50 \text{ m/s})]$$

$$f = (420 \text{ Hz}) \cdot [(350 \text{ m/s}) / (400 \text{ m/s})]$$

$$f = (147{,}000 \text{ Hz·m/s}) / (400 \text{ m/s})$$

$$f = 368 \text{ Hz}$$

7. A is correct.

$$F_B = V\rho g$$

where F_B = buoyant force upward, V = volume of fluid that the object displaces, ρ = density of the fluid, g = acceleration due to gravity

The lead weight starts in the air above the water, so initially, it experiences zero buoyant force from the water.

As the weight breaks the surface of the water and sinks, the buoyant force increases until the weight is fully submerged, after which the force remains constant since the density of water is uniform.

8. A is correct.

A moving charge experiences a magnetic force from a magnetic field, but the force is perpendicular to the velocity (and the magnetic field), so the speed does not change.

$$F_B = qvB \sin\theta$$

F_B is perpendicular to both v and B

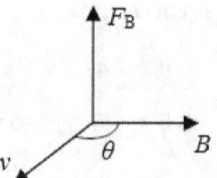

9. D is correct.

If one mass is halved, then the gravitational attraction between them is halved.

For equilibrium, the electrostatic repulsion must be halved.

$$F_e = F_g$$
$$F_e = kQ_1Q_2 / r^2$$
$$F_g = Gm_1m_2 / r^2$$
$$kQ_1Q_2 / r^2 = Gm_1m_2 / r^2$$
$$kQ_1Q_2 = Gm_1m_2$$

k and G are constants and cannot be manipulated

If the mass of object 1 is halved, while equilibrium is maintained:

$$G(m_1 / 2)m_2 = (Gm_1m_2) / 2$$

If the electrostatic force is halved, the charge of one of the objects must be halved:

$$(kQ_1Q_2) / 2 = kQ_1(Q_2 / 2)$$

10. E is correct.

Calculate the focal length:

$$1 / f = 1 / d_i + 1 / d_o$$

where f is focal length d_o is distance to the object and d_i is distance to the image

$$1 / f = 1 / 2 \text{ m} + 1 / 4 \text{ m}$$
$$1 / f = 2 / 4 \text{ m} + 1 / 4 \text{ m}$$
$$1 / f = 3 / 4 \text{ m}$$
$$f = 4 / 3 \text{ m}$$

11. B is correct.

$$^A_Z n + e^- \rightarrow ^A_{Z-1}(n-1) + v_e$$

A proton captures an electron and transforms it into a neutron.

Therefore, the atomic number Z decreases by 1 because the nucleus contains one less proton.

The $(n-1)$ signifies the new element name (since the atomic number changed), and v_e is the release of an electron neutrino.

12. B is correct.

To balance the torques due to the weight, the fulcrum must be placed 4 times farther from the son than the man because the father weighs 4 times more.

Since the total length of the seesaw is 10 m, the fulcrum must be placed 8 m from the son and 2 m from the father, who is on the heavier end.

$x + 4x = 10$ m

$5x = 10$ m

$x = 2$ m

Another method to solve the problem:

200 N		800 N
$10 - x$	Δ	x

$(200 \text{ N}) \cdot (10 - x) = (800 \text{ N})x$

$x = 2$ m

13. E is correct.

Energy before release and at the top of each bounce equals gravitational PE:

$PE = mgh$

Gravitational potential energy is proportional to height, and mass and g stay constant.

Multiply by 0.8 (80%) to determine the height after a bounce if 20% of the energy is lost.

$h_{initial} = 250$ cm

250 cm × (0.8 × 0.8 × 0.8), equals h after 3 bounces

$h_3 = (250 \text{ cm}) \cdot (0.8)^3$

$h_3 = 128$ cm

14. A is correct.

$T = 1 / f$

$T = 1 / (10 \text{ Hz})$

$T = 0.1$ s

15. A is correct.

$f = v / \lambda$

$f = (1{,}600 \text{ m/s}) / (2.5 \text{ m})$

$f = 640$ Hz

16. B is correct.

Formula for *rotational kinematics*:

$$\omega_f = \omega_i + \alpha \Delta t$$

Solve for Δt:

$$\Delta t = (\omega_f - \omega_i) / \alpha$$

$$\Delta t = (33.3 \text{ rad/s} - 15.0 \text{ rad/s}) / 3.45 \text{ rad/s}^2$$

$$\Delta t = 5.30 \text{ s}$$

17. C is correct.

Perpendicular magnetic and electric fields are required for a mass spectrometer.

18. A is correct.

Find Capacitive Reactance:

$$X_c = 1 / 2\pi C f$$

$$X_c = 1 / (2\pi) \cdot (26 \times 10^{-6} \text{ F}) \cdot (60 \text{ Hz})$$

$$X_c = 102 \text{ } \Omega$$

Find rms current:

$$I = V_{rms} / X_c$$

$$I = 120 \text{ V} / 102 \text{ } \Omega$$

$$I = 1.2 \text{ A}$$

19. C is correct.

Snell's Law:

$$n_g / n_w = (\sin \phi) / (\sin \theta)$$

Find the index of refraction for glass:

$$n_g / 1.33 = (\sin 61°) / (\sin 48°)$$

$$n_g / 1.33 = (0.875) / (0.743)$$

$$n_g / 1.33 = 1.18$$

$$n_g = (1.18) \cdot (1.33)$$

$$n_g = 1.57$$

continued…

Solve for the new angle of refraction after the angle of incidence has changed:

$$1.57 / 1.33 = (\sin \phi) / (\sin 25°)$$

$$1.18 = (\sin \phi) / (0.423)$$

$$(1.18)·(0.423) = \sin \phi$$

$$\sin \phi = 0.5$$

$$\phi = 30°$$

20. C is correct.

$$f = nv / 2L$$

n = 1 for the frequency of the fundamental

$$f = v / 2L$$

$$f = (340 \text{ m/s}) / 2(0.1 \text{ m})$$

$$f = 1{,}700 \text{ Hz}$$

21. B is correct.

Enthalpy of fusion.

As a solid undergoes a phase change, the temperature stays constant until the phase change is complete.

To calculate the amount of heat absorbed to completely melt the solid, multiply the heat of fusion by the mass undergoing the phase change.

Heat needed to melt a solid:

$$q = m\Delta H_f$$

22. E is correct.

$$a = g \sin \theta$$

An object's acceleration down a frictionless ramp (with an incline angle) is constant.

23. D is correct.

Sound cannot travel through a vacuum because there is no medium to propagate the wave.

In the air, sound waves travel through gas; in the ocean, they travel through liquid; and in the Earth, they travel through solids. These are all mediums in which sound waves can propagate.

However, vacuums are devoid of matter; there is no medium, and the wave cannot pass.

24. C is correct.

The *lowest* harmonic (i.e., fundamental) frequency (f_1) corresponds to the *longest* harmonic (i.e., fundamental) wavelength (λ_1).

$f_1 = v / \lambda_1$

$f_1 = (8 \text{ m/s}) / 4 \text{ m}$

$f_1 = 2 \text{ Hz}$

25. B is correct.

By Newton's Third Law,

F_1 and F_2 form an *action-reaction* pair.

The ratio of their magnitudes equals 1.

26. E is correct.

$P = IV$

$P = (2 \text{ A}) \cdot (120 \text{ V})$

$P = 240 \text{ W}$

An ampere (A) is a rate of electric charge flowing in a circuit in coulombs per second (C/s), where 1 A = 1 C/s.

The volt (V) measures the difference in electric potential between two points, where 1 V is defined as the electric potential difference when 1 ampere consumes 1 watt (W) of power.

Power is a measure of energy per unit time:

1 W = 1 A·V

1 W = 1 J / s

1 W = 1 N·m/s

1 W = 1 kg·m^2/s^3

27. A is correct.

When an atom absorbs energy, its valence electrons move to higher "orbits."

As the electrons fall back to their original (ground state), they emit the absorbed energy as light.

28. C is correct.

The de Broglie wavelength of a matter wave is:

$\lambda = h / p$

$\lambda = h / (mv)$

$\lambda = (6.626 \times 10^{-34} \text{ J·s}) / [(1.30 \text{ kg})·(28.10 \text{ m/s})]$

$\lambda = (6.626 \times 10^{-34} \text{ J·s}) / [(1.30 \text{ kg})·(28.10 \text{ m/s})]$

$\lambda = 1.81 \times 10^{-35}$ m

29. B is correct.

$d = ½gt^2$

$t^2 = 2d / g$

$t^2 = 2(42 \text{ m}) / 10 \text{ m/s}^2$

$t^2 = 8.4 \text{ s}^2$

$t \approx 2.9$ s

30. D is correct.

$F = ma$

$W = mg$

$m = W / g$

$F = (W / g)a$

$a = F / m$

The $F_{\text{friction}} = 8.8$ N, and the mass is known from the box's weight.

$8.8 \text{ N} = (40 \text{ N} / 10 \text{ m/s}^2)a$

$a = (8.8 \text{ N}) / (4 \text{ N/m/s}^2)$

$a = 2.2 \text{ m/s}^2$

Since the box moves at constant velocity when F is applied, F = force due to kinetic friction.

Once the force F is removed, the net force that causes its deceleration is the frictional force.

31. B is correct.

A longer barrel gives the propellant a longer time to impart a force upon a bullet and thus a higher velocity.

This is characterized by impulse.

$J = F\Delta t$

32. A is correct.

$KE_{final} = 0$ since $v_f = 0$

The length of the skid marks is irrelevant.

$\Delta Energy = KE_{final} - KE_{initial}$

$\Delta E = \frac{1}{2}mv_f^2 - \frac{1}{2}mv_i^2$

$\Delta E = 0 \text{ J} - \frac{1}{2}(1{,}000 \text{ kg}) \cdot (30 \text{ m/s})^2$

$\Delta E = -4.5 \times 10^5 \text{ J}$

33. B is correct.

The position of an object in SHM is represented as a function of time using sine or cosine:

$x = A \sin(\omega t - \theta)$

where x = position, A = amplitude (i.e., max displacement of object from equilibrium position), ω = angular velocity in radians/sec (or degrees/sec), t = time elapsed, θ = phase

Here, $\theta = 0$ since the graph matches the phase of the standard sine graph, so there is no need for phase correction. $A = 1$ is used for simplicity.

$x = \sin(\omega t)$

The object's velocity in SHM is represented by the derivative of the position function:

$v = \omega \cos(\omega t)$

The object's acceleration in SHM is represented by the derivative of the velocity function:

$a = -\omega^2 \sin(\omega t)$

The acceleration of objects in simple harmonic motion is represented as the opposite value of the position, multiplied by the square of angular velocity.

ω is constant, so the graphs keep the same wavelengths.

34. D is correct.

$KE_{avg} = (3/2)kT$

$KE_{avg} = (3/2) \cdot (1.38 \times 10^{-23} \text{ J/K}) \cdot (740 \text{ K})$

$KE_{avg} = 1.5 \times 10^{-20} \text{ J}$

35. B is correct.

The power of the combination is:

$$P = P_1 + P_2$$

Since power is the reciprocal of the focal length (in meters),

$f_1 = 10$ cm $= 1 / 10$ m

$P_1 = 1 / f_1$

$P_1 = 1 / (1 / 10$ m$)$

$P_1 = 10$ D

$f_2 = 20$ cm $= 1 / 5$ m

$P_2 = 1 / f_2$

$P_2 = 1 / (1 / 5$ m$)$

$P_2 = 5$ D

$P = P_1 + P_2$

$P = 10$ D $+ 5$ D $= 15$ D

36. E is correct.

$C = 1 / (2\pi R f)$

$C = 1 / (2\pi \times 4{,}000 \ \Omega \times 600$ Hz$)$

$C = 6.6 \times 10^{-8}$ F

Because the answers are in the micro-Faradays, divide by 10^{-6} to find proper units (μFaradays)

$C = 6.6 \times 10^{-8}$ F $/ (10^{-6})$

$C = 0.066$ μF

37. C is correct.

The voltage drop across 3 Ω resistor is 2 V, the current through the 3 Ω resistor is:

$I = V / R$

$I = 2$ V $/ 3 \ \Omega$

$I = 2/3$ amps

Since the 1.5 Ω resistor is connected in parallel with the 3 Ω resistor, voltage drop = 2 V (parallel resistors share the same voltage drop).

continued...

The current through the 1.5 Ω resistor is:

$I = 2 \text{ V} / 1.5 \text{ Ω}$

$I = 4/3$ amps

Then, sum the currents:

$I_{\text{total}} = 2/3$ amps $+ 4/3$ amps

$I_{\text{total}} = 2$ amps

38. A is correct.

mass = density × volume

$m = \rho V$

$m = (1 \times 10^{18} \text{ kg/m}^3) \cdot (1.76 \times 10^{-6} \text{ m})^3$

$m = 5.45$ kg ≈ 5.5 kg

39. A is correct.

An *isothermal process* does more work than an adiabatic process.

Thus, there is more area under the curve of the isothermal process than the adiabatic.

40. E is correct.

$\lambda = vt$

$\lambda = (4.6 \text{ m/s}) \cdot (10 \text{ s})$

$\lambda = 46$ m

41. D is correct.

$W = Fd \cos \theta$

$W = (20 \text{ N}) \cdot (2 \text{ m})$

$W = 40$ J

42. C is correct.

The total momentum of the system is always conserved.

Before the ball was thrown, the momentum was zero because all mass on the canoe was stationary.

After the ball is thrown and caught on the canoe, the momentum must still equal zero, so the canoe remains stationary.

$p = mv$

43. C is correct.

In Joseph's frame, the two events are separated by $\Delta x = 0$ and $\Delta t = (61 - 21)$ years, and in the Earth's frame $\Delta x' = 50$ light-years.

The *Lorentz transformation* is:

$$\Delta x' = \gamma (\Delta x + v \Delta t)$$

where $\gamma = 1/\sqrt{1 - \beta^2}$ $\quad \beta = v / c$

$$50 \text{ light-years} = [1/\sqrt{1 - \beta^2}] \cdot (0 + \beta \cdot 40 \text{ light-years})$$

Squaring both sides:

$$2500 = 1600 \cdot \beta^2 / (1 - \beta^2)$$

Solving for β^2:

$$\beta^2 = (25/16) / (1 + 25/16)$$

$$\beta^2 = 0.610$$

and:

$$v = 0.781c$$

44. A is correct.

$$f = 1 / T$$

Period $= (60 \text{ s}) / (10 \text{ oscillations})$

$T = 6 \text{ s}$

$f = 1 / 6 \text{ s}$

$f = 0.17 \text{ Hz}$

45. D is correct.

For materials with a positive coefficient of thermal expansion, a hole drilled in the material expands as temperature increases.

Regardless of the surrounding metal's expansion, the hole's diameter increases with higher temperatures.

46. C is correct.

The kinetic energy of a falling object is directly proportional to the height from which it falls.

This is because mass and gravity are constants, so only the height varies the kinetic energy of a dropped object.

$\text{KE} = \text{PE}$

$\frac{1}{2}mv^2 = mgh$

Diagnostic Test 1 – Detailed Explanations

47. B is correct.

If the two sound sources are in phase, then there is no destructive interference.

The point can be related to the wavelength of the sound wave.

$$0.5 \text{ m} = x\lambda$$
$$0.5 \text{ m} = x(1 \text{ m})$$
$$x = \tfrac{1}{2}$$

The microphone is located one-half wavelength from the speaker.

The speaker is a source of the sound pressure wave, so it is an antinode.

One-half wavelength from an antinode is an antinode.

48. C is correct.

An object becomes electrostatically charged when a charge imbalance exists.

A charge can only be transferred by electrons because protons are not mobile.

Thus, electron transfer creates an electrostatic charge.

49. D is correct.

Volume strain:

$$\Delta V / V$$
$$B = -\Delta P / (\Delta V / V)$$
$$\Delta P = -B(\Delta V / V)$$
$$\Delta P = (-6.3 \times 10^9 \text{ Pa}) \cdot (-3 \times 10^{-4})$$
$$\Delta P = 19 \times 10^5 \text{ Pa}$$
$$\Delta P = (19 \times 10^5 \text{ Pa} / 1) \cdot (1 \text{ atm} / 10^5 \text{ Pa})$$
$$\Delta P = 19 \text{ atm}$$

The negative sign indicates that the final volume is lower than the initial volume; since the volume has decreased, the pressure increased.

50. D is correct.

Velocity is in the direction of the current:

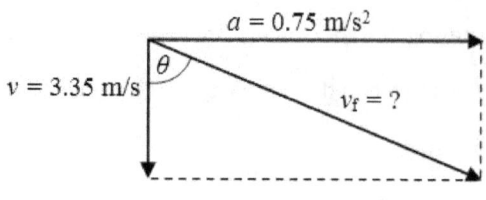

$v_c = at$

$v_c = (0.75 \text{ m/s}^2) \cdot (33.5 \text{ s})$

$v_c = 25 \text{ m/s}$

Final velocity: $v_f^2 = v^2 + v_c^2$

$v_f^2 = (3.35 \text{ m/s})^2 + (25 \text{ m/s})^2$

$v_f^2 = 636.2 \text{ m}^2/\text{s}^2$

$v_f = 25 \text{ m/s}$

The angle of motion with respect to the initial velocity:

$\theta = \tan^{-1}(v_c / v)$

$\theta = \tan^{-1}[(25 \text{ m/s}) / (3.35 \text{ m/s})]$

$\theta = \tan^{-1} 7.5$

$\theta = 82.4°$

51. D is correct.

Divide into 3 parts: initial acceleration, constant velocity, final deceleration.

1) Initial acceleration: determine α, then solve for the displacement during the acceleration. The initial velocity is zero.

Convert the displacement in radians to revolutions.

$\alpha = (\omega_f - \omega_i) / t$

$\alpha = (58 \text{ radians/s} - 0) / 10 \text{ s}$

$\alpha = 5.8 \text{ radians/s}^2$

$\theta = \frac{1}{2}\alpha t^2$

$\theta = \frac{1}{2}(5.8 \text{ radians/s}^2) \cdot (10 \text{ s})^2$

$\theta = 290 \text{ radians}$

$Rev = \theta / 2\pi$

$Rev = 290 \text{ radians} / 2\pi$

$Rev = 46 \text{ revolutions}$

2) Constant velocity: solve θ using constant angular velocity.

continued...

Convert radians to revolutions.

$\theta = \omega t$

$\theta = (58 \text{ radians/s}) \cdot (30 \text{ s})$

$\theta = 1{,}740$ radians

$Rev = \theta / 2\pi$

$Rev = 1{,}740 \text{ radians} / 2\pi$

$Rev = 277$ revolutions

3) Final deceleration: determine t for the period of deceleration using the final velocity as zero.

Solve for the displacement during this constant deceleration and convert to revolutions.

$\alpha = (\omega_f - \omega_i) / t$

$t = (\omega_f - \omega_i) / \alpha$

$t = (0 - 58 \text{ radians/s}) / (-1.4 \text{ radians/s}^2)$

$t = 41$ s

$\theta = \omega_i t + \frac{1}{2}\alpha t^2$

$\theta = [(58 \text{ radians/s}) \cdot (41 \text{ s})] + [\frac{1}{2}(-1.4 \text{ radians/s}^2) \cdot (41 \text{ s})^2]$

$\theta = 1{,}201$ radians

$Rev = 1{,}201 \text{ radians} / 2\pi$

$Rev = 191$ revolutions

Add the revolutions:

$Rev_{total} = 46 \text{ rev} + 277 \text{ rev} + 191 \text{ rev}$

$Rev_{total} = 514 \approx 510$ revolutions

52. E is correct.

Snell's Law:

$n_1 \sin \theta_1 = n_2 \sin \theta_2$

$n_1 = n_2 (\sin \theta_2 / \sin \theta_1)$

$n_1 = (1.33) \cdot [\sin (72°) / \sin (48°)]$

$n_1 = 1.7$

$(1.7) \sin (37°) = (1.33) \sin \theta_2$

$\sin \theta_2 = 0.769$

$\theta_2 = 50°$

53. C is correct.

Label the tuning forks I, II, III, and IV.

Beats: I & II, I & III, I & IV, II & III, II & IV and III & IV

From six pairs, there is the possibility of six different beat frequencies.

This is a combination problem since the order does not matter.

The formula for combinations is:

$$C(n, k) = n! / (n – k)!k!$$

where n is the given sample size, and k is the number of tuning forks per pair.

When solving combination problems, choose n and k by asking: "how do I find all the ways to pick n and k?"

$$C(4,2) = 4! / (4 – 2)!2!$$
$$C(4,2) = 4! / 2!2!$$
$$C(4,2) = 6$$

54. E is correct.

Capacitance:

$$C = Q / V$$
$$Q = CV$$
$$V = IR$$
$$Q = C \times (IR)$$
$$Q = (12 \times 10^{-6}\,F) \cdot (33 \times 10^{-6}\,A) \cdot (8.5 \times 10^{6}\,\Omega)$$
$$Q = 0.0034\ C$$

Divide by 10^{-6} to determine micro-coulombs:

$$Q = (0.0034\ C) / 10^{-6}$$
$$Q = 3{,}400\ \mu C$$

55. E is correct.

Use the conservation of energy.

Note that the ball is at rest in the initial state, so the initial kinetic energy is zero.

Take the zero of gravitational potential energy to be the height of the ball at its final position.

continued…

The final kinetic energy has a translational part and a rotational part:

$$E_f = E_i$$

$$KE_f = PE_i$$

$$KE_{translation} + KE_{rotation} = PE_i$$

$$\tfrac{1}{2}mv^2 + \tfrac{1}{2}I\omega^2 = mgh$$

The ball is rolling without slipping, so the relationship between translational and rotational velocity is:

$$v = r\omega$$

The above becomes:

$$\tfrac{1}{2}mr^2\omega^2 + \tfrac{1}{2}(2/5)mr^2\omega^2 = mgh$$

$$(7/10)mr^2\omega^2 = mgh$$

Solving for the *angular speed*:

$$\omega = \sqrt{(10gh / 7r^2)}$$

$$\omega = \sqrt{\{[10 \cdot (9.8 \text{ m/s}^2) \cdot (5.3 \text{ m})] / [7 \cdot (1.7 \text{ m})^2]\}}$$

$$\omega = 5.1 \text{ rad/s}$$

56. E is correct.

A: it is possible to convert work entirely into heat, but heat cannot be converted entirely into work.

B: heat can be transferred from a colder body to a hotter body (e.g., refrigeration), but the process is not spontaneous and requires work.

C: the second law of thermodynamics is not a consequence of the first law and goes beyond the limitations imposed by the first law.

57. C is correct.

$$y = v_i t + \tfrac{1}{2}at^2$$

$$50 \text{ m} = 0 + \tfrac{1}{2}(10 \text{ m/s}^2)t^2$$

$$50 \text{ m} = \tfrac{1}{2}(10 \text{ m/s}^2)t^2$$

$$t^2 = 50 \text{ m} / 5 \text{ m/s}^2$$

$$t^2 = 10 \text{ s}^2$$

$$t = 3.2 \text{ s}$$

Solve for speed:

$$v_f = v_i + at$$

$$v_f = 0 + (10 \text{ m/s}^2) \cdot (3.2 \text{ s})$$

$$v_f = 32 \text{ m/s}$$

58. D is correct.

Newton's Second Law for each block:

$ma = F_{net}$ acting on the object

The tension and acceleration on each block are equal in magnitude but act in different directions.

The only nonzero net forces will be horizontal for the 15 kg block and vertical for the 60 kg block.

For the 15 kg block:

ma = tension acting to the right

$(15 \text{ kg})a = F_T$

For the 60 kg block:

ma = (weight acting downward) − (tension acting upward)

$(60 \text{ kg})a = (60 \text{ kg}) \cdot (10 \text{ m/s}^2) − F_T$

Substitute F_T from the first equation into the second:

$(60 \text{ kg})a = (60 \text{ kg}) \cdot (10 \text{ m/s}^2) − (15 \text{ kg})a$

$(60 \text{ kg})a + (15 \text{ kg})a = (60 \text{ kg}) \cdot (10 \text{ m/s}^2)$

$(75 \text{ kg})a = (60 \text{ kg}) \cdot (10 \text{ m/s}^2)$

$a = [(60 \text{ kg}) \cdot (10 \text{ m/s}^2)] / (75 \text{ kg})$

$a = 8 \text{ m/s}^2$

59. C is correct.

charge = (# electrons)·(electron charge)

$Q = n(e^-)$

$n = Q / e^-$

$n = (-9 \times 10^{-6} \text{ C}) / (-1.6 \times 10^{-19} \text{ C})$

$n = 5.6 \times 10^{13}$ electrons

60. B is correct.

$^0_0\gamma$ is a *gamma particle*, so the atomic mass and atomic number do not change.

Alpha decay: the parent nuclide sheds two protons and two neutrons, identical to the nucleus of ^{4}He.

$^A_Z X \rightarrow \,^{A-4}_{Z-2} Y + \,^4_2 \alpha$

continued…

Beta-decay (*minus*): the parent nuclide sheds an electron and electron antineutrino.

However, a neutron converts to a proton in the process, so the mass number remains the same, but the atomic number increases by 1.

$$^A_Z X \rightarrow ^A_{Z+1} Y + ^0_{-1}e^- + ^0_0\bar{\nu}_e$$

Beta-decay (*plus*): the parent nuclide sheds a positron and neutrino.

However, a proton converts to a neutron in the process, so the mass number remains the same, but the atomic number decreases by 1.

$$^A_Z X \rightarrow ^A_{Z-1} Y + ^0_{+1}e^+ + ^0_0\nu_e$$

61. A is correct.

Coulomb's Law:

$F = kQ_1Q_2 / r^2$

When charges are doubled:

$F_2 = k(2Q_1)(2Q_2) / r^2$

$F_2 = 4kQ_1Q_2 / r^2$

$F_2 = 4F$

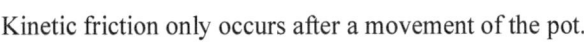

F increases by a factor of 4 (quadruples).

62. B is correct.

The angle the board makes before the pot slides depends on the static friction coefficient, as static friction influences the force of friction before the pot slides.

Kinetic friction only occurs after a movement of the pot.

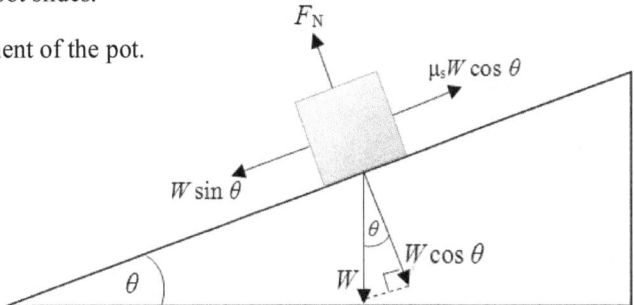

63. E is correct.

$Q = mc\Delta T$

where ΔT is constant

If m = 4 times increase and c = 3 times increase

$Q = (4)\cdot(3)\Delta T$

$Q = (12)\Delta T$

$Q_1 / 12 = \Delta T$

64. C is correct.

Ignoring air resistance, energy is conserved.

Loss in PE = gain in KE

$$KE = \tfrac{1}{2}mv^2$$

$$KE = \tfrac{1}{2}(20 \text{ kg}) \cdot (30 \text{ m/s})^2$$

$$KE = 9{,}000 \text{ J}$$

This equals the amount of PE that is lost (i.e., converted into KE).

65. C is correct.

velocity = frequency × wavelength

$$v = f\lambda$$

$$\lambda = v / f$$

$$f = 1 / T$$

$$\lambda = v \times T$$

$$\lambda = 360 \text{ m/s} \times 4.2 \text{ s}$$

$$\lambda \approx 1{,}512 \text{ m}$$

66. B is correct.

$$PE = \tfrac{1}{2}kx^2$$

$$PE = \tfrac{1}{2}k(2x)^2$$

$$PE = 4(\tfrac{1}{2}kx^2)$$

67. C is correct.

Young's Modulus is expressed as:

$$E = \sigma \text{ (stress)} / \varepsilon \text{ (strain)}$$

$$E = (F / A) / (\Delta L / L)$$

$$E = (FL) / (\Delta L A)$$

Solve for *E*:

$$E = (8.8 \text{ kg}) \cdot (9.8 \text{ m/s}^2) \cdot (4.4 \text{ m}) / (0.0033 \text{ m}) \cdot (\pi / 4) \cdot (0.0016 \text{ m})^2$$

$$E = 5.7 \times 10^{10} \text{ N/m}^2$$

68. D is correct.

Coulomb's Law, the electrostatic force is *inversely proportional* to the square of the distance between charges.

$$F = kq_1q_2 / r^2$$

If the distance increases by a factor of 2, then the force decreases by a factor of $2^2 = 4$.

69. C is correct.

$$P = IV$$

$$I = P / V$$

$$I = (1 \times 10^{-3} \text{ W}) / (9 \text{ V})$$

$$P = 0.00011 \text{ A}$$

$$P = 0.11 \text{ mA}$$

70. C is correct.

The plane mirror is double the distance from an object, so $\frac{1}{2}h$ is required for the minimum length.

Law of reflection:

$$\theta_1 = \theta_2$$

$$h = 2x$$

$$x = \frac{1}{2}h$$

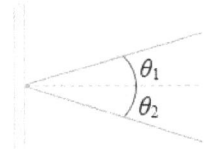
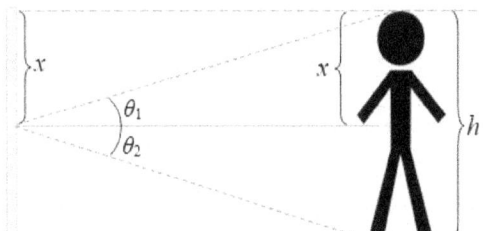

71. C is correct.

Fission occurs when an atom with a larger atomic number is struck by a free neutron and splits.

For example: $^{235}_{92}\text{U} + ^{1}_{0}\text{n} \rightarrow ^{92}_{36}\text{Kr} + ^{141}_{56}\text{Ba} + 3^{1}_{0}\text{n}$

72. B is correct.

Momentum is conserved:

$$m_1v_1 + m_2v_2 = m_3v_3$$

momentum before = momentum after

$$p_i = p_f$$

$$p_{total} = m_1v_1 + m_2v_2$$

$$p_{total} = (1 \text{ kg}) \cdot (1 \text{ m/s}) + (6 \text{ kg}) \cdot (0 \text{ m/s})$$

$$p_{total} = 1 \text{ kg} \cdot \text{m/s}$$

73. C is correct.

The work done by the force can be related to kinetic energy.

6 kg mass:

$$KE = W$$

$$\tfrac{1}{2}(6 \text{ kg})\cdot(2 \text{ m/s})^2 = Fd_1$$

$$d_1 = 12 / F$$

3 kg mass:

$$KE = W$$

$$\tfrac{1}{2}(3 \text{ kg})\cdot(4 \text{ m/s})^2 = Fd_2$$

$$d_2 = 24 / F$$

$$d_2 = 2(12 / F)$$

Therefore:

$$2d_1 = d_2$$

74. A is correct.

Separation between maxima in a double-slit interference pattern is given by:

$$\Delta y = \lambda D / d$$

where y = the maximum separation, λ = wavelength, D = the distance from the slit to the diffraction pattern, and d = the slit separation

Red light has the highest value wavelength and has the largest maximum separation.

75. B is correct.

The λ of a stretched string of length L fixed at both ends is:

$$\lambda = 2L / n$$

where n = 1 for the fundamental frequency

Therefore, if the fundamental frequency (f_1) is 860 Hz:

$$\lambda = 2(0.25 \text{ m})$$

$$\lambda = 0.5 \text{ m}$$

$$v = f\lambda$$

$$v = (860 \text{ Hz})\cdot(0.5 \text{ m})$$

$$v = 430 \text{ m/s}$$

76. A is correct.

The volume flow rate (Q) is how much water is flowing per second.

Find the cross-sectional area of the pipe and multiply the area with the velocity of the water.

$Q = A_{pipe} v$

$Q = \pi(0.03 \text{ m})^2 \times (4 \text{ m/s})$

$Q = 0.0113 \text{ m}^3/\text{s}$

$Q = 1.1 \times 10^{-2} \text{ m}^3/\text{s}$

77. E is correct.

No statement about how a conductor differs from an insulator is correct.

78. C is correct.

When a copper wire joins two conductors, they have the same potential because the wire allows a charge to flow.

Any potential difference is neutralized by charge flow.

79. C is correct.

$d = 130$ m

$f = 3.6$ MHz

$\theta = ?$

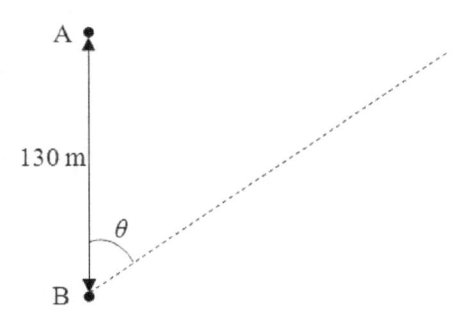

velocity = frequency × wavelength

$\lambda = c / f$

$\lambda = (3 \times 10^8 \text{ m/s}) / (3.6 \times 10^6 \text{ Hz})$

$\lambda = 83.3$ m

The condition for constructive interference is:

$m\lambda = d \cos \theta$

or $\cos \theta = m\lambda / d$

The value of cosine gets closer to 1.0 as the angle gets *smaller*.

To find the smallest angle, find the largest possible value of $\cos \theta$, as long as the value does not exceed 1.0.

With $m = 1$:

$\theta = \cos^{-1}(83.3 \text{ m} / 130 \text{ m})$

$\theta = \cos^{-1}(0.641)$

$\theta = 50°$

continued…

With $m = 2$:

$\theta = \cos^{-1}(2 \times 83.3 \text{ m} / 130 \text{ m})$

$\theta = \cos^{-1}(1.281)$, which has no solution.

Therefore, the smallest angle for which constructive interference occurs is 50°.

80. A is correct.

The magnetic quantum number is an interval from $-\ell$ to $+\ell$ (includes zero and an allowable value).

The total amount of magnetic quantum numbers possible is:

$2(\ell) + 1$

For example:

$\ell = 2$

$m_\ell = -2, -1, 0, 1, 2$

$m_{\ell \text{ total possible}} = 2(2) + 1$

$m_{\ell \text{ total possible}} = 5$

In the range from $-\ell$ to $+\ell$, there are 5 possible magnetic quantum numbers.

Verification through the equation:

$m_{\ell \text{ tot possible}} = 2(\ell) + 1$

81. D is correct.

In the frame of the Earth:

$d = vt$

$t = d / v$

$t = (4.367 \text{ years} \cdot c) / [0.8c]$

$t = 5.49 \text{ years}$

82. C is correct.

Within the nucleus, the Coulomb repulsion is overcome by the strong nuclear forces.

The strong nuclear force counteracts the repulsion and binds the nucleus.

83. B is correct.

Sound intensity is defined as power per unit area and is usually expressed as W/m^2.

Thus, sound intensity is directly proportional to power.

84. E is correct.

Output pressure (P_2) = input pressure (P_1)

$P_2 = P_1$

$(F_2 / A_2) = (F_1 / A_1)$

$F_1 = (A_1 / A_2)F_2$

$F_1 = [\pi(3 \text{ cm})^2 / \pi(12.5 \text{ cm})^2] \cdot (12{,}000 \text{ N})$

$F_1 = (9 \text{ cm} / 156.25 \text{ cm}) \cdot (12{,}000 \text{ N})$

$F_1 = 691 \text{ N}$

85. D is correct.

$E = F / q$

$F = Eq_{proton}$

$F = (4 \times 10^4 \text{ N/C}) \cdot (1.6 \times 10^{-19} \text{ C})$

$F = 6.4 \times 10^{-15} \text{ N}$

86. A is correct.

$Q = mc\Delta T$

$c = Q / m\Delta T$

$c = (200 \text{ kcal}) / [(3 \text{ kg}) \cdot (90 \text{ °C})]$

$c = 0.74 \text{ kcal/kg·°C}$

87. A is correct.

Optical density is related to the index of refraction of a material and describes how electromagnetic waves travel in a medium.

The optical density of a material is not related to its mass (physical) density.

88. B is correct.

Beta radiation is more powerful than alpha radiation but less powerful than gamma rays.

Beta radiation can penetrate the skin, paper, or even a light layer of clothing.

89. D is correct.

Increasing the viscosity decreases Reynold's number, decreasing the likelihood of turbulence occurring.

Increasing the flow rate or increasing the radius of the pipe increases the likelihood of turbulence.

Reynold's number does not include a term for temperature (i.e., no bearing on the value of Reynold's number).

90. D is correct.

$$T = W \pm ma$$

$$F = W \pm ma$$

Positive (add) if a body is moving upward and negative (subtract) if moving downward.

$$T = mg + ma$$

$$T = m(g + a)$$

$$T = (900 \text{ kg}) \cdot (9.8 \text{ m/s}^2 + 0.6 \text{ m/s}^2)$$

$$T = 9{,}360 \text{ N}$$

91. B is correct.

Check if KE is conserved:

$$KE_{before} = KE_{after} \text{ if the collision is elastic}$$

Before:

$$(\tfrac{1}{2}) \cdot (4 \text{ kg}) \cdot (1.8 \text{ m/s})^2 + (\tfrac{1}{2}) \cdot (6 \text{ kg}) \cdot (0.2 \text{ m/s})^2 = KE_{before}$$

$$KE_{before} = 6.6 \text{ J}$$

After:

$$(\tfrac{1}{2}) \cdot (4 \text{ kg}) \cdot (0.6 \text{ m/s})^2 + (\tfrac{1}{2}) \cdot (6 \text{ kg}) \cdot (1.4 \text{ m/s})^2 = KE_{after}$$

$$KE_{after} = 6.6 \text{ J}$$

Therefore:

$$KE_{before} = KE_{after}$$

The collision was completely elastic because kinetic energy was conserved.

92. B is correct.

Assuming that the swing was at rest at the beginning and end of Steve's pull:

$W = mgh$

Referring to the figure:

$h = L(1 - \cos \theta)$

where L is the length of the chain.

Hence:

$W = mgL(1 - \cos \theta)$

Therefore:

$m = W / [gL(1 - \cos \theta)]$

$m = (174 \text{ J}) / [(9.8 \text{ m/s}^2) \cdot (5.1 \text{ m}) \cdot (1 - \cos (32°))]$

$m = (174 \text{ J}) / [(9.8 \text{ m/s}^2) \cdot (5.1 \text{ m}) \cdot (1 - 0.848)]$

$m = 22.9 \text{ kg}$

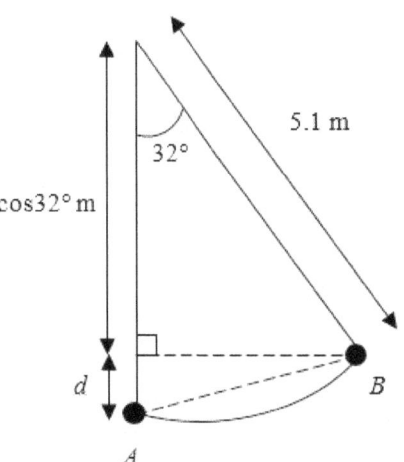

93. A is correct.

It takes 2 s for the object to travel from one end of its displacement to the other (½ a cycle).

The time for a complete cycle is 2 s + 2 s = 4 s.

$f = 1 / T$

$f = 1 / (4 \text{ s})$

$f = 0.25 \text{ s}^{-1} = 0.25 \text{ Hz}$

94. D is correct.

2 (10 cm aluminum rods) + 5 (8 cm steel rods) = 20 cm + 40 cm = 60 cm

$\Delta L = L \alpha \Delta T$

where α is the coefficient of linear expansion

$\Delta L_{alum} = (20 \text{ cm}) \cdot (2.4 \times 10^{-5} \text{ K}^{-1}) \cdot (80 \text{ °C} - 5 \text{ °C})$

$\Delta L_{alum} = 3.6 \times 10^{-2} \text{ cm}$

$\Delta L_{alum} = 0.36 \text{ mm}$

$\Delta L_{steel} = (40 \text{ cm}) \cdot (1.2 \times 10^{-5} \text{ K}^{-1}) \cdot (80 \text{ °C} - 5 \text{ °C})$

$\Delta L_{steel} = 3.6 \times 10^{-2} \text{ cm}$

$\Delta L_{steel} = 0.36 \text{ mm}$

The change in length of the composite rod is the sum of these: 0.72 mm

95. D is correct.

Diverging mirrors (i.e., convex mirrors) are curved outward toward the light source and have a focal point *behind* the mirror, so the focal length f is a negative value.

Using the equation for focal length:

$$1/f = 1/d_o + 1/d_i$$

where d_o is the distance to the light source, and d_i is the distance to the image.

$$-1/6 \text{ m} = 1/12 \text{ m} + 1/d_i$$

$$1/d_i = -1/4 \text{ m}$$

$$d_i = -4 \text{ m}$$

96. E is correct.

By doubling the frequency of the light, the energy of each photon doubles, but not the number of photons (hence B is wrong).

Although doubling the energy of each photon would cause more electrons to be ejected, it does not necessarily, double that number since the number is not linearly proportional to the incident energy (A is not always true).

The kinetic energy of the ejected electrons would more than double since the initial kinetic energy is less than the energy of the incident photons - the work function of the surface reduced it (hence C is wrong).

The kinetic energy would increase by at least 2 but not necessarily by 4 (hence D is wrong).

97. B is correct.

Efficiency of a heat engine:

$$n = Q_H - Q_C / Q_H$$

$$n = (8{,}500 \text{ J} - 4{,}500 \text{ J}) / (8{,}500 \text{ J})$$

$$n = 0.47$$

$$n = 47\%$$

98. B is correct.

Buoyancy force:

$$F_B = \rho g h A$$

$$F_B = \rho g V$$

Because the cube is lowered at a constant rate, the volume of the cube underwater increases linearly, and thus the F_B increases linearly.

After the cube is submerged, F_B stays constant because the volume of displaced water is constant.

Thus, a linearly increasing line then a steady flat slope of zero describes the buoyant force *vs.* time graph.

99. C is correct.

The source is stationary, and the detector is traveling towards the source at $v_d = 50$ m/s.

Since they are moving toward each other, the velocity of the detector is positive, and the detected frequency will be higher than the emitted frequency.

For the frequency perceived when the source of noises is approaching; the Doppler calculation takes the form:

$f_{observed} = f_{source} (v + v_{observer}) / (v - v_{source})$

$f_{observed} = (420 \text{ Hz}) \times [(350 \text{ m/s} + 50 \text{ m/s}) / (350 \text{ m/s} - 0 \text{ m/s})]$

$f_{observed} = 420 \text{ Hz} \times 1.14$

$f_{observed} = 480 \text{ Hz}$

100. A is correct.

Beats are observed when two sound waves of different frequency approach and alternating constructive/destructive causes the sound to be soft and loud.

$f_{beat} = |f_1 - f_2|$

Notes for active learning

Diagnostic Test 2 – Detailed Explanations

Answer Key

1	B	Work & energy	26	D	Sound
2	E	Equilibrium & momentum	27	C	Fluids & solids
3	D	Force, motion, gravitation	28	B	Electrostatics & electromagnetism
4	A	Kinematics & dynamics	29	D	Special relativity
5	B	Rotational motion	30	C	Light & optics
6	E	Work & energy	31	E	Atomic & nuclear structure
7	D	Sound	32	C	Equilibrium & momentum
8	A	Electrostatics & electromagnetism	33	C	Work & energy
9	B	Work & energy	34	B	Waves & periodic motion
10	B	Kinematics & dynamics	35	C	Sound
11	E	Equilibrium & momentum	36	B	Electrostatics & electromagnetism
12	D	Waves & periodic motion	37	E	Electrostatics & electromagnetism
13	C	Fluids & solids	38	B	Electric circuits
14	D	Electric circuits	39	B	Light & optics
15	D	Atomic & nuclear structure	40	C	Atomic & nuclear structure
16	E	Heat & thermodynamics	41	E	Heat & thermodynamics
17	C	Kinematics & dynamics	42	C	Force, motion, gravitation
18	A	Force, motion, gravitation	43	D	Sound
19	C	Equilibrium & momentum	44	D	Fluids & solids
20	B	Atomic & nuclear structure	45	B	Electrostatics & electromagnetism
21	D	Kinematics & dynamics	46	C	Electric circuits
22	A	Force, motion, gravitation	47	E	Light & optics
23	D	Heat & thermodynamics	48	B	Atomic & nuclear structure
24	E	Work & energy	49	E	Quantum mechanics
25	E	Waves & periodic motion	50	B	Force, motion, gravitation

#	Ans	Topic	#	Ans	Topic
51	E	Equilibrium & momentum	76	A	Heat & thermodynamics
52	A	Work & energy	77	B	Kinematics & dynamics
53	D	Waves & periodic motion	78	C	Force, motion, gravitation
54	C	Heat & thermodynamics	79	B	Equilibrium & momentum
55	D	Light & optics	80	E	Atomic & nuclear structure
56	A	Electric circuits	81	E	Kinematics & dynamics
57	B	Electrostatics & electromagnetism	82	B	Force, motion, gravitation
58	E	Sound	83	B	Heat & thermodynamics
59	A	Waves & periodic motion	84	B	Work & energy
60	B	Waves & periodic motion	85	E	Waves & periodic motion
61	A	Work & energy	86	D	Sound
62	C	Equilibrium & momentum	87	A	Fluids & solids
63	D	Rotational motion	88	D	Electrostatics & electromagnetism
64	A	Kinematics & dynamics	89	C	Electric circuits
65	B	Light & optics	90	B	Light & optics
66	A	Work & energy	91	C	Atomic & nuclear structure
67	A	Light & optics	92	D	Equilibrium & momentum
68	A	Electrostatics & electromagnetism	93	A	Work & energy
69	B	Special relativity	94	D	Waves & periodic motion
70	A	Kinematics & dynamics	95	E	Sound
71	C	Equilibrium & momentum	96	C	Fluids & solids
72	C	Quantum mechanics	97	B	Electrostatics & electromagnetism
73	E	Fluids & solids	98	A	Electric circuits
74	A	Electric circuits	99	A	Light & optics
75	B	Atomic & nuclear structure	100	C	Atomic & nuclear structure

1. B is correct.

The extension x beyond the resting length is given by:

$x = F_{spring} / k$

spring force ← M → centripetal force

$F_{spring} = F_{centripetal}$

$x = (12 \text{ N}) / (40 \text{ N/m})$

$x = 0.3 \text{ m}$

Since the radius of the circle of revolution is 2 m and the spring is pulling, the resting length of the spring is:

$(2 \text{ m} - 0.3 \text{ m}) = 1.7 \text{ m}$

2. E is correct.

The acceleration of a point on a rotating circle is the centripetal acceleration:

$a_c = v^2 / r$

where $r = 7.2 \text{ cm} = 0.072 \text{ m}$

Convert rpm to rps:

$(2{,}640 \text{ rotation/min}) \cdot (1 \text{ min} / 60 \text{ s}) = 44.0 \text{ rps}$

The speed of a point on the edge of the motor is:

$v = C / T$

where C is the circumference and T is the period of revolution

Calculate the circumference:

$C = 2\pi r$

$C = 2\pi(0.072 \text{ m}) = 0.452 \text{ m}$

Calculate the speed:

$v = (0.452 \text{ m}) / (1 \text{ s} / 44 \text{ rps})$

$v = 19.89 \text{ m/s}$

$a = v^2 / r$

$a = (19.89 \text{ m/s})^2 / (0.072 \text{ m})$

$a = 5{,}494.6 \approx 5{,}495 \text{ m/s}^2$

3. **D is correct.**

$$x = x_0 + v_0 t + \tfrac{1}{2}at^2$$

where $t = 0.51$ s

$$x = 0 + 0 + \tfrac{1}{2}at^2$$

$$x = \tfrac{1}{2}at^2$$

$$2x / t^2 = a$$

$$2(1 \text{ m}) / (0.51 \text{ s})^2 = a$$

$$a = 2 \text{ m} / 0.26 \text{ s}^2$$

$$a = 7.7 \text{ m/s}^2$$

$$(m_1 + m_2)a = m_1 g - m_2 g$$

$$m_1 a + m_2 a = m_1 g - m_2 g$$

$$m_2(a + g) = m_1(g - a)$$

$$m_2 = [m_1(g - a)] / (a + g)$$

$$m_2 = [(100 \text{ kg}) \cdot (9.8 \text{ m/s}^2 - 7.7 \text{ m/s}^2)] / (7.7 \text{ m/s}^2 + 9.8 \text{ m/s}^2)$$

$$m_2 = [(100 \text{ kg}) \cdot (2.1 \text{ m/s}^2)] / (17.5 \text{ m/s}^2)$$

$$m_2 = 12 \text{ kg}$$

4. **A is correct.**

The projectile experiences a net force downward due to gravity, so the acceleration is negative.

Since deceleration upward is equivalent to acceleration downward, the rock is constantly accelerating downward.

5. **B is correct.**

Three rotating bodies – total angular momentum is the sum of angular momenta.

$$L_{total} = L_1 + L_2 + L_3$$

The angular momentum of a rotating object is:

$$L = I\omega$$

The flywheels are identical, so they each have the same rotational inertia I, so:

$$L = I(\omega_1 + \omega_2 + \omega_3)$$

The moment of inertia of the flywheels is:

$$I = \tfrac{1}{2}mr^2$$

$$I = 0.5(65.0 \text{ kg}) \cdot (1.47 \text{ m})^2$$

continued…

$I = 70.23$ kg m^2

Thus:

$L = (70.23$ kg m$^2) \cdot (3.83$ rad/s $+ 3.83$ rad/s $- 3.42$ rad/s$)$

$L = 298$ kg m^2/s

6. E is correct.

The object has a constant velocity upward, and the force necessary to propel the object is constant and upward.

Power = Watts / time

$W = Fd$

$P = Fd / t$

$d / t = v$

$P = Fv$

$P = (50$ N$) \cdot (10$ m/s$)$

$P = 500$ W

7. D is correct.

For a pipe closed at one end, the harmonic frequencies are odd multiples of the fundamental frequency, so:

n = 1, 3, 5….

The harmonic frequency has a wavelength $\lambda_n = 4L / n$

The wavelength of the fundamental frequency is:

$\lambda_1 = 4L / 1$

$\lambda_1 = 4L$

Since the wavelength is four times greater than the length of the pipe, the pipe accommodates ¼λ, and therefore it has a displacement node at the closed end and an antinode at the open end.

8. A is correct.

Charged objects interact with other charged objects.

If the electric charge is conserved, by definition, the charge cannot be created or destroyed.

Electrons and protons are the fundamental particles that carry a charge.

Therefore, the charge of an object is a whole-number multiple of an electron ($Q = $ ne$^-$).

The electric charge of an object can be infinitely large, so there are infinite whole-number multiples of an electron's charge.

Because the charge is a whole-number multiple of an electron's charge, it cannot have a value that is not a whole-number multiple, and therefore it occurs in restricted quantities.

9. B is correct.

Hooke's Law:

$F = -kx$, so force and distance stretched are inversely proportional.

The negative sign indicates that the force is in a direction opposite to the direction of the displacement.

10. B is correct.

The area under a curve is the same as taking the integral of velocity with respect to time.

The integral of velocity with respect to time gives the displacement.

11. E is correct.

Momentum is conserved in the collision, so it neither increases nor decreases.

Kinetic energy is conserved in the collision since the collision is elastic; the pucks are rubber and bounce off each other without losing speed due to the frictionless surface.

If these pucks have identical masses and travel towards each other with identical speeds, each puck has a final velocity equal in magnitude but opposite in direction to its initial velocity.

12. D is correct.

$c = \lambda f$

$\lambda = c / f$

$\lambda = (3 \times 10^8 \text{ m/s}) / (2.4 \times 10^{20} \text{ Hz})$

$\lambda = 1.25 \times 10^{-12} \text{ m}$

$r = 5 \times 10^{-13} \text{ cm}$

$r = 5 \times 10^{-15} \text{ m}$

$\lambda / r = (1.25 \times 10^{-12} \text{ m}) / (5 \times 10^{-15} \text{ m})$

$\lambda / r = 250$

13. C is correct.

The pressure in the hose depends on the height of the water above it and the pressure in the atmosphere.

So, the pressure in the hose is given by:

$P = \rho g h + P_{atm}$

where h = height of water

The 12 m water tank is 25 m above the ground, so:

$h = 37 \text{ m}$

continued…

So, the water pressure in the hose is:

$P = (1{,}000 \text{ kg/m}^3) \cdot (9.8 \text{ m/s}^2) \cdot (37 \text{ m}) + 101{,}325 \text{ N/m}^2$

$P = 362{,}600 \text{ N/m}^2 + 101{,}325 \text{ N/m}^2$

$P = 4.6 \times 10^5 \text{ N/m}^2$

14. D is correct.

Series: [diagram of R_1, R_2, R_3 in series] $R_{tot} = R_1 + R_2 + R_3$

Parallel: [diagram of R_1, R_2, R_3 in parallel] $1/R_{tot} = 1/R_1 + 1/R_2 + 1/R_3$

15. D is correct.

In the photoelectric effect, photons from a light source are absorbed by electrons on a metal surface and cause them to be ejected.

The energy of the ejected electrons is only dependent upon photon frequency and is found by:

$KE = hf - \phi$

where h = Planck's constant, f = frequency and ϕ = stopping potential

Increasing the intensity of the light only increases the number of photons incident upon the metal and, thus, the number of ejected electrons, but their KE does not change.

16. E is correct.

The Carnot cycle is an idealized thermodynamic cycle consisting of two isothermal processes and two adiabatic processes. It is the most efficient heat engine operating between two temperatures.

17. C is correct.

$\Delta d = v \Delta t$

$\Delta d = (-3 \text{ m/s}) \cdot (4 \text{ s})$

$\Delta d = -12 \text{ m}$

$d - d_0 = -12 \text{ m}$

$d - 4 \text{ m} = -12 \text{ m}$

$d = -8 \text{ m}$

18. A is correct.

Horizontal velocity does not affect the pebble's downward trajectory, so it is effectively in free fall as the second pebble.

19. C is correct.

Efficiency is defined as $KE_{final} / KE_{initial}$.

Kinetic energy:

$$KE = \tfrac{1}{2}mv^2$$

$$KE = p^2/2m$$

Note that momentum is conserved in a collision:

$$p_{initial} = p_{final} = p$$

Therefore:

$$\text{Efficiency} = KE_{final}/KE_{initial}$$

$$\text{Efficiency} = (p^2/2m_2) / (p^2/2m_1)$$

$$\text{Efficiency} = m_1 / m_2$$

$$\text{Efficiency} = (2.0 \text{ kg}) / (2.5 \text{ kg})$$

$$\text{Efficiency} = 0.8 \times 100\% = 80\%$$

20. B is correct.

The nucleus of an atom consists of protons and neutrons held by the strong nuclear force.

This force counteracts the electrostatic force of repulsion between the protons in the nucleus.

The gravitational and weak nuclear forces are negligible when discussing the nucleus and the force acting within it.

21. D is correct.

There is no acceleration in the horizontal direction, so velocity is constant.

$$v_{0x} = v_x$$

$$d = v_x \times t$$

$$d = (30 \text{ m/s}) \cdot (75 \text{ s})$$

$$d = 2{,}250 \text{ m}$$

22. A is correct.

The friction described in the scenario is between the tires and the road because the problem asks for the force of friction *on the car*.

Note that the car is skidding, meaning that the wheels are locked and are being dragged along the road; therefore, there is relative motion between the tires and the road; the condition for *kinetic friction*.

Static friction applies when there is no relative motion between the tires and the road at the point of contact, such as when the wheels rotate normally.

$$F_{friction} = \mu_k N$$

Because the angle is described as *slight*, the incline can be ignored.

23. D is correct.

The rate of heat transfer:

$$Q/t = kA\Delta T / d$$

where k is the thermal conductivity of the wall material, A is the surface area of the wall, d is the wall's thickness, and ΔT is the temperature difference on either side

Therefore, if thickness d is doubled, the rate is halved.

24. E is correct.

Energy can exist as PE, KE, heat, waves, etc.

Energy in any of its forms can be defined as the *ability to do work*.

The various mathematical expressions for energy specify the amount of work that can be done.

The conversion between energy and work goes both ways: work can generate any form of energy.

25. E is correct.

The displacement of the tuning fork's tines from their resting positions measures the amplitude of the resulting sound wave.

26. D is correct.

$$I \text{ (dB)} = 10 \log_{10}(I / I_o)$$

27. C is correct.

Hydrostatic equilibrium:

$$P_{bottom} = P_{atm} + \rho g(h_3 - h_1)$$

Absolute pressure:

$$P = P_{atm} + \rho g(h_3 - h_1)$$

Gauge pressure: $P = \rho g(h_3 - h_1)$

28. B is correct.

Voltage results in a current but not vice versa.

Voltage is a potential difference across a circuit but does not flow through it.

29. D is correct.

In the stationary observer's time frame, the time between clock ticks ($\Delta t'$) appears twice as measured in the clock's frame (Δt).

The Lorentz transformation for the time interval from the clock's frame (in which $\Delta x = 0$) to the observer's frame is:

$$\Delta t' = \gamma (\Delta t - v\, \Delta x / c^2)$$

where $\gamma = 1 / \sqrt{(1 - \beta^2)}$; $\beta = v / c$

$$\Delta t' = \gamma\, \Delta x$$

Expression desired:

$$\Delta t' / \Delta t = 2.0$$

Therefore:

$$\gamma = 2.0$$

Solving for β:

$$\beta = 0.866c$$

30. C is correct.

Ultraviolet radiation has the highest frequency among the choices.

Energy is related to frequency by:

$$E = hf$$

Thus, ultraviolet light has the most energy per photon because it has the highest frequency, directly proportional to the energy.

31. E is correct.

This is an example of a β^- decay.

A *neutron* converts to a proton and an electron (e^-) along with an electron neutrino (v_e).

32. C is correct.

Moment of inertia I is defined as the ratio of the angular momentum L of a system to its angular velocity ω around a principal axis.

Moment of inertia:

$I = L / \omega$

Angular acceleration around a fixed axis:

$\tau = \alpha I$

Mass moment of inertia of a thin disk:

$I = \frac{1}{2}mr^2$

$\tau = \alpha(\frac{1}{2}mr^2)$

$m = (2\tau) / \alpha r^2$

$m = [(2)\cdot(14 \text{ N·m})] / [(5.3 \text{ rad/s}^2)\cdot(0.6 \text{ m})^2]$

$m = 14.7 \text{ kg}$

33. C is correct.

$v = v_0 + at$

$29 \text{ m/s} = 0 + (10 \text{ m/s}^2)t$

$v = at$

$t = v / a$

$t = (29 \text{ m/s}) / (10 \text{ m/s}^2)$

$t = 2.9 \text{ s}$

$y = \frac{1}{2}at^2$

$y = \frac{1}{2}(10 \text{ m/s}^2)\cdot(2.9 \text{ s})^2 + 1 \text{ m}$

$y = \frac{1}{2}(10 \text{ m/s}^2)\cdot(8.41 \text{ s}^2) + 1 \text{ m}$

$y = 42 \text{ m} + 1 \text{ m}$

$y = 43 \text{ m}$

34. B is correct.

The *resonant frequency of a spring-mass system*:

$\omega = \sqrt{k / m}$

Increasing the spring constant k results in a higher resonant frequency.

35. C is correct.

Sound velocity in an ideal gas:

$$v_{sound} = \sqrt{yRT / M}$$

where y = adiabatic constant, R = gas constant, T = temperature and M = molecular mass of gas

Increasing the temperature increases the velocity of sound in air.

36. B is correct.

The work is path-independent.

$$W = q\Delta V$$
$$W = q(V_D - V_A)$$
$$W = (10^{-14} \text{ C}) \cdot (750 \text{ V} - 750 \text{ V})$$
$$W = 0 \text{ J}$$

The force does positive work during the first part of the motion and negative work in the final part. The total work is zero.

37. E is correct.

The Earth's magnetic field is thought to be created by circulating electric currents in the Earth's mantle (liquid portion).

These charges move slowly due to the convection currents in the mantle and create the magnetic field through many charges present.

38. B is correct.

Polarity switches twice per wavelength.

There are 60 λ per second.

$$2 \times f = \text{\# polarity switches}$$
$$2 \times 60 = 120 \text{ times/s}$$

39. B is correct.

A blue object illuminated with yellow light appears black because it absorbs the yellow light and reflects none.

40. C is correct

$$^A_Z X \rightarrow {^{A-4}_{Z-2}} Y + {^4_2}\alpha$$

Alpha decay: the parent nuclide ejects two protons and two neutrons as ^4_2He (essentially a helium nucleus).

The daughter nucleus has an atomic number of two less than the parent nucleus and an atomic weight of four less than the parent nucleus.

41. E is correct.

From the Second Law of Thermodynamics: it is not possible to extract heat from a hot reservoir and convert it all into useful work.

The *maximum efficiency* is that of a Carnot cycle given as:

$\eta = (Q_H - Q_C) / Q_H$

42. C is correct.

Centripetal acceleration:

$F_c = (m) \cdot (v^2 / r)$

$F_c = (1{,}200 \text{ kg}) \cdot [(3.5 \text{ m/s})^2 / 4 \text{ m}]$

$F_c = (1{,}200 \text{ kg}) \cdot [(12.25 \text{ m}^2/\text{s}^2) / 4 \text{ m}]$

$F_c = 3{,}675 \text{ N}$

43. D is correct.

Frequency, length, and velocity are related by:

$f = v / 2L$

$v = f \times 2L$

$v = (440 \text{ Hz}) \cdot (2 \times 0.14 \text{ m})$

$v = 123.2 \text{ m/s} \approx 123 \text{ m/s}$

$L = v / 2f$

$L = (123 \text{ m/s}) / (2) \cdot (520 \text{ Hz})$

$L = 0.118 \text{ m}$

$\Delta L = 0.14 \text{ m} - 0.118 \text{ m}$

$\Delta L = 0.022 \text{ m}$

$\Delta L = 2.2 \text{ cm}$

44. D is correct.

gauge pressure $= \rho g h$

gauge pressure $= (1{,}000 \text{ kg/m}^3) \cdot (10 \text{ m/s}^2) \cdot (110 \text{ m})$

gauge pressure $= 1.1 \times 10^6 \text{ Pa}$

45. B is correct.

Magnetic moment of a circular loop:

$$\mu = IA$$

$$A = (\pi r^2)$$

$$\mu = I(\pi r^2)$$

If r is doubled:

$$\mu = I\pi(2r)^2$$

$$\mu = I\pi(4r^2)$$

The magnetic moment increases by a factor of 4.

46. C is correct.

First, find the total resistance of each set of resistors in parallel.

Resistors in parallel:

$$1 / R_{total} = 1 / R_1 + 1 / R_2 \ldots + 1 / R_n$$

$$1 / R_{total} = 1 / 600\ \Omega + 1 / 600\ \Omega$$

$$R_{total} = 300\ \Omega$$

The two sets of parallel resistors are in series. Resistors in series:

$$R_{total} = R_1 + R_2 \ldots + R_n$$

$$R_{total} = 300\ \Omega + 300\ \Omega$$

$$R_{total} = 600\ \Omega$$

47. E is correct.

A simple compound microscope usually uses a short focal length objective and a long focal length eyepiece.

48. B is correct.

De Broglie equation:

$$\lambda = h / mv$$

If velocity increases, then λ decreases because they are inversely proportional.

Diagnostic Test 2 – Detailed Explanations

49. E is correct.

The Balmer formula for hydrogen is:

$1/\lambda = (1/91.2 \text{ nm}) \cdot (1/m^2 - 1/n^2)$

From the n = 3 level, the transition to the n = 2 level has the lowest energy and therefore the longest wavelength.

The Balmer formula for n = 3, m = 2 is:

$1/\lambda = (1/91.2 \text{ nm}) \cdot (1/2^2 - 1/3^2)$

$1/\lambda = (1/91.2 \text{ nm}) \cdot (5/36)$

$\lambda = 656$ nm

50. B is correct.

Newton's Third Law: when two objects interact, the force exerted on one object is equal in strength and opposite direction to the force exerted on the other object.

51. E is correct.

$F = ma$

$m = F/a$

$m_1 = (69 \text{ N}) / (9.8 \text{ m/s}^2)$

$m_1 = 7.04$ kg

$m_2 = (94 \text{ N}) / (9.8 \text{ m/s}^2)$

$m_2 = 9.59$ kg

$m_1 r_1 = m_2 r_2$

$m_1 / m_2 = r_2 / r_1$

$m_1 / m_2 = (7.04 \text{ kg}) / (9.59 \text{ kg})$

$m_1 / m_2 = 0.734$

$r_2 / r_1 = 0.734$

$r_2 + r_1 = 10$ m

Two equations, two unknowns:

Eq$_1$: $r_2 + r_1 = 10$ m

Eq$_2$: $r_2 - (0.734) \cdot (r_1) = 0$

Multiply Eq$_2$ by −1 and add to Eq$_1$:

$(1.734) r_1 = 10$ m

$r_1 = 5.8$ m

continued…

Alternatively:

For the object to be in equilibrium, the torques due to the two forces must sum to zero:

$\tau_L + \tau_R = 0$

Taking a counterclockwise torque to be positive:

$F_L x_L - F_R x_R = 0$

Let L be the length of the object, 10 m:

$x_R = (L - x_L)$

and:

$F_L x_L - F_R (L - x_L) = 0$

Solve for x_L:

$x_L = L F_R / (F_L + F_R)$

$x_L = (10 \text{ m}) \cdot (94 \text{ N}) / (69 \text{ N} + 94 \text{ N})$

$x_L = 5.77 \text{ m} \approx 5.8 \text{ m}$

52. A is correct.

The potential energy of a system can be zero because PE is defined against an arbitrary reference point.

In a gravitational potential problem, if the reference point is ground level and the object is below ground level, it will have negative potential energy relative to the reference point.

53. D is correct.

$\lambda = 2 \text{ m}$ and $T = 1 \text{ s}$

$f = 1 / T$

$f = 1 / 1 \text{ s}$

$f = 1 \text{ Hz}$

$v = f\lambda$

$v = (1 \text{ Hz}) \cdot (2 \text{ m})$

$v = 2 \text{ m/s}$

54. C is correct.

$\Delta E = E_2 - E_1$

$\Delta E = 110 \text{ J} - 40 \text{ J}$

$\Delta E = 70 \text{ J}$

55. D is correct.

A plane mirror has a magnification of m = 1.

$$m = -d_i / d_o$$

$$m = h_i / h_o$$

$$1 = -d_i / d_o$$

$$1 = h_i / h_o$$

$$-d_i = d_o$$

$$h_i = h_o$$

The negative image distance indicates that the image is virtual, and the positive image height indicates that the image is erect.

56. A is correct.

The voltage through the 8 Ω resistor is:

$$V = IR$$

$$V = (8\ \Omega) \cdot (0.8\ A)$$

$$V = 6.4\ V$$

Since the 8 Ω resistor is in parallel with the 16 Ω resistor, the voltage across the 16 Ω resistor is 6.4 V, and the current through it is:

$$I = V / R$$

$$I = (6.4\ V) / (16\ \Omega)$$

$$I = 0.4\ A$$

The total current in the upper branch is the sum of these:

$$I_{upper} = 0.4\ A + 0.8\ A$$

$$I_{upper} = 1.2\ A$$

The voltage across 20 Ω resistor:

$$V = IR$$

$$V = (1.2\ A) \cdot (20\ \Omega)$$

$$V = 24\ V$$

The total voltage across the upper branch is:

$$V_{upper} = 6.4\ V + 24\ V$$

$$V_{upper} = 30.4\ V$$

This is the power supply voltage and the voltage across the lower branch. *continued…*

The 2 Ω and 6 Ω resistors are in parallel, so it is the voltage across the 2 Ω resistor.

Therefore, the current in the 2 Ω resistor is:

$I = V / R$

$I = (30.4 \text{ V}) / (2 \text{ Ω})$

$I = 15.2 \text{ A}$

57. B is correct.

Force exerted on a particle of charge q:

$F = qE$

The acceleration of the proton is to the right, so the force is to the right.

Therefore, the electric field must be to the right.

58. E is correct.

Since a beat of frequency 4 Hz is produced, the violin string vibrates at either:

(340 Hz – 4 Hz) = 336 Hz

or

(340 Hz + 4 Hz) = 344 Hz

Since the string is too taut, the perceived f is too high.

Therefore, the string vibrates at 344 Hz.

The period is the reciprocal of the frequency.

$T = 1 / f$

$T = 1 / 344 \text{ sec}$

59. A is correct.

Hooke's Law describes simple harmonic motion (SHM):

$F = -kx$

Combining it with Newton's Second Law ($F = ma$), find:

$a = -(k/m)x$

Acceleration is proportional to displacement.

60. B is correct. Period of a pendulum:

$T = 2\pi\sqrt{(L / g)}$

The period does not depend on mass, so changes to M do not affect the period.

61. A is correct.

Work equation:

$$W = Fd$$

$$W = (70 \text{ N}) \cdot (45 \text{ m})$$

$$W = 3{,}150 \text{ J}$$

Power equation:

$$P = W / t$$

$$P = (3{,}150 \text{ J}) / (60 \times 30 \text{ s})$$

$$P = (3{,}150 \text{ J}) / (180 \text{ s})$$

$$P = 18 \text{ W}$$

62. C is correct.

To balance the seesaw, the total torque about the center must be zero.

Let the subscripts S, M, and J represent Shawn, Mark, and John, respectively.

Then:

$$\tau_S + \tau_M + \tau_J = 0$$

Take the positive sense of torque to be counterclockwise.

Then:

$$F_S x_S - F_M x_M - F_J x_J = 0$$

$$(m_S g) x_S - (m_M g) x_M - (m_J g) x_J = 0$$

$$m_S x_S - m_M x_M - m_J x_J = 0$$

Solve for John's position, x_J:

$$x_J = (m_S x_S - m_M x_M) / m_J$$

$$x_J = [(105 \text{ kg}) \cdot (5.5 \text{ m}) - (20 \text{ kg})(10 \text{ m})] / (20 \text{ kg})$$

$$x_J = [(105 \text{ kg}) \cdot (5.5 \text{ m}) - (20 \text{ kg})(10 \text{ m})] / (20 \text{ kg})$$

$$x_J = (377.5 \text{ kg} \cdot \text{m}) / (20 \text{ kg})$$

$$x_J = 18.9 \text{ m} \approx 19 \text{ m}$$

63. D is correct.

There is not enough information because neither the masses nor the radii of the sphere and cylinder are known. However, various parameters often cancel.

Try to find the acceleration of the two objects.

The object with the larger acceleration reaches the bottom first.

An object rolling down an incline experiences three forces, and hence three torques.

The forces are the force of gravity acting on the center of mass of the object, the normal force between the incline and the object, and the force of friction between the incline and the object.

If the origin is taken to be the center of the object, the force of gravity provides zero torque.

This can be seen by noting that the distance between the origin and the point of application of the force is zero.

$$\tau_{gravity} = F_{gravity} r = mg(0) = 0$$

Similarly, the normal force contributes zero torque because the direction of the force is directed through the origin (pivot point).

$$\tau_{normal} = F_{normal} \, r \sin \theta = F_{normal}(R) \cdot (\sin 180°) = F_{normal}(R) \cdot (0) = 0$$

Use a coordinate system in which the x-axis is parallel to the incline, and the y-axis is perpendicular.

The object is rolling in the positive x-direction.

The dynamical equation for linear motion along the x-direction is:

$$F_{net} = ma$$

$$(mg \sin \theta - f) = ma$$

Note that the normal force is only in the y-direction and thus does not directly contribute to the acceleration in the x-direction. The dynamical equation for rotational motion is:

$$\tau_{net} = I\alpha$$

$$fR = I\alpha$$

(Note that the frictional force is perpendicular to the r vector, and $\sin 90° = 1$)

where R is the radius of the object, f is the force of friction, and I is the moment of inertia

A relation is needed to couple these two dynamical equations.

This is the equation of constraint imposed by the restriction that the object rolls without slipping:

$$\alpha = a / R$$

To find the linear acceleration, use the equation of constraint to eliminate α from the rotational equation by replacing it with a / R:

$$fR = I(a / R)$$

continued…

The force of friction is of no interest, so rearrange this last expression:

$$f = Ia / R^2$$

Substitute this into the linear dynamic equation from above in place of f:

$$mg \sin \theta - (Ia / R^2) = ma$$

Solving this for a:

$$a = mg \sin \theta / [(m + (I / R^2)]$$

$$a = g \sin \theta / [1 + (I / mR^2)]$$

For a sphere,

$$I = (2/5)mR^2$$

so:

$$a_{sphere} = g \sin \theta / (1 + 2/5)$$

$$a_{sphere} = (5/7)g \sin \theta$$

For the cylinder,

$$I = \tfrac{1}{2}mR^2$$

so:

$$a_{cylinder} = g \sin \theta / (1 + \tfrac{1}{2})$$

$$a_{cylinder} = (2/3)g \sin \theta$$

Since 5/7 > 2/3, the acceleration of the sphere is greater than the acceleration of the cylinder so that the sphere will reach the bottom first.

Interestingly, neither the mass nor the size of the sphere or cylinder enters the result.

Indeed, the mass and radius cancel.

Since neither the masses nor the radii were given in the statement of the problem, it would not be possible to solve this problem by brute force numerical calculation.

64. A is correct.

$v_y = 3.13 \sin 30°$

$v_y = 1.6$ m/s

$v_f = v_o + at$

$0 = (1.6$ m/s$) + (-9.8$ m/s$^2)t$

$(9.8$ m/s$^2)t = (1.6$ m/s$)$

$t = (1.6$ m/s$) / (9.8$ m/s$^2)$

$t = 0.16$ s

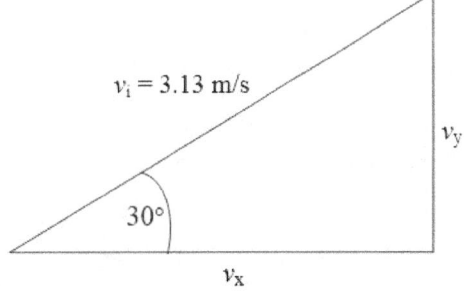

65. B is correct.

For a concave mirror with an object located at a distance equal to the focal length:

$$d_o = f$$

An object located at the focal point of a concave mirror will form no image.

66. A is correct.

Before it is released, the hammer has zero velocity and a gravitational PE of mgh.

This PE is converted entirely into KE when it reaches the ground.

$$PE\ (top) = KE\ (bottom)$$

$$mgh_0 = \tfrac{1}{2}m(v_0^2)$$

cancel m from each side of the expression

$$v_0 = \sqrt{2gh_0}$$

If h_0 increases by a factor of 2, substitute $2h_0$ for h_0

$$v = \sqrt{[2g(2h_0)]}$$

$$v = \sqrt{2} \times \sqrt{2gh_0}$$

$$v = \sqrt{2} \times (v_0)$$

The new velocity is $\sqrt{2}$ times faster.

67. A is correct.

There is a common notion that a light's color is dependent on its wavelength (ranging from 700 nm at the red end of the spectrum to 400 nm at the violet end for visible light).

However, this is true *only* when the light travels through one medium (e.g., air).

When light enters another medium (e.g., water), the wavelength changes, even though the color remains the same.

If the color depended on wavelength, it would change when the light entered a medium with a different refraction index.

The color of light is *always* dependent on frequency – the number of cycles per second (i.e., the number of waves of the light per second passing a given plane in space per second).

To illustrate this concept, visualize a ball floating in a swimming pool. If one were to dive into the pool and view the ball while underwater, all the ball colors would appear the same as they did when they were observed in the air.

When the light changes medium, the speed of the light and the wavelength of the light change.

However, the frequency of the series of light waves does not.

68. A is correct.

An electric current measures the amount of charge passing a point in the circuit per unit of time.

In a flow of water, the analogous parameter is the volume (i.e., amount) of water passing a point per unit of time (i.e., volume flow rate).

By analogy, the current is the same for resistors in series, and the volume flow rate is constant (absent any branching) along with a flow, which is not true of flow velocity.

69. B is correct.

Because of time dilation, clocks moving with respect to Earth run more slowly than clocks at rest on Earth.

So, fewer days will have passed for the astronaut during the length of the mission than for observers on Earth.

70. A is correct.

For constant acceleration, the velocity increases with time.

If velocity increases with time, the position *vs.* timeline of the graph is curved over each time interval.

71. C is correct.

Use conservation of momentum for momenta in the x coordinate to solve the x component of the second ball's final velocity.

Use m as the mass for the first ball and $1.4m$ as the mass of the second ball.

$p_{before} = p_{after}$

$m(4 \text{ m/s}) \cos 60° = 1.4 m v_x$

$v_x = (4 \text{ m/s}) \cdot (\cos 60°) / 1.4$

$v_x = (4 \text{ m/s}) \cdot (0.5) / 1.4$

$v_x = 1.4 \text{ m/s}$

72. C is correct.

Photons are light, and the speed of light is invariant.

Therefore, I is wrong. The energy of the photons in a beam of light determines the color of the light.

If the color is unchanged, then the average energy of the photons is unchanged, making II wrong.

Changing the brightness of a beam of light increases the number of photons in the beam of light.

73. E is correct.

The flow rate must be equal in both parts of the pipe:

$A_1 v_1 = A_2 v_2$

$(\pi / 4) D_1^2 v_1 = (\pi / 4) D_2^2 v_2$

cancel $(\pi / 4)$ from each side of the expression

$D_1^2 v_1 = D_2^2 v_2$

$v_2 = (D_1^2 v_1) / D_2^2$

$v_2 = [(0.2 \text{ m})^2 / (0.004 \text{ m})^2] \cdot (0.03 \text{ m/s})$

$v_2 = (2,500) \cdot (0.03 \text{ m/s})$

$v_2 = 75 \text{ m/s}$

74. A is correct.

The magnetic force on a charged particle can change the velocity and direction of the particle but cannot change its speed.

Kinetic energy is calculated using the square of speed:

$KE = \frac{1}{2} m v^2$

Thus, if speed does not change, the energy of the charge does not change.

75. B is correct.

Beta (β^-) decay: parent nuclide ejects an electron and electron antineutrino.

However, a neutron converts to a proton in the process, so the mass number remains the same, but the atomic number increases by 1.

$$^A_Z X \rightarrow \,^A_{Z+1} Y + \,^0_{-1} e^- + \,^0_0 \bar{\nu}_e$$

76. A is correct.

$Q = c_p m \Delta T$

$c_{p1} m_w \Delta T = c_{p2} m \Delta T$

$c_{p2} = (c_{p1} m_w \Delta T_w) / (m_2 \Delta T_2)$

$c_{p2} = (1 \text{ kcal/kg·°C}) \cdot (0.2 \text{ kg}) \cdot (40 \text{ °C}) / (0.06 \text{ kg}) \cdot (60 \text{ °C})$

$c_{p2} = 2.2 \text{ kcal/kg·°C}$

77. B is correct.

First, determine how long it takes the ball to drop 50 m:

$$PE = KE$$

$$mgh = \tfrac{1}{2}mv_{yf}^2$$

cancel m from both sides of the expression

$$gh = \tfrac{1}{2}v_{yf}^2$$

$$v_{yf}^2 = 2gh$$

$$v_{yf}^2 = (2)\cdot(10 \text{ m/s}^2)\cdot(50 \text{ m})$$

$$v_{yf}^2 = 1{,}000 \text{ m}^2/\text{s}^2$$

$$v_{yf} \approx 32 \text{ m/s}$$

$$t = (v_{yf} - v_{yi})/a$$

$$t = (32 \text{ m/s} - 0)/(10 \text{ m/s}^2)$$

$$t = 3.2 \text{ s}$$

Calculate the distance traveled horizontally in 3.2 s:

$$d_x = v_x \times t$$

$$d_x = (5 \text{ m/s})\cdot(3.2 \text{ s})$$

$$d_x = 16 \text{ m}$$

78. C is correct.

$$F_{tot} = F_{gravity} + F_{friction}$$

$$ma_{tot} = mg\sin\theta + \mu_k mg\cos\theta$$

cancel m from each side

$$a_{tot} = g(\sin\theta + \mu_k \cos\theta)$$

$$a_{tot} = -9.8 \text{ m/s}^2(\sin 30° + 0.3\cos 30°)$$

$$a_{tot} = -7.44 \text{ m/s}^2$$

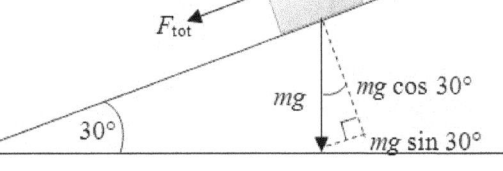

Find time taken to reach 0 m/s:

$$v_f = v_0 + at$$

$$0 \text{ m/s} = 14 \text{ m/s} + (-7.44 \text{ m/s}^2)t$$

$$t = 1.88 \text{ s}$$

$$x = x_0 + v_0 t + \tfrac{1}{2}at^2$$

continued…

$x = 0$ m $+ (14$ m/s$) \cdot (1.88$ s$) + \frac{1}{2}(-7.44$ m/s$^2) \cdot (1.88$ s$)^2$

$x = 13.2$ m

Find vertical component of x:

$y = x \sin \theta$

$y = (13.2$ m$) \sin 30°$

$y = 6.6$ m

79. B is correct.

Forces on the block (with the bullet) are gravity and the tension of the string.

The tension is perpendicular to the direction of travel, so the tension does no work.

Solve by using the conservation of energy, assuming a complete transfer of KE into gravitational PE.

KE (block with bullet at bottom) = PE (block with bullet at top)

$\frac{1}{2}mv^2 = mgh$

cancel m from each side of the expression

$\frac{1}{2}v^2 = gh$

$\frac{1}{2}(2$ m/s$)^2 = (9.8$ m/s$^2)h$

$\frac{1}{2}(4$ m^2/s$^2) = (9.8$ m/s$^2)h$

$(2$ m^2/s$^2) = (9.8$ m/s$^2)h$

$h = (2$ m^2/s$^2) / (9.8$ m/s$^2)$

$h = 0.20$ m

$h = 20$ cm

80. E is correct.

Photon energy:

$E = hf$

$f = c / \lambda$

$f = (3 \times 10^8$ m/s$) / (6.5 \times 10^{-6}$ m$)$

$f = 4.6 \times 10^{13}$ Hz

$E = (4.136 \times 10^{-15}$ eV·s$) \cdot (4.6 \times 10^{13}$ Hz$)$

$E = 0.19$ eV

81. E is correct.

$$v_f^2 = v_0^2 + 2ad$$

where $v_0 = 0$

$$v_f^2 = 0 + 2ad$$
$$v_f^2 = 2ad$$

Since a is constant, d is proportional to v_f^2

If v_f increases by a factor of 4, then d increases by a factor of $4^2 = 16$.

82. B is correct.

The period of a satellite is found through Kepler's Third Law:

$$T = 2\pi\sqrt{(r^3 / GM)}$$

where T = period, r = distance from Earth's center, G = gravitational constant and M = mass of Earth

The period does not depend on the mass of the satellite, so the period remains the same.

83. B is correct.

Heat conduction follows the equation:

$$\Delta Q / \Delta t = kA\Delta T / d$$

Assuming all other values are constant, the equation can be written as:

$$\Delta Q / \Delta t = (1 / d)x$$

where x is a constant

The rate of heat loss is inversely proportional to thickness, so by increasing d, the slope of the curve is negative.

84. B is correct.

The arrows experience the same stopping force when they hit the hay bales.

The kinetic energy can be related to the work done by the force:

Arrow 1: $KE_1 = W$

$KE_1 = Fd_1$

$d_1 = KE_1 / F$

Arrow 2: $KE_2 = 2KE_1$

$2KE_1 = W$

$2KE_1 = Fd_2$

$d_2 = 2KE_1 / F$

$d_2 = 2d_1$

85. E is correct.

The period of a pendulum:

$$T = 2\pi\sqrt{L/g}$$

where L is the length of the pendulum, and g is the acceleration due to gravity

Use $g/6$ for g.

$$T = 2\pi\sqrt{L/(g/6)}$$

$$T = 2\pi\sqrt{6L/g}$$

$$T = 2\pi\sqrt{L/g} \times \sqrt{6}$$

New period = $T\sqrt{6}$

86. D is correct.

Decibels use a logarithmic scale.

$$\text{Intensity (dB)} = 10\log_{10}[I/I_0]$$

where I_0 is the intensity at the threshold of hearing (10^{-12} W/m²)

$$I = 10\log_{10}[10^{-7} \text{ W/m}^2 / 10^{-12} \text{ W/m}^2]$$

$$I = 10\log_{10}[10^5]$$

$$I = 50 \text{ dB}$$

87. A is correct.

The buoyant force is given by:

$$F_B = \rho V g$$

ρ = density of fluid, V = volume of fluid displaced, g = acceleration due to gravity

To calculate how much the brick *appears* to weigh, subtract the buoyant force from the force due to gravity, where m is the *actual* mass of the brick:

$$F_{apparent} = F_{gravity} - F_{buoyant}$$

$$F_{apparent} = mg - \rho V g$$

The *actual* mass of the brick can be calculated as density × volume

$$V_{brick} = (2 \text{ in} \times 4 \text{ in} \times 6 \text{ in})$$

$$V_{brick} = 48 \text{ in}^3$$

$$V_{brick} = (48 \text{ in}^3) \cdot (16.4 \text{ cm}^3/1 \text{ in}^3)$$

$$V_{brick} = 787 \text{ cm}^3$$

continued…

mass = density × volume

$m = (11.4 \text{ g/cm}^3) \cdot (787 \text{ cm}^3)$

$m = 8{,}972 \text{ g}$

To get the apparent *mass*, divide the apparent force by *g*:

$m_{apparent} = F_{apparent} / g$

$m_{apparent} = (m_{brick}g - \rho V g) / g$

$m_{apparent} = m_{brick} - \rho V$

$m_{apparent} = 8{,}972 \text{ g} - [(0.92 \text{ g/cm}^3) \cdot (787 \text{ cm}^3)]$

$m_{apparent} = (8{,}972 \text{ g} - 724 \text{ g})$

$m_{apparent} = 8.2 \text{ kg}$

88. D is correct.

Acceleration is always positive and away from charge Q.

Therefore, velocity increases (no opposing force of friction).

The energy of the system starts as electrical PE.

$PE_{elec} = (kQq) / r$

where r is the initial distance between the point charges

Electrical PE is the energy required to bring a system together from the charges starting at infinity.

After charge Q has moved far away, the energy of the system is only $KE = \frac{1}{2}mv^2$

v has a limit because KE cannot exceed kQq / r

89. C is correct.

Resistance in series experience equal current because there is only one path for the current to travel.

90. B is correct.

For the critical angle, the refracted angle is 90°

$n_{water} \sin \theta_{crit} = n_{air} \sin 90°$

$\sin \theta_{crit} = (n_{air} / n_{water}) \sin 90°$

$\theta_{crit} = \sin^{-1} [(1 / 1.33) \cdot (1)]$

$\theta_{crit} = \sin^{-1} (3/4)$

91. C is correct.

The energy needed to change hydrogen from one state to another:

$$E = -13.6 \text{ eV}[(1/n_1^2) - (1/n_2^2)]$$

To ionize hydrogen, the electron must be removed to the n = ∞ state.

The energy needed to change from the ground state:

$$E = -13.6[(1/1) - (1/\infty)]$$

$$E = -13.6 \text{ eV}$$

Energy is expressed as a negative number to indicate that this much energy is needed to be input to the atom.

92. D is correct.

Find the perimeter (i.e., circumference) of the carousel: distance traveled in one revolution.

$$\text{Perimeter} = \pi \times d$$

$$\text{Perimeter} = \pi \times 18 \text{ m}$$

$$\text{Perimeter} = 56.5 \text{ m}$$

Convert to rev/min, to rev/s

$$v = (5 \text{ rev/min}) \cdot (1 \text{ min}/60 \text{ s})$$

$$v = 0.083 \text{ rev/s}$$

Convert rev/s to m/s:

where 1 rev = 56.5 m

$$v = (0.083 \text{ rev/s}) \cdot (56.5 \text{ m}/1 \text{ rev})$$

$$v = 4.7 \text{ m/s}$$

93. A is correct.

$$\text{Power} = \text{work} / \text{time}$$

$$\text{Power} = (\text{force} \times \text{distance}) / \text{time}$$

Newton's First Law: no force is required to keep the object moving with constant velocity.

The projectile maintains horizontal v since no forces are acting on the horizontal axis.

The vertical forces must be balanced since it maintains its elevation (only moving horizontally).

Since there is no net force, no work is done, and therefore no power is required.

94. D is correct.

Solve for spring constant k:

$$PE = \tfrac{1}{2}kx^2$$

$$k = 2(PE) / x^2$$

where x is the amplitude (maximum distance traveled from rest).

$$k = 2(10 \text{ J}) / (0.2 \text{ m})^2$$

$$k = 500 \text{ N/m}$$

Solve for the period:

$$T = 2\pi[\sqrt{(m/k)}]$$

$$T = 2\pi[\sqrt{(0.4 \text{ kg} / 500 \text{ N/m})}]$$

$$T = 0.18 \text{ s}$$

Convert the period to frequency:

$$f = 1/T$$

$$f = 1/(0.18 \text{ s})$$

$$f = 5.6 \text{ Hz}$$

95. E is correct.

Electromagnetic waves propagate at the speed of light, oscillations of electric and magnetic fields that propagate at the speed of light.

The oscillations of the two fields form a transverse wave perpendicular to each other and perpendicular to energy and wave propagation.

96. C is correct.

Under one meter of water, the gauge pressure is:

$$P = \rho g h$$

Jack's lungs are open to atmospheric pressure due to the snorkel, so he only needs to overcome the gauge pressure.

The force needed is:

$$P = F/A$$

$$F = PA$$

$$F = (\rho g h)A$$

The area is the area of his chest, as this is what the pressure acts against.

$$F = (\text{gauge pressure}) \cdot (\text{chest area})$$

97. B is correct.

Before the switch is closed, there is no current and no magnetic field in either solenoid.

After the switch is closed, the current in the first solenoid begins to flow from positive to negative, and a magnetic field is created, pointing to the right (according to the solenoid right-hand rule).

According to Lenz's Law, an EMF is induced in the second solenoid to oppose the change in magnetic flux.

Since the field from the first solenoid points to the right, the induced current in the second solenoid must induce a current pointing to the left to oppose it.

By the right-hand solenoid rule, the current in the galvanometer must flow from right to left to create a temporary magnetic field in the direction opposite to the original field.

98. A is correct.

Resistance in a wire:

$R = (\rho L) / A$

where ρ = resistivity, L = length of wire and A = cross-sectional area of wire

Cross-sectional area of a wire:

$A = \pi r^2$

$A = \pi D^2 / 4$

If D is doubled:

$A = \pi (2D)^2 / 4$

$A = \pi (4D^2) / 4$

The area is increased by a factor of 4.

The new resistance if D is doubled and L is doubled:

$R = (\rho \times 2L) / (4A)$

$R = \frac{1}{2}(\rho L / A)$

99. A is correct.

Magnification is defined as:

$m = -d_i / d_o$

For a mirror:

$1 / f = 1 / d_i + 1 / d_o$

continued…

Multiplying both sides by d_o gives:

$d_o / f = d_o / d_i + 1$

$d_o / f = -(1 / m) + 1$

or:

$1 / m = 1 - d_o / f$

$m = 1 / (1 - d_o / f)$

$m = 1 / (1 - 5 \text{ m} / 10 \text{ m})$

$m = +2$

The magnification is positive, so the image is upright and twice as large.

100. C is correct.

Background radiation is the energy source that provides most of a person's annual exposure to radiation.

Notes for active learning

Diagnostic Test 3 – Answer Key and Detailed Explanations

Answer Key

1	D	Electric circuits	26	D	Work & energy
2	C	Force, motion, gravitation	27	D	Sound
3	E	Sound	28	B	Electrostatics & electromagnetism
4	A	Fluids & solids	29	B	Light & optics
5	D	Rotational motion	30	D	Heat & thermodynamics
6	C	Electric circuits	31	C	Equilibrium & momentum
7	D	Light & optics	32	B	Waves & periodic motion
8	D	Atomic & nuclear structure	33	A	Fluids & solids
9	E	Kinematics & dynamics	34	B	Electric circuits
10	A	Force, motion, gravitation	35	C	Atomic & nuclear structure
11	B	Equilibrium & momentum	36	D	Heat & thermodynamics
12	E	Work & energy	37	B	Kinematics & dynamics
13	C	Waves & periodic motion	38	D	Force, motion, gravitation
14	C	Heat & thermodynamics	39	B	Equilibrium & momentum
15	B	Light & optics	40	E	Atomic & nuclear structure
16	A	Atomic & nuclear structure	41	B	Quantum mechanics
17	B	Electrostatics & electromagnetism	42	A	Force, motion, gravitation
18	A	Fluids & solids	43	B	Heat & thermodynamics
19	A	Sound	44	A	Work & energy
20	D	Waves & periodic motion	45	B	Waves & periodic motion
21	C	Work & energy	46	E	Sound
22	C	Equilibrium & momentum	47	D	Fluids & solids
23	D	Special relativity	48	D	Electrostatics & electromagnetism
24	E	Kinematics & dynamics	49	B	Electric circuits
25	A	Heat & thermodynamics	50	B	Light & optics

#	Ans	Topic	#	Ans	Topic
51	D	Atomic & nuclear structure	76	B	Electric circuits
52	C	Equilibrium & momentum	77	D	Electrostatics & electromagnetism
53	E	Work & energy	78	A	Fluids & solids
54	A	Waves & periodic motion	79	C	Sound
55	E	Fluids & solids	80	E	Waves & periodic motion
56	E	Fluids & solids	81	A	Work & energy
57	A	Electric circuits	82	B	Equilibrium & momentum
58	B	Electric circuits	83	C	Force, motion, gravitation
59	C	Light & optics	84	B	Kinematics & dynamics
60	B	Rotational motion	85	C	Heat & thermodynamics
61	D	Atomic & nuclear structure	86	A	Work & energy
62	C	Force, motion, gravitation	87	E	Sound
63	B	Sound	88	A	Electrostatics & electromagnetism
64	A	Fluids & solids	89	D	Light & optics
65	D	Electrostatics & electromagnetism	90	C	Electric circuits
66	E	Electric circuits	91	C	Equilibrium & momentum
67	B	Fluids & solids	92	C	Waves & periodic motion
68	D	Electrostatics & electromagnetism	93	E	Fluids & solids
69	B	Kinematics & dynamics	94	C	Electric circuits
70	D	Force, motion, gravitation	95	C	Atomic & nuclear structure
71	A	Equilibrium & momentum	96	D	Heat & thermodynamics
72	B	Work & energy	97	B	Kinematics & dynamics
73	B	Waves & periodic motion	98	D	Quantum mechanics
74	B	Special relativity	99	E	Equilibrium & momentum
75	A	Light & optics	100	A	Atomic & nuclear structure

1. D is correct.

The magnitude of the acceleration is given by:

$F = ma$

$a = F / m$

$F = qE_0$

$a = qE_0 / m$

$|a_p / a_e| = |(q_p E_0 / m_p) / (q_e E_0 / m_e)|$

$a_p / a_e = (m_e / m_p)$

$a_p = (m_e / m_p) a_e$

$a_p = (9 \times 10^{-31} \text{ kg} / 1.67 \times 10^{-27} \text{ kg}) a_e$

$a_p \approx (1 / 1,850) a_e$

The electric field is constant.

The charges of the proton and the electron are equal in magnitude and are opposite in sign.

1.67×10^{-27} kg divided by 9×10^{-31} kg $\approx 1,850$.

The proton is about 1,850 times more massive, so its acceleration is 1,850 times smaller.

2. C is correct.

By Newton's second law, the acceleration is:

$a = F_{net} / m$

The object has two forces: gravity (F_g), pointing down the slope, and friction (F_f), opposing the motion, pointing up the slope.

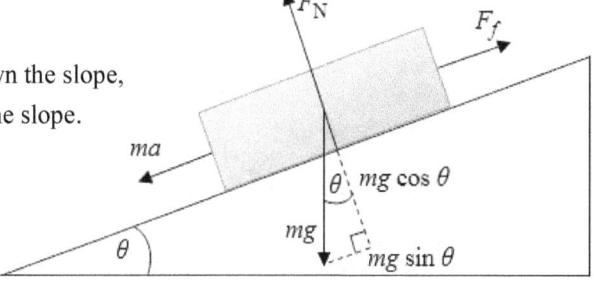

Therefore:

$F_{net} = F_g - F_f$

where the positive direction points down the slope in the direction of motion.

From the diagram, the force due to gravity is:

$F_g = mg \sin \theta$

The force of friction is:

$F_f = \mu_k F_{normal}$

$F_f = \mu_k mg \cos \theta$

continued...

Therefore:

$$a = (mg \sin \theta - \mu_k mg \cos \theta) / m$$

cancel m from each side of the expression

$$a = g (\sin \theta - \mu_k \cos \theta)$$

$$a = (9.8 \text{ m/s}^2) \cdot (\sin 40° - 0.19 \times \cos 40°)$$

$$a = (9.8 \text{ m/s}^2) \cdot (0.49)$$

$$a = 4.9 \text{ m/s}^2$$

3. **E is correct.**

The difference in intensity between a shout and a whisper is given by:

$$90 \text{ dB} - 20 \text{ dB} = 70 \text{ dB}$$

A change of 10 decibels corresponds to a factor change of 10 in the intensity I in W/m².

An increase of 70 dB corresponds to an intensity increase of seven factors of 10, which equals a 10^7 or a ten-million-fold increase in the intensity of the sound.

4. **A is correct.**

$$F_B = mg$$

$$F_B = \rho V g$$

$$\rho V g = mg$$

cancel g from each side of the expression

$$m = \rho V$$

$$m = (\pi r^2 h) \cdot (1 \text{ g/cm}^3)$$

$$m = \pi (1 \text{ cm})^2 \cdot (14 \text{ cm}) \cdot (1 \text{ g/cm}^3)$$

$$m = 14\pi \text{ g}$$

$$m = 44 \text{ g}$$

5. **D is correct.**

$$813.0 \text{ rpm} \cdot (1 \text{ min} / 60 \text{ s}) \cdot (2\pi \text{ rad/rev}) = 85.14 \text{ rad/s}$$

6. C is correct.

$$V = IR$$

Resistivity is the measure of resistance along the length of a given material:

$$R = \rho L / A$$

where ρ is the resistivity

$$\rho = RA / L$$

$$\rho = (\Omega)\cdot(m^2) / (m)$$

$$\rho = \Omega\cdot m$$

7. D is correct.

The *magnification*:

$$m = -d_i / d_o$$

$$m = -3\ m / 6\ m$$

$$m = -½$$

where the negative sign indicates that the image is inverted

8. D is correct.

Uranium decays because the electromagnetic repulsion of the protons overcomes the strong nuclear force due to its limited range and the massive size of the uranium nucleus.

9. E is correct.

To achieve a due north bearing, the east-west velocity must be made to be zero.

Find the horizontal component of the NE drift:

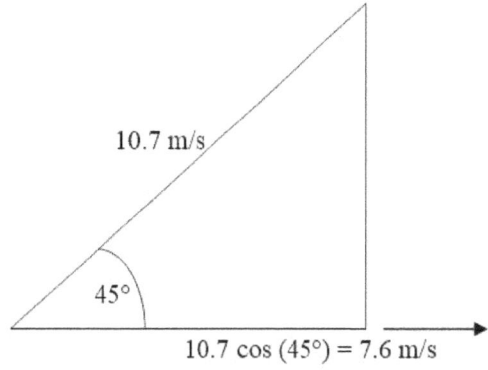

$$10.7 \cos (45°) = 7.6\ m/s$$

continued...

Find the horizontal component of the NW acceleration:

The drift is fully corrected when there is zero east-west velocity.

$$v_f = v_i + a_x t$$

$$0 = (7.6 \text{ m/s}) + (3.1 \text{ m/s}^2) \cdot (t)$$

$$t = (7.6 \text{ m/s}) / (3.1 \text{ m/s}^2)$$

$$t = 2.4 \text{ s}$$

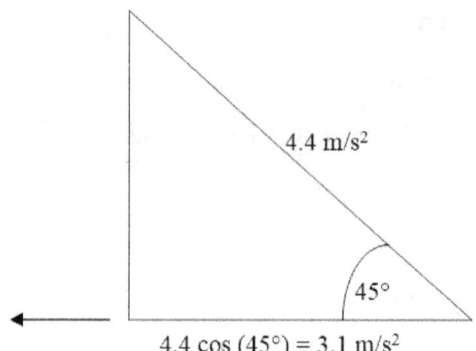

10. A is correct.

For constant acceleration, average speed is:

$$v_{avg} = (v_{final} + v_{initial}) / 2$$

In this example:

$$v_{initial} = 0 \text{ m/s}$$

so:

$$v_{avg} = v_{final} / 2$$

With that:

$$d = v_{avg} t$$

$$d = v_{avg} t / 2$$

$$d = (20 \text{ m/s}) \cdot (10 \text{ s}) / 2$$

$$d = 100 \text{ m}$$

11. B is correct.

The relationship between the final speed and initial height can be found using the conservation of energy:

$$PE_{initial} = KE_{final}$$

$$mgh = \tfrac{1}{2} mv^2$$

Therefore:

$$v^2 = gh$$

To double v, h increases by a factor of 4.

12. E is correct.

There are two forces on the ball: tension and the force of gravity.

When the ball is at the bottom of its circular path, tension points up, and the force of gravity points down.

The tension at the bottom is related to the net force on the ball and the force of gravity:

$$F_{net} = T - F_{gravity}$$

$$F_{net} = T - mg$$

or

$$T = F_{net} + mg$$

The ball executes circular motion, but its speed changes.

Consider the speed to be approximately constant if observed for short intervals, so the motion is uniform circular motion during a short interval.

For an object executing uniform circular motion, the net force is the centripetal force:

$$F_{net} = F_{centripetal}$$

$$F_{net} = mv^2/r$$

Putting this into the previous equation:

$$T = mv^2/r + mg$$

The velocity at the bottom can be related to the initial height using conservation of energy:

$$KE_{final} = PE_{initial}$$

$$\tfrac{1}{2} mv^2 = mgh$$

The height change is equal to twice the radius:

$$\tfrac{1}{2} mv^2 = mg(2r)$$

Therefore:

$$mv^2 = 4mgr$$

Putting this into the equation for the tension:

$$T = (4mgr)/r + mg$$

$$T = 5mg$$

$$T = 5(4 \text{ kg}) \cdot (9.8 \text{ m/s}^2)$$

$$T = 196.0 \text{ N}$$

13. C is correct.

$T = 12$ s

$f = 1 / T$

$f = (1 / 12$ s$)$

$f = 0.083$ Hz

$\lambda = v / f$

$\lambda = (4.5$ m/s $/ 0.083$ Hz$)$

$\lambda = 54$ m

14. C is correct.

Although an iceberg has a much lower temperature than hot coffee, it contains far more thermal energy due to its much greater mass.

For example, assume coffee at 90 °C goes to 80 °C:

$Q = mc\Delta T$

$Q = (1$ kg$) \cdot (4.2$ kJ/kg·K$) \cdot (90$ °C $- 80$ °C$)$, where specific heat for water $= 4.2$ kJ/kg·K

$Q = 42$ kJ were released during temperature change

If a 10,000 kg iceberg (tiny iceberg) were to go from 0 °C to −10 °C:

$Q = mc\Delta T$

$Q = (10{,}000$ kg$) \cdot (2.05$ kJ/kg·K$) \cdot (0$ °C $- (-10$ °C$))$

where specific heat for ice $= 2.05$ kJ/kg·K

$Q = 205{,}000$ kJ were released during temperature change

Even a small iceberg at a much lower temperature contains more thermal energy due to its far greater mass.

15. B is correct.

The equation for focal length is:

$1 / f = 1 / d_o + 1 / d_i$

where d_o is object distance and d_i is image distance

$1 / f = 1 / d_o + 1 / d_i$

$1 / f = 1 / 24$ cm $+ 1 / 3$ cm

$1 / f = 1 / 24$ cm $+ 8 / 24$ cm

$1 / f = 0.375$ cm

$f = 2.7$ cm

16. A is correct.

When an electron is emitted, the parent nucleus is undergoing β^- decay:

$$^A_Z X \rightarrow \,^A_{Z+1} Y + \,^0_{-1} e^- + \,^0_0 \bar{\nu}_e$$

The parent nuclide ejects an electron and electron antineutrino.

A neutron converts to a proton in the process, so the mass number remains the same, but the atomic number increases by 1.

The daughter nuclei should have the same mass number but an atomic number 1 higher than the parent nuclei.

This causes the daughter nuclei to be a different element; in this example, an atomic number of 91 is Pa.

17. B is correct.

The equation for the magnetic field of an infinitely long straight wire is given as:

$B = \mu I / 2\pi r$

where μ = permittivity of free space, I = current, and r is the radial distance from the wire

When r doubles, B decreases by a factor of ½.

18. A is correct.

Flow velocity:

$v = f / A$

where flow rate $f = 0.04$ m³/s

$v = f / (\pi r^2)$

$v = (0.04 \text{ m}^3/\text{s}) / [\pi (0.06 \text{ m})^2]$

$v = 3.5$ m/s

19. A is correct.

Intensity related to decibels may be expressed as:

$I \text{ (dB)} = 10 \log_{10}(I / I_0)$

If intensity is increased by 100:

$I \text{ (dB)} = 10 \log_{10}(100 / 1)$

$I \text{ (dB)} = 20$

20. D is correct.

Velocity in a wire:

$$v = \sqrt{T / (m / L)}$$

$$T = v^2(m / L)$$

volume = $A_{\text{cross-section}} \times$ length

$$V = (\pi D^2 / 4) \cdot (L)$$

$$m = V\rho$$

$$m = (\rho) \cdot (\pi D^2 / 4) \cdot (L)$$

Rewrite tension formula with mass:

$$T = v^2[(\rho) \cdot (\pi D^2 / 4)(L) / (L)]$$

$$T = v^2(\rho \times \pi D^2 / 4)$$

$$T = (42 \text{ m/s})^2 \times (\pi / 4) \cdot (2{,}600 \text{ kg/m}^3) \cdot (0.0044 \text{ m})^2$$

$$T = (1{,}764 \text{ m}^2/\text{s}^2) \cdot (\pi / 4) \cdot (2{,}600 \text{ kg/m}^3) \cdot (0.00001936 \text{ m}^2)$$

$$T = 69.74 \approx 70 \text{ N}$$

21. C is correct.

Force along the incline:

$$F = mg \sin \theta$$

$$F = (5 \text{ kg}) \cdot (9.8 \text{ m/s}^2) \sin 30°$$

$$F = (5 \text{ kg}) \cdot (9.8 \text{ m/s}^2) \cdot (0.5)$$

$$F = 24.5 \text{ N}$$

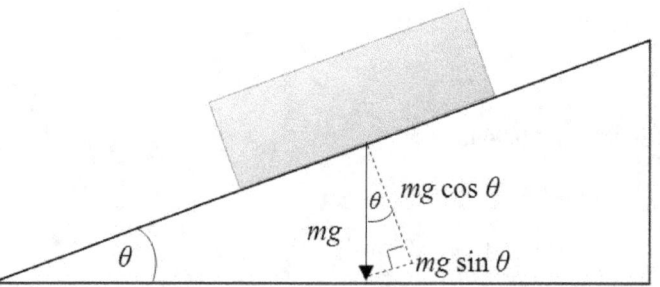

Solving for work:

Work = force × distance

$$W = (24.5 \text{ N}) \cdot (10 \text{ m}) = 245 \text{ J}$$

22. C is correct.

Momentum is:

$$p = mv$$

$$v = 400 \text{ m} / 50 \text{ s}$$

$$v = 8 \text{ m/s}$$

$$p = (65 \text{ kg}) \cdot (8 \text{ m/s})$$

$$p = 520 \text{ kg·m/s}$$

23. D is correct.

The speed of light has the same value for an observer regardless of the state of motion of the source or observer. This is a fundamental principle of special relativity.

24. E is correct.

$$\Delta v = at$$

$$a = \Delta v / t$$

$$a = (v_f - v_i) / t$$

$$(v_f - v_i) = at$$

$$v_f = at + v_i$$

$$v_f = (2 \text{ m/s}^2) \cdot (6 \text{ s}) + (5 \text{ m/s})$$

$$v_f = 17 \text{ m/s}$$

25. A is correct.

Stay in SI units for consistency; express the specific heat in SI units:

$$c = 108 \text{ cal/kg/°C}$$

$$c = 108 \text{ cal/kg/°C } (4.186 \text{ J/cal})$$

$$c = 450.09 \text{ J/kg/°C}$$

Calculate ΔT:

$$\Delta T = Q/(mc) = (1/2 \, mv^2) / (mc) = v^2/(4c)$$

$$\Delta T = (1{,}250 \text{ m/s})^2 / [4 \, (450.09 \text{ J/kg/°C})]$$

$$\Delta T = 1728 \text{ °C}$$

Note that the temperature increase is independent of the mass of the meteorite.

26. D is correct.

Relate KE to PE:

$KE_B = PE_B$

$½mv_B^2 = mgh_B$

cancel m from each side of the expression

$½v_B^2 = gh_B$

$h_B = v_B^2 / 2g$

If $v_M = 2v_B$:

$h_M = (2v_B)^2 / 2g$

$h_M = 4v_B^2 / 2g$

$h_M = 4h_B$

Mary's ball travels four times as high as Brittany's ball.

27. D is correct.

For tubes open at both ends:

$\lambda = 2L / n$

where n = 1 is the fundamental

$\lambda = 2 (0.2 \text{ m}) / 1$

$\lambda = 0.4$ m

28. B is correct.

According to the *Lorentz force law* for a charged particle in a magnetic field, the magnetic force on a particle acts in a direction given by the right-hand rule.

For particles with velocity perpendicular to the field (as in this case), the magnitude of the force is:

$F_B = qvB$

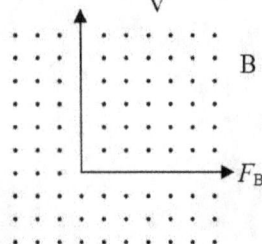

continued…

The centripetal force F_C on the proton is the magnetic force F_B:

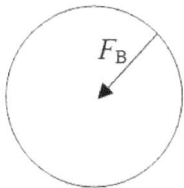

$F_B = F_C$

$qvB = (mv^2) / r$

Therefore:

$r = (mv) / (qB)$

The values for m, q and B are fixed.

If the speed is increased by a factor of three, so is the radius.

Thus, the time needed to complete one circular path is:

$t = 2\pi(3r) / 3v$

$t = 2\pi r / v$, unchanged from the original time.

29. **B is correct.**

$1/f = 1/d_i + 1/d_o$

$1/2 \text{ m} = 1/d_i + 1/6 \text{ m}$

$1/3 \text{ m} = 1/d_i$

$d_i = 3 \text{ m}$

$m = -d_i / d_o$

$m = h_i / h_o$

$-3 \text{ m} / 6 \text{ m} = h_i / h_o$

$-½ = h_i / h_o$

For a converging lens, a positive image distance indicates a real image.

The image is inverted because $h_i / h_o = -½$ and h_o cannot be negative, so h_i must be negative.

A negative image height indicates an inverted image.

30. D is correct.

Break into three parts: heating the ice, melting the ice, heating the water.

Heating ice:

$$\Delta Q_1 = mc\Delta T$$

$$\Delta Q_1 = (10 \text{ kg}) \cdot (0.5 \text{ kcal/kg} \cdot °C) \cdot [0 \text{ °C} - (-8 \text{ °C})]$$

$$\Delta Q_1 = 40 \text{ kcal}$$

Melting ice:

$$\Delta Q_2 = mL_f$$

$$\Delta Q_2 = (10 \text{ kg}) \cdot (80 \text{ kcal/kg})$$

$$\Delta Q_2 = 800 \text{ kcal}$$

Heating water:

$$\Delta Q_3 = mc\Delta T$$

$$\Delta Q_3 = (10 \text{ kg}) \cdot (1 \text{ kcal/kg} \cdot °C) \cdot (14 \text{ °C} - 0 \text{ °C})$$

$$\Delta Q_3 = 140 \text{ kcal}$$

Combine values:

$$\Delta Q_{total} = (40 \text{ kcal} + 800 \text{ kcal} + 140 \text{ kcal})$$

$$\Delta Q_{total} = 980 \text{ kcal}$$

31. C is correct.

Conservation of energy states that momentum before and after the collision is consistent.

$$p_i = p_f$$

$$p = mv$$

$$m_1 v_1 = m_2 v_2$$

$$m_1 \Delta v_1 = m_2 \Delta v_2$$

$$(6 \text{ kg}) \cdot (v_f - v_i)_1 = (8 \text{ kg}) \cdot (v_f - v_i)_2$$

$(v_f - v_i)_1$ must be larger

32. B is correct.

The period of a pendulum is:

$T = 2\pi\sqrt{L/g}$

If the elevator accelerates upwards, the constant acceleration adds to apparent acceleration due to gravity, and the period decreases.

$g < g + a$

The period decreases as the gravitational acceleration increases.

33. A is correct.

The 400 m distance between the aircraft (i.e., source) and the observer remains constant.

Since there is no relative motion between the source and the detector, f_d equals f_s.

34. B is correct.

The magnetic force on an object changes the direction of the charge and thus the velocity; speed does not change.

Kinetic energy is calculated using speed; thus, if the magnetic force does not change the speed, it does not change its kinetic energy.

35. C is correct.

Heavy nuclides with atomic numbers greater than 83 almost always undergo alpha decay to reduce the number of neutrons and protons in the nucleus.

36. D is correct.

A larger coefficient of thermal expansion causes a greater size increase than materials with smaller thermal expansion coefficients.

If the pin were removed easily while hot, it did not expand as much as material X and must have a smaller coefficient.

37. B is correct.

According to Newton's First Law: every object remains at rest or in motion in a straight line unless acted upon by an unbalanced force.

Thus, if the object's motion changes, an unbalanced force is being applied.

38. D is correct.

Objects in orbit around Earth still experience the force of gravity.

The astronauts feel weightless in space because they are in free fall and cannot feel the force of gravity.

39. B is correct.

Find the amount of time it takes the ball to fall 1 m from the apex:

$\Delta x = v_0 t + \frac{1}{2}at^2$

$1 \text{ m} = (0 \text{ m/s})t + \frac{1}{2}(9.8 \text{ m/s}^2)t^2$

$1 \text{ m} = \frac{1}{2}(9.8 \text{ m/s}^2)t^2$

$1 \text{ m} = 4.9 \text{ m/s}^2 t^2$

$t = \sqrt{1 \text{ m} / 4.9 \text{ m/s}^2}$

$t = 0.452 \text{ s}$

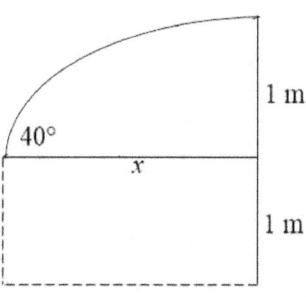

Find initial upward velocity of the ball:

$v^2 = v_0^2 + 2a\Delta x$

$0 \text{ m/s} = v_0^2 + 2(-9.8 \text{ m/s}^2)\cdot(1 \text{ m})$

$0 \text{ m/s} = v_0^2 + (-19.6 \text{ m}^2/\text{s}^2)$

$19.6 \text{ m}^2/\text{s}^2 = v_0^2$

$v_0 = \sqrt{19.6 \text{ m}^2/\text{s}^2}$

$v_0 = 4.427 \text{ m/s}$

Calculate the horizontal velocity:

$\tan(40°) = (4.427 \text{ m/s}) / v_h$

$0.839 = (4.427 \text{ m/s}) / v_h$

$v_h = (4.427 \text{ m/s}) / 0.839$

$v_h = 5.277 \text{ m/s}$

Calculate the distance from the fence using elapsed time and horizontal velocity:

$d = v_h t$

$d = 5.277 \text{ m/s} \times 0.452 \text{ s}$

$d = 2.39 \text{ m} \approx 2.4 \text{ m}$

40. E is correct.

During beta minus decay, the parent nuclide ejects an electron and electron antineutrino.

However, in the process, a neutron converts to a proton, so the mass number remains the same, but the atomic number increases by 1.

$^{A}_{Z}X \rightarrow\ ^{A}_{Z+1}Y +\ ^{0}_{-1}e^- +\ ^{0}_{0}\overline{\nu}_e$

41. B is correct.

Using the Compton Equation:

$$\Delta\lambda = h / m_0 c \, (1 - \cos\theta)$$

$$\Delta\lambda = \lambda_{Compton} (1 - \cos\theta)$$

where $\lambda_{Compton} = 2.43 \times 10^{-12}$ m

At $\theta = 180°$, $\Delta\lambda = \lambda$

Thus:

$$\lambda = 2\lambda_{Compton}$$

$$\lambda = 4.86 \times 10^{-12} \text{ m}$$

42. A is correct.

An object in motion with constant nonzero velocity experiences no acceleration and thus cannot have a net force upon it.

If v = constant, then:

$$a = 0$$

$$F = ma$$

$$F = m(0 \text{ m/s}^2)$$

$$F = 0 \text{ N}$$

43. B is correct.

Heat conduction equation:

$$Q/t = (kA\Delta T)/d$$

where k is the thermal conductivity of the wall material, A is the surface area of the wall, d is the wall's thickness, and ΔT is the temperature difference on either side

Assume heat flow is lengthwise so barrier distance:

$$d = 2d_0$$

The important value here is A (A = πr^2), which is the surface area of the object with respect to the direction of heat flow, and d, which is barrier thickness.

The surface area with respect to heat flow:

$$A_0 = (\pi/4)D_0^2$$

$$A = (\pi/4)\cdot(2D_0)^2$$

$$A = 4(\pi/4)D_0^2$$

$$A = 4A_0$$

continued...

Substitute values into original equation:

$Q / t = (k\Delta T) \cdot (4A_0 / 2d_0)$

$Q / t = 2(kA_0\Delta T / d_0)$

$Q / t = 2(Q_0 / t)$

$Q / t = 2(30 \text{ W})$

$Q / t = 60 \text{ W}$

44. A is correct.

The spring has the 0.9 kg mass attached to it, so its original equilibrium length is the length needed to counteract the force of gravity.

If the spring is stretched further, then the net force only includes the component from the spring force:

$F_{spring} = k\Delta x$

$F_{spring} = (3 \text{ N/m}) \cdot (0.18 \text{ m})$

$F_{spring} = 0.54 \text{ N}$

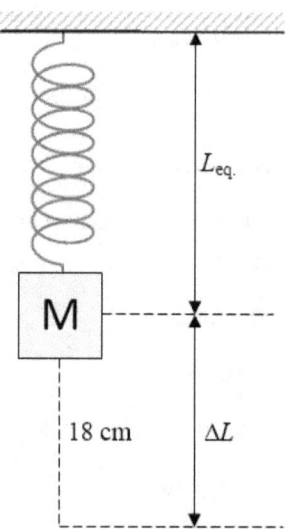

45. B is correct.

Convert weight to mass:

$F = ma$

$m = F / a$

$m = 30 \text{ N} / (9.8 \text{ m/s}^2)$

$m = 3.061 \text{ kg}$

Frequency of a spring system:

$\omega = \sqrt{k / m}$

$\omega = \sqrt{40.0 \text{ N/m} / 3.061 \text{ kg}}$

$\omega = 3.615 \text{ rad/s}$

$f = \omega / 2\pi$

$f = (3.615 \text{ rad/s}) / 2\pi$

$f = 0.58 \text{ Hz}$

46. E is correct.

$$f_{beat} = |f_2 - f_1|$$

$$f_{beat} = |786.3 \text{ Hz} - 785.8 \text{ Hz}|$$

$$f_{beat} = 0.5 \text{ Hz}$$

47. D is correct.

The weight of the piston surface area is a ratio:

$$F_1 / A_1 = F_2 / A_2$$

$$F_2 = F_1 \times A_2 / A_1$$

$$F_2 = (600 \text{ N}) \cdot (50 \text{ cm}^2) / 5 \text{ cm}^2$$

$$F_2 = 6,000 \text{ N}$$

48. D is correct.

Equation for potential energy:

$$\Delta PE = q\Delta V$$

Note: q is negative because electrons are negatively charged

$$\Delta PE = (-1.6 \times 10^{-19} \text{ C}) \cdot (-500 \text{ V} - 500 \text{ V})$$

$$\Delta PE = (-1.6 \times 10^{-19} \text{ C}) \cdot (-1,000 \text{ V})$$

$$\Delta PE = 1.6 \times 10^{-16} \text{ J}$$

Moving this electron increases its PE energy.

49. B is correct.

$$P = VI$$

$$V = IR$$

substituting into the equation

$$P = (IR) \times I$$

$$P = I^2 \times R$$

If P is on y-axis and R is on x-axis,

$$\text{slope} = P / R$$

$$\text{slope} = I^2$$

50. B is correct.

The resolution is the smallest distance between two objects, where they can be distinguished as separate objects.

Therefore, the light with a shorter wavelength gives a smaller distance between peaks and a higher resolution.

51. D is correct.

The photoelectric effect is when light incident upon a metallic surface causes electrons to be emitted.

The photoelectric effect is described by:

$$KE_{max} = hf - \phi$$

where h = Planck's constant, f = frequency and ϕ = work function

If the light of a threshold frequency of $hf > \phi$ shines upon a metallic surface, electrons are ejected with KE.

Increasing frequency increases KE.

Note: hf is the energy carried by one incident photon.

Since the incident intensity remains constant, but the energy per photon has increased, the rate of incident photons must decrease, resulting in a decrease in the rate of ejection events.

52. C is correct.

First calculate the stone's speed at impact:

$$v^2 = v_0^2 + 2ad$$

where $v_0 = 0$ and $a = g$

$$v^2 = 2ad$$
$$v^2 = 2(10 \text{ m/s}^2) \cdot (5 \text{ m})$$
$$v^2 = 100 \text{ m}^2/\text{s}^2$$
$$v = 10 \text{ m/s}$$

Use the speed of impact to calculate momentum p:

$$p = mv$$
$$p = (3 \text{ kg}) \cdot (10 \text{ m/s})$$
$$p = 30 \text{ kg·m/s}$$

Another method to solve this problem:

$$PE = KE$$
$$mgh = \tfrac{1}{2}mv^2$$

cancel m from each side of the expression

continued…

$gh = \frac{1}{2}v^2$

$2(gh) = v^2$

$v^2 = 2(10 \text{ m/s}^2 \times 5 \text{ m})$

$v^2 = 100 \text{ m}^2/\text{s}^2$

$v = 10 \text{ m/s}$

$p = mv$

$p = (3 \text{ kg}) \cdot (10 \text{ m/s})$

$p = 30 \text{ kg} \cdot \text{m/s}$

53. E is correct.

Find kinetic energy and set it equal to the work done by friction:

$\text{KE} = W_f$

$\frac{1}{2}mv^2 = F_f \times d$

$\frac{1}{2}m / F_f = d / v^2$

Because the mass is constant,

$d / v^2 = $ constant regardless of velocity.

Solve for the new skid distance:

$d_1 / v_1^2 = d_2 / v_2^2$

$d_2 = (d_1) \cdot (v_2^2) / (v_1^2)$

$d_2 = (30 \text{ m}) \cdot (150 \text{ km/h})^2 / (45 \text{ km/h})^2$

$d_2 = 333 \text{ m}$

54. A is correct.

The time it takes to complete one cycle is period T.

$T = 1 / f$

The period is measured in seconds.

Frequency = s^{-1} or Hz

55. E is correct.

Volume strain = Δ Volume / Volume

$$\Delta \text{Volume} = \text{Volume strain} \times \text{Volume}$$

$$\Delta V = (-3 \times 10^{-4}) \cdot (12 \text{ L})$$

$$\Delta V = -3.6 \times 10^{-3} \text{ L}$$

$$\Delta V = -3.6 \text{ mL}$$

The value of ΔV is negative, meaning the final volume is smaller than the initial volume, confirming a reduction.

56. E is correct.

First, find the total mass of the mixture once the ethanol has been added to the chloroform:

$$\text{Total Mass} = x + 5 \text{ grams}$$

where x is the mass of the ethanol added.

Then, find the volume of the resulting mixture.

Volume is found using the specific gravity formula:

$$\text{Volume (mL)} = \text{Mass (g)} / SG$$

Therefore:

$$V_e = x / 0.8$$

$$V_c = 5 \text{ g} / 1.5$$

$$\text{Total Volume} = V_e + V_c$$

$$\text{Total Volume} = (x / 0.8) + (5 \text{ g} / 1.5)$$

$$\text{Total Volume} = 1.25x + 3.33 \text{ mL}$$

Find the mass of added ethanol using the given specific gravity of the mixture:

$$SG_{\text{mixture}} = \text{Total Mass} / \text{Total Volume}$$

$$1.2 = (x + 5 \text{ g}) / (1.25x + 3.33 \text{ mL})$$

$$(1.2) \cdot (1.25x + 3.33 \text{ mL}) = (x + 5 \text{ g})$$

$$1.5x + 3.96 \text{ g} = x + 5 \text{ g}$$

$$0.5x = 1.04 \text{ g}$$

$$x = 2.08 \text{ g} \approx 2 \text{ g}$$

57. A is correct.

Batteries in series add voltage:

$$V_{eq} = 4(8 \text{ V})$$

$$V_{eq} = 32 \text{ V}$$

Resistors in series add resistance:

$$R_{eq} = 45 \text{ } \Omega + 25 \text{ } \Omega$$

$$R_{eq} = 70 \text{ } \Omega$$

Ohm's law:

$$V = IR$$

$$I = V / R$$

Current through series circuit is constant:

$$I = 32 \text{ V} / 70 \text{ } \Omega$$

$$I = 0.46 \text{ A}$$

The point at which the current is measured does not matter because the circuit is connected in series.

58. B is correct.

The equivalent resistance of resistors in parallel:

$$R_{eq} = 1 / (1 / R_1 + 1 / R_2 + 1 / R_3 \ldots)$$

The equivalent resistance is smaller than the smallest resistance:

For example:

$$R_1 = 20 \text{ } \Omega, R_2 = 30 \text{ } \Omega, R_3 = 30 \text{ } \Omega$$

$$R_{eq} = 1 / (1 / 20 \text{ } \Omega + 1 / 30 \text{ } \Omega + 1 / 30 \text{ } \Omega \ldots)$$

$$R_{eq} = 8.75 \text{ } \Omega$$

$$R_{eq} < R_1$$

59. C is correct.

First, assume that the distance of the Moon from the lens is ∞.

$d_o = \infty$

Next, assume d_i is the distance from the lens where the image forms.

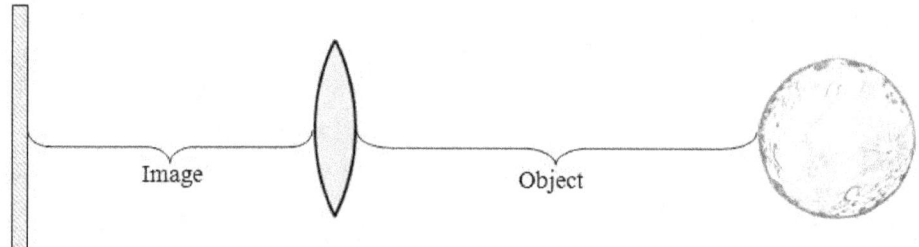

By the thin lens equation:

$1/d_i + 1/d_o = 1/f$

$1/d_i + 1/\infty = 1/f$

$1/d_i + 0 = 1/f$

$1/d_i = 1/f$

$d_i = f$

60. B is correct.

The angular speed changes according to the kinematic relation:

$\Delta\omega = \alpha \Delta t$

$\Delta t = \Delta\omega / \alpha$

$\Delta t = (0 \text{ rad/s} - 96.0 \text{ rad/s}) / (-1.5 \text{ rad/s}^2)$

The angular acceleration is negative because the wheel is slowing down, and the initial ω is in the positive direction.

$\Delta t = 64.0 \text{ s}$

61. D is correct.

If ^{23}Na absorbs a proton and releases a neutron, the atomic number increases by one, and the mass number remains constant.

Thus, ^{23}Mg has the same mass number but has an atomic number one higher than ^{23}Na.

62. C is correct.

The *coefficient of static friction* (object at rest) is larger than the coefficient of kinetic friction (object in motion).

The coefficient of static friction is proportional to the force needed to take a stationary object out of static equilibrium and accelerate it.

The *coefficient of kinetic friction* is proportional to the force needed to maintain dynamic equilibrium in an object moving at a constant speed.

Therefore, the force required to take a stationary object out of static equilibrium and accelerate it is greater than that required to keep a moving object in dynamic equilibrium.

It is more difficult to set an object in motion than it is to keep it in motion.

63. B is correct.

The two ends count as nodes.

From a standing wave with four nodes, there are three antinodes.

Therefore, there are three half-wavelengths.

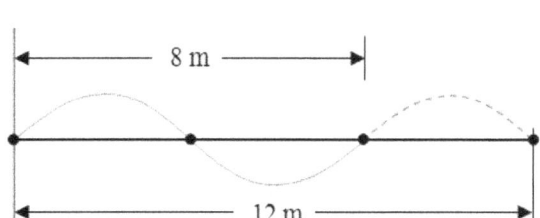

Each dot on the curve represents one of the four nodes.

One complete wave (the solid line) includes three nodes.

    1 wave = (2/3) entire string

    1 wave = (2/3)·(12 m)

    1 wave = 8 m

64. A is correct.

The reading on the meter is the net force between the necklace weight and buoyant force.

$F_{total} = F_O - F_B$

$F_{total} = mg - \rho V g$

$F_{total} = g(m - \rho V)$

$F_{total} = (9.8 \text{ m/s}^2) \cdot [(0.06 \text{ kg}) - (1 \text{ g/cm}^3) \cdot (5.7 \text{ cm}^3) \cdot (1 \text{ kg}/1{,}000 \text{ g})]$

$F_{total} = 0.53 \text{ N}$

65. D is correct.

The centripetal force F_C on a charged particle in a magnetic field is the magnetic Lorentz force:

$$F_M = qvB$$

Therefore:

$$F_C = F_M$$

$$mv^2 / r = qvB$$

or

$$r = mv / qB$$

Increasing the speed by a factor of two increases the radius by a factor of two.

66. E is correct.

The equation for force in a magnetic field (B):

$$F = qv \times B$$

The direction of B is given by the right-hand rule.

If the thumb is oriented upward, as shown for F, the magnetic field B points into the page.

67. B is correct.

$$PV = nRT$$

$$T_0 = PV / nR$$

$$T_1 = (2P) \cdot (2V) / (nR)$$

$$T_1 = 4(PV / nR)$$

68. D is correct.

Electric field lines go from positive charge to negative charge.

The electron will go against the field lines due to attraction to the positive charge emitting the field lines.

They do not follow path X because momentum in the horizontal direction must be conserved.

69. B is correct.

The maximum and minimum of position *vs.* time always equal zero velocity.

70. D is correct.

$F_f = \mu_k F_N$

$F_f = \mu_k mg \cos\theta$

Find the length of travel:

$L = 50 \text{ m} / \sin 10°$

$L = 50 \text{ m} / 0.174$

$L = 288 \text{ m}$

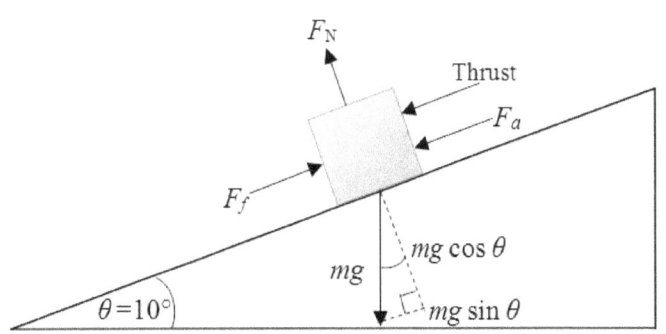

Find acceleration to reach 40 m/s:

$v_f^2 = v_i^2 + 2ad$

$(40 \text{ m/s})^2 = 0 + 2a(288 \text{ m})$

$1{,}600 \text{ m}^2/\text{s}^2 = a(576 \text{ m})$

$a = (1{,}600 \text{ m}^2/\text{s}^2) / (576 \text{ m})$

$a = 2.8 \text{ m/s}^2$

Find normal (F_N) and gravitational (F_G) forces:

$F_N = mg \cos\theta$

$F_N = (50 \text{ kg})\cdot(9.8 \text{ m/s}^2) \cos 10°$

$F_N = (50 \text{ kg})\cdot(9.8 \text{ m/s}^2)\cdot(0.985)$

$F_N = 483 \text{ N}$

$F_G = mg \sin\theta$

$F_G = (50 \text{ kg})\cdot(9.8 \text{ m/s}^2) \sin 10°$

$F_G = (50 \text{ kg})\cdot(9.8 \text{ m/s}^2)\cdot(0.174)$

$F_G = 85 \text{ N}$

$F_{total} = (260 \text{ N} + 85 \text{ N})$

$F_{total} = 345 \text{ N}$, the total force experienced by the skier.

The skier experiences acceleration down the slope that was reduced by friction.

The coefficient of kinetic friction can be calculated by:

$F_{total} - F_{friction} = F_{experienced}$

$F_{total} - \mu F_N = F_{experienced}$

$345 \text{ N} - \mu_k 483 \text{ N} = (2.8 \text{ m/s}^2 \times 50 \text{ kg})$

$345 \text{ N} - \mu_k 483 \text{ N} = 140 \text{ N}$

continued...

$$345 \text{ N} - 140 \text{ N} = \mu_k 483 \text{ N}$$

$$205 \text{ N} = \mu_k 483 \text{ N}$$

$$205 \text{ N} / 483 \text{ N} = \mu_k$$

$$\mu_k = 0.42$$

71. A is correct.

Conservation of momentum:

$$m_1 v_1 = m_2 v_2$$

$$m_1 = \text{putty}$$

$$m_2 = \text{putty} + \text{bowling ball}$$

$$m_2 = (1 \text{ kg} + 7 \text{ kg})$$

$$m_2 = 8 \text{ kg}$$

$$v_2 = (m_1 v_1) / m_2$$

$$v_2 = (1 \text{ kg}) \cdot (1 \text{ m/s}) / 8 \text{ kg}$$

$$v_2 = 1/8 \text{ m/s}$$

72. B is correct.

Upward force due to the spring (Hooke's law):

$$F = k\Delta x$$

where k is the spring constant and Δx is the distance the spring is stretched

The downward force due to gravity:

$$F = mg$$

System is in equilibrium so set the expressions equal:

$$k\Delta x = mg$$

$$\Delta x = mg / k$$

$$\Delta x = (4 \text{ kg}) \cdot (10 \text{ m/s}^2) / (10 \text{ N/m})$$

$$\Delta x = 4 \text{ m}$$

73. B is correct.

The overtone or harmonic can be found by the following:

harmonic	overtone
n^{th} harmonic	$(n^{th} - 1)$ overtone

Thus, the third harmonic has the 2nd $(3 - 1)$ overtone

74. B is correct.

Proper time is the elapsed time between two events in the frame in which the events happen at the same location.

The two events here are the spaceship passing Earth and the spaceship passing Mars.

In the frame of the spaceship, both events occur in the same place: at the spaceship.

75. A is correct.

$$m = -d_i / d_o$$

$$m = -(-6 \text{ m}) / (2 \text{ m})$$

$$m = 3$$

Positive magnification means an upright image.

76. B is correct.

Calculate impedance for each circuit element.

Resistor:

$$Z_r = 30 \text{ }\Omega$$

Inductor:

$$Z_i = j2\pi f L$$

where j is an imaginary number used in calculating complex impedance

$$Z_i = j2\pi(50 \text{ Hz})\cdot(0.4 \text{ H})$$

$$Z_i = j125.6 \text{ }\Omega$$

Capacitor:

$$Z_c = -j / (2\pi f C)$$

$$Z_c = -j / [2\pi(50 \text{ Hz})\cdot(50 \times 10^{-6} \text{ F})]$$

$$Z_c = -j63.7 \text{ }\Omega$$

Adding in series:

$$Z_{total} = 30 \text{ }\Omega + j(125.6 \text{ }\Omega - 63.7 \text{ }\Omega)$$

To find magnitude, use Pythagorean Theorem:

$$|Z| = \sqrt{[(30 \text{ }\Omega)^2 + (61.9 \text{ }\Omega)^2]}$$

$$|Z| = 68.79 \text{ }\Omega$$

$$V = IR$$

continued...

Using impedance (Z) as resistance:

$V = (1.8 \text{ A}) \cdot (68.79 \text{ } \Omega)$

$V = 124 \text{ V}$

77. D is correct.

Coulomb's Law, which describes the repulsive force between two particles, is given as:

$F = kq_1q_2 / r^2$

The expression does not include mass, so the repulsive force remains the same when m changes.

Note: gravitational (attractive) forces do rely on the masses of the objects.

78. A is correct.

The gauge pressure is referenced at ambient air pressure thus:

$P = \rho g h$

$P = (1{,}000 \text{ kg/m}^3) \cdot (9.8 \text{ m/s}^2) \cdot (6 \text{ m} + 22 \text{ m})$

$P = 2.7 \times 10^5 \text{ N/m}^2$

79. C is correct.

$f_{beat} = |f_2 - f_1|$

$\pm f_{beat} = f_2 - f_1$

$f_2 = \pm f_{beat} + f_1$

$f_2 = \pm 5 \text{ Hz} + 822 \text{ Hz}$

$f_2 = 817 \text{ Hz}, 827 \text{ Hz}$

Only 827 Hz is an answer choice.

80. E is correct.

A phase change occurs when waves reflect from the surface of a medium with a higher refractive index than the medium they are traveling in.

Glass has a higher refractive index than air.

Therefore, when the light ray moves from glass to air, no change occurs.

81. A is correct.

Convert PE (before release) into KE (as it is about to strike the ground):

$mgh = KE$

KE is proportional to h.

82. B is correct.

Determine distance in one revolution (the perimeter or circumference):

$P = \pi d$

$P = \pi(18 \text{ m})$

$P = 56.55 \text{ m}$

Convert rev/min to rev/s:

$v = (5.3 \text{ rev/min}) \cdot (1 \text{ min}/60 \text{ s})$

$v = 0.0883 \text{ rev/s}$

Convert rev/s to m/s

where 1 rev = 56.55 m

$v = (0.0883 \text{ rev/s}) \cdot (56.55 \text{ m}/1 \text{ rev})$

$v = 5 \text{ m/s}$

83. C is correct.

$F_{net} = ma$

The only acceleration is centripetal:

$a_{cent} = v^2 / r$

$a_{cent} = (4 \text{ m/s})^2 / 16 \text{ m}$

$a_{cent} = (16 \text{ m}^2/\text{s}^2) / 16 \text{ m}$

$a_{cent} = 1 \text{ m/s}^2$

$F_{net} = ma$

$F_{net} = (40 \text{ kg}) \cdot (1 \text{ m/s}^2)$

$F_{net} = 40 \text{ kg} \cdot \text{m/s}^2$

$F_{net} = 40 \text{ N}$

84. B is correct.

When an object is accelerating, its velocity must change in speed or direction.

The speed or direction does not always change, but velocity does.

85. C is correct.

Solve for heat required:

$$Q = cm\Delta T$$

$$Q = (113 \text{ cal/kg·°C})·(1.14 \text{ kg})·(90 \text{ °C} - 18 \text{ °C})$$

$$Q = 9,275 \text{ cal}$$

Convert to Joules:

$$Q = (9,275 \text{ cal} / 1)·(4.186 \text{ J} / \text{cal})$$

$$Q = 38,825 \text{ J}$$

86. A is correct.

Include the term for work done by air resistance in the conservation of energy equation.

$$KE_i + PE_i + W_{\text{air resis}} = KE_f + PE_f$$

$$0 + mgh + (-F_{\text{air}} \times d) = \tfrac{1}{2}mv_f^2 + 0$$

$$mgh + (-mad) = \tfrac{1}{2}mv_f^2$$

$$(1.2 \text{ kg})·(10 \text{ m/s}^2)·(6 \text{ m}) + (-3.4 \text{ kg·m/s}^2)·(6 \text{ m}) = \tfrac{1}{2}(1.2 \text{ kg})v_f^2$$

$$v_f^2 = 86 \text{ m}^2/\text{s}^2$$

$$v_f = 9.2 \text{ m/s}$$

87. E is correct.

Sound intensity is expressed as power / area.

Units of intensity:

$$W/m^2 = J/s/m^2 = J/m^2/s$$

which is the unit of energy per unit area per unit time.

88. A is correct.

Unit of watt = work / time

Multiply by time, time cancels, and work is left.

Alternatively, 1 kilowatt-hour:

$$1 \text{ watt} = 1 \text{ J/s}$$

$$1 \text{ watt·second} = 1 \text{ J}$$

$$(1 \text{ hr.}/60 \text{ s})·(60 \text{ min}/1 \text{ hr.})·(60 \text{ s}/1\text{min}) = 60^2 \text{ s}$$

$$1 \times 10^3 \text{ Watt} \times (1 \text{ hour})·(60^2 \text{ s}/1 \text{ hour}) = 36 \times 10^5 \text{ J}$$

Work = force × distance

Joule is a unit of work.

89. D is correct.

Critical angle = $\sin^{-1}(n_2/n_1)$

Critical angle = $\sin^{-1}(1.3/1.6)$

Critical angle = $\sin^{-1}(0.81)$

Critical angle = $54°$

90. C is correct.

$PE_e = PE_1 + PE_2 + PE_3$

$PE_e = (kQ_1Q_2)/r_1 + (kQ_2Q_3)/r_2 + (kQ_1Q_3)/r_3$

The distance between the bottom charge and the right charge can be found using the Pythagorean Theorem.

$d^2 = (0.06 \text{ m})^2 + (0.04 \text{ m})^2$

$d^2 = 0.0036 \text{ m}^2 + 0.0016 \text{ m}^2$

$d^2 = 0.0052 \text{ m}^2$

$d = 0.072 \text{ m}$

$Q_1 = Q_2 = Q_3$

So:

$PE_e = kQ^2[(1/0.06 \text{ m}) + (1/0.04 \text{ m}) + (1/0.072 \text{ m})]$

$PE_e = (9 \times 10^9 \text{ Nm}^2/\text{C}^2) \cdot (6.2 \times 10^{-9} \text{ C})^2 \cdot [(1/0.06 \text{ m}) + (1/0.04 \text{ m}) + (1/0.072 \text{ m})]$

$PE_e = (9 \times 10^9 \text{ Nm}^2/\text{C}^2) \cdot (6.2 \times 10^{-9} \text{ C})^2 \times (16.7 \text{ m}^{-1} + 25 \text{ m}^{-1} + 13.9 \text{ m}^{-1})$

$PE_e = 1.9 \times 10^{-5}$ J

91. C is correct.

If the cylinder's center of mass does not move, it must be stationary, and the forces must balance:

$F = mg \sin \theta$

The force produced by torque is:

$\tau = I\alpha$

where I is the mass moment of inertia, and α is angular acceleration.

The moment of inertia for a solid cylinder is:

$I = \tfrac{1}{2}mr^2$

$\tau = I\alpha$

$\tau = Fr$

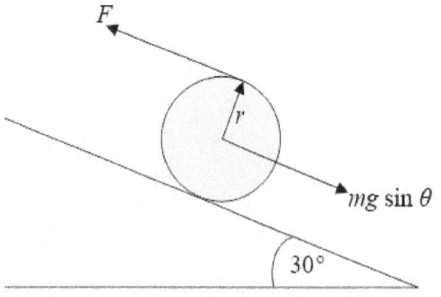

continued...

$$I\alpha = Fr$$

$$(½mr^2)\alpha = (mg \sin \theta)r$$

$$\alpha = (2g \sin \theta) / r$$

$$\alpha = (2)·(10 \text{ m/s}^2) \sin 30° / (0.8 \text{ m})$$

$$\alpha = (2)·(10 \text{ m/s}^2)·(0.5) / (0.8 \text{ m})$$

$$\alpha = 12.5 \text{ rad/s}^2$$

92. C is correct.

Frequency = # cycles / time

$$f = 2 \text{ cycles} / 1 \text{ s}$$

$$f = 2 \text{ s}^{-1}$$

$$v = \lambda f$$

where λ is wavelength

$$v = (12 \text{ m})·(2 \text{ s}^{-1})$$

$$v = 24 \text{ m/s}$$

93. E is correct.

Atmospheric pressure is not considered because it acts at the water's surface and around Mike's finger, so it cancels.

$$P_{water} = \rho g h$$

$$P_{water} = (10^3 \text{ kg/m}^3)·(10 \text{ m/s}^2)·(1 \text{ m})$$

$$P_{water} = 10^4 \text{ N/m}^2$$

Area of hole:

$$A = (0.01 \text{ m})·(0.01 \text{ m})$$

$$A = 10^{-4} \text{ m}^2$$

$$F = PA$$

$$F = (10^4 \text{ N/m}^2)·(10^{-4} \text{ m}^2)$$

$$F = 1 \text{ N}$$

94. C is correct.

$P = IV$

Voltage is considered potential in a DC circuit.

Slope = Voltage / Power

Slope = V / IV

Slope = 1 / I

Slope = 1 / current

95. C is correct.

The energy needed to change hydrogen from one state to another:

$E = -13.6 \text{ eV}[(1/n_1^2) - (1/n_2^2)]$

To ionize hydrogen, the electron must be removed to the n = ∞ state.

The energy needed to change from the ground state:

$E = -13.6[(1/1) - (1/\infty)]$

$E = -13.6 \text{ eV}$

Energy is expressed as a negative number to indicate that this much energy is needed to be input to the atom.

96. D is correct.

The heat required to melt a solid:

$Q = mL_f$

$Q = (70 \text{ kg}) \cdot (334 \times 10^3 \text{ J/kg})$

$Q = 2.3 \times 10^4 \text{ kJ}$

97. B is correct.

Distance traveled is represented by the area under the velocity *vs.* time curve.

At the point where each of those curves intersects on this plot, there is more area under the curve of the truck velocity than there is under the curve of the car velocity.

98. D is correct.

The energy of the photons in the beam depends linearly on the frequency of the beam.

Since the frequency of beam B is twice the frequency of beam A, the photons in beam B have twice the energy as the photons in beam A.

The intensity depends on the energy of the photons and the number of photons per second carried by the beam. Nothing is known about the intensity or the number of photons per second.

99. E is correct.

$\tau = 120$ N·m

$\tau = I\alpha$

Mass moment of inertia: mass moment of inertia of disk about z-axis

$I = \frac{1}{2}mr^2$

$\tau = [\frac{1}{2}mr^2]\alpha$

120 N $= \frac{1}{2}(12$ kg$)\cdot(4$ m$)^2\,\alpha$

$\alpha = (120$ N$) / [\frac{1}{2}(12$ kg$)\cdot(4$ m$)^2]$

$\alpha = 1.25$ rad/s^2

$\omega = \alpha t$

7.35 rad/s $= (1.25$ rad/s$^2)t$

$t = (7.35$ rad/s$) / (1.25$ rad/s$^2)$

$t = 5.9$ s

100. A is correct.

Beam A is due to high energy electrons because it deflects towards the positive plate, indicating attraction.

Notes for active learning

Notes for active learning

Glossary of Physics Terms

A

Absolute humidity (or saturation value) − the maximum amount of water vapor that could be present in 1 m³ of air at a given temperature.

Absolute magnitude − a classification scheme that compensates for the differences in the distance to stars; calculates the brightness that stars would appear to have if they were at a defined, standard distance of 10 parsecs.

Absolute scale − temperature scale set so that zero is the theoretically lowest temperature possible (this would occur when random motion of molecules has ceased).

Absolute zero − the theoretically lowest temperature possible, at which temperature the molecular motion vanishes; −273.16 °C or 0 K.

Absorptance − the ratio of the total absorbed radiation to the total incident radiation.

Acceleration − the rate of change of velocity of a moving object with respect to time; the SI units are m/s²; by definition, this change in velocity can result from a change in speed, a change in direction, or a combination of changes in speed and direction.

Acceleration due to gravity − the rate of change in velocity produced in a body due to the Earth's attraction; denoted by the letter g (SI unit − m/s²); on the surface of the Earth, its average value is 9.8 m/s²; increases when going towards the poles from the equator; decreases with altitude and with depth inside the Earth; the value of g at the center of the Earth is zero.

Achromatic − capable of transmitting light without decomposing it into its constituent colors.

Acid rain − rainwater with a pH of less than 5.7 is acid rain. It is caused by the gases NO_2 (from car exhaust fumes) and SO_2 (from the burning of fossil fuels) dissolving in the rain. Acid rain kills fish, kills trees, and destroys buildings and lakes.

Acoustics − the science of the production, transmission, and effects of sound.

Acoustic shielding − a sound barrier that prevents the transmission of acoustic energy.

Adiabatic − any change in which there is no gain or loss of heat.

Adiabatic cooling − the decrease in temperature of an expanding gas that involves no additional heat flowing out of the gas; the cooling from the energy lost by expansion.

Adiabatic heating − the increase in temperature of the compressed gas that involves no additional heat flowing into the gas; the heating from the energy gained by compression.

Afocal lens − a lens of zero convergent power whose focal points are infinitely distant.

Air mass − a large, uniform body of air with nearly identical temperature and moisture conditions throughout.

Albedo − the fraction of the total light incident on a reflecting surface, especially a celestial body, which is reflected in all directions.

Allotropic forms − elements with several structures with different physical properties (e.g., graphite and diamond).

Alpha (α) particle – the nucleus of a helium atom (i.e., two protons and two neutrons) emitted as radiation from a decaying heavy nucleus (α-decay).

Alternating current – the flow of charge (i.e., current) traveling in one direction for one-hundredth of a second but the opposite direction for the next hundredth of a second. An electric current that first moves in one direction, then in the opposite direction with a regular frequency.

Amorphous – term that describes solids that have neither definite form nor structure.

Amp – unit of electric current; equivalent to coulomb/second.

Ampere – the unit amp; the SI unit of electric current; one ampere is the flow of one coulomb of charge per second.

Amplitude – the maximum absolute value attained by the disturbance of a wave or by any quantity that varies periodically. The height of the wave crest above the average position.

Amplitude (of an oscillation) – the maximum displacement of a body from its mean position during an oscillatory motion.

Amplitude (of waves) – the maximum displacement of particles of the medium from their mean positions during the propagation of a wave.

Angle of contact – the angle between tangents to the liquid surface and the solid surface inside the liquid; both tangents are drawn at the point of contact.

Angle of incidence – the angle of an incident (arriving) ray or particle to a surface; measured from a line perpendicular to the surface (the normal).

Angle of reflection – the angle of a reflected ray or particle from a surface; measured from a line perpendicular to the surface (the normal).

Angle of refraction – the angle between the refracted ray and the normal.

Angle of repose – the degree of inclination of a plane with the horizontal such that a body placed on the plane is on the verge of sliding but does not.

Angstrom – a unit of length; $1 = 10^{-10}$ m.

Angular acceleration – the rate of change of angular velocity of a body moving along a circular path; denoted by a.

Angular displacement – the angle described at the center of the circle by a moving body along a circular path. It is measured in radians.

Angular momentum (or *moment of momentum*) – the cross-product of position vector and momentum.

Angular momentum quantum number – from the quantum mechanics model of the atom, one of four descriptions of the energy state of an electron wave; describes the energy sublevels of electrons within the principal energy levels of an atom.

Angular velocity – the rate of change of angular displacement per unit of time.

Annihilation – when a particle and an antiparticle combine and release their rest energies in other particles.

Antineutrino – the antiparticle of neutrino; has zero mass and spin ½.

Archimedes principle – a body immersed in a fluid experiences an apparent loss of weight equal to the weight of the fluid displaced by the body.

Area – the amount of surface enclosed within the boundary lines.

Astronomical unit – the radius of the Earth's orbit is defined as one astronomical unit (A.U.).

Atom – the smallest unit of an element that can exist alone or in combination with other elements.

Atomic mass unit – relative mass unit (amu) of an isotope based on the standard of the ^{12}carbon isotope; one atomic mass unit (1 amu) = 1/12 the mass of a ^{12}C atom = 1.66×10^{-27} Kg.

Atomic number – the number of protons in the nucleus of an atom.

Atomic weight – weighted average of the masses of stable isotopes of an element as they occur in nature; based on the abundance of each isotope of the element and the atomic mass of the isotope compared to ^{12}carbon.

Avogadro's number – the number of ^{12}carbon atoms in exactly 12.00 g of C is 6.02×10^{23} atoms or other chemical units; the number of chemical units in one mole of a substance.

Avogadro's Law – under the same temperature and pressure conditions, equal volumes of gases contain an equal number of molecules.

Axis – the imaginary line about which a planet or other object rotates.

B

Background radiation – ionizing radiation (e.g., alpha, beta, gamma rays) from natural sources.

Balanced forces – when some forces act on a body and the resultant force is zero; see *Resultant forces*.

Balmer lines – lines in the spectrum of the hydrogen atom in the visible range; produced by the transition between a high energy level to n = 2, with n being the principal quantum number.

Balmer series – a set of four-line spectra; narrow lines of color emitted by hydrogen atom electrons as they drop from excited states to the ground state.

Bar – a unit of pressure; equal to 10^5 Pascal.

Barometer – an instrument that measures atmospheric pressure; used in weather forecasting and determining elevation above sea level.

Baryon – subatomic particle composed of three quarks.

Beat – a phenomenon of the periodic variation in sound intensity due to the superposition of waves differing slightly in frequency; rhythmic increases and decreases of volume from constructive and destructive interference between two sound waves of slightly different frequencies.

Bernoulli's theorem – states that the total energy per unit volume of a non-viscous, incompressible fluid in a streamline flow will remain constant.

Beta (β) particle – high-energy electron emitted as ionizing radiation from a decaying nucleus (β-decay); also a *beta ray*.

Big bang theory – the current model of galactic evolution in which the universe is assumed to have been created by an intense and brilliant explosion from a primeval fireball.

Binding energy – the net energy required to break a nucleus into its constituent protons and neutrons; also, the energy equivalent released when a nucleus is formed.

Biomass – the chemical energy stored in fast growing plants.

Black body – an ideal body which would absorb all incident radiation and reflect none.

Black body radiation – electromagnetic radiation emitted by an ideal material (the black body) perfectly absorbs and perfectly emits radiation.

Black hole – the remaining theoretical core of a supernova that is so dense that even light cannot escape.

Bohr model – model of the structure of the atom that attempted to correct the deficiencies of the solar system model and account for the Balmer series.

Boiling point – the temperature at which a phase change of liquid to gas occurs through boiling; the same temperature as the condensation point.

Boundary – the division between two regions of differing physical properties.

Boyle's Law – for a given mass of a gas at a constant temperature, the volume of the gas is inversely proportional to the pressure.

Brewster's Law – states that the refractive index of a material is equal to the tangent of the polarizing angle for the material.

British thermal unit (Btu) – the amount of energy or heat needed to increase the temperature of one pound of water one-degree Fahrenheit.

Brownian motion – the continuous random motion of solid microscopic particles suspended in a fluid medium due to their ongoing bombardment by atoms and molecules.

Bulk modulus of elasticity – the ratio of normal stress to the volumetric strain produced in a body.

Buoyant force – the upward force on an object immersed in a fluid.

C

Calorie – a unit of heat; 1 Calorie = 4.186 joule.

Candela – the SI unit of luminous intensity defined as the luminous intensity in each direction of a source that emits monochromatic photons of frequency 540×10^{12} Hz and has a radiant intensity in that direction of 1/683 W/sr.

Capacitance – the ratio of the charge stored per increase in potential difference.

Capacitor – an electrical device used to store charge and energy in the electrical field.

Capillarity – the rise or fall of a liquid in a tube of a fine bore.

Carnot's theorem – no engine operating between two temperatures can be more efficient than a reversible engine working between the same two temperatures.

Cathode rays – negatively charged particles (electrons) emitted from a negative terminal in an evacuated glass tube.

Celsius scale of temperature − the ice-point is taken as the lower fixed point (0 °C), and the steam-point is taken as the upper fixed point (100 °C); the interval between the ice-point and the steam-point is divided into 100 equal divisions; the unit division on this scale is 1 °C; previously called the centigrade scale; the relationship relates the temperatures on the Celsius scale and the Fahrenheit scale, $C/100 = (F - 32) / 180$; the temperature of a healthy person is 37 °C or 98.6 °F.

Center of gravity – the point through which all the weight of an object appears to act.

Centrifugal force − an apparent outward force on an object in circular motion; a consequence of the third law of motion.

Centripetal force − the radial force required to keep an object moving in a circular path.

Chain reaction − a self-sustaining reaction where some of the products can produce more reactions of the same kind (e.g., in a nuclear chain reaction, neutrons are the products that produce more nuclear reactions in a self-sustaining series).

Charles' Law − for a given mass of a gas at constant pressure, the volume is directly proportional to the temperature.

Chromatic aberration − an optical lens defect causing color fringes due to the lens bringing different colors of light to focus at different points.

Circular motion − the motion of a body along a circular path.

Closed system − the system which cannot exchange heat or matter with the surroundings.

Coefficient of areal expansion − the fractional change in surface area per degree of temperature change; see *Coefficient of thermal expansion.*

Coefficient of linear expansion − the fractional change in length per degree of temperature change; see *Coefficient of thermal expansion.*

Coefficient of thermal expansion − the fractional change in the size of an object per degree of change in temperature at constant pressure; the SI unit is K^{-1}.

Coefficient of volumetric expansion − the fractional change in volume per degree of temperature change; see *Coefficient of thermal expansion.*

Coherent source − when there is a constant phase difference between waves emitted from different parts of the source.

Combustion (or *burning*) – the combining of a substance with oxygen.

Compass – a magnet, which is free to rotate and indicate direction.

Complementary colors – two colors which, when mixed, give white. For examples: blue + yellow = white; red + cyan = white and green + magenta = white.

Compound – a substance made up of two or more elements chemically combined.

Compression − a part of a longitudinal wave in which the density of the particles of the medium is higher than the typical density.

Compressive stress − a force that tends to compress the surface as the Earth's plates move into each other.

Concave lens – a lens that spreads out light rays.

Condensation – the changing of a gas to a liquid state.

Condensation (sound) – a compression of gas molecules; a pulse of increased density and pressure that moves through the air at the speed of sound.

Condensation (water vapor) – where more vapor or gas molecules are returning to the liquid state than are evaporating.

Condensation nuclei – tiny particles such as dust, soot, or salt crystals suspended in the air on which water condenses.

Condensation point – the temperature at which a gas or vapor changes back to liquid; see *Boiling point*.

Conduction – the transfer of heat from a region of higher temperature to a region of lower temperature by increased kinetic energy moving from molecule to molecule, without the movement of the solid.

Constructive interference – the condition in which two waves are arriving at the same place at the same time and in phase add amplitudes to create a new wave.

Control rods – material inserted between fuel rods in a nuclear reactor to absorb neutrons and control the rate of the nuclear chain reaction.

Convection – transfer of heat from a region of higher temperature to a region of lower temperature by the displacement of high-energy molecules. For example, the displacement of warmer, less dense air (higher kinetic energy) by cooler, denser air (lower kinetic energy).

Conventional current – the opposite of electron current; considers an electric current to consist of a drift of positive charges that flow from the positive terminal to the negative terminal of a battery.

Convex lens – a lens that brings light rays together.

Corrosion – an undesired process where a metal is converted to one of its compounds (e.g., rusting).

Coulomb – a unit used to measure the quantity of electric charge; equivalent to the charge resulting from the transfer of 6.24 billion particles such as the electron.

Coulomb's Law – the relationship between charge, distance, and magnitude of the electrical force between two bodies; the force between two charges is directly proportional to the product of charges and inversely proportional to the square of the distance between the charges.

Covalent bond – a chemical bond formed by the sharing of a pair of electrons.

Covalent compound – chemical compound held by covalent bonds.

Crest – the point of maximum positive displacement on a transverse wave.

Critical angle – the limit to the angle of incidence when all light rays are reflected internally.

Critical mass – the mass of fissionable material needed to sustain a chain reaction.

Current – a flow of charge. Unit is Ampere (A).

Curvilinear motion – the motion of a body along a curved path.

Cycle – a complete vibration.

Cyclotron – a device used to accelerate the charged particles.

D

De-acceleration − negative acceleration when the velocity of a body decreases with time.

Decibel − unit of sound level; if P_1 & P_2 are two amounts of power, the first is said to be n decibels greater, where $n = 10 \log 10 (P_1/P_2)$.

Decibel scale − a nonlinear scale of loudness based on the ratio of the intensity level of a sound to the intensity at the hearing threshold.

Density − the mass of a substance per unit volume.

Destructive interference − the condition in which two waves arriving at the same point at the same time out of phase add amplitudes that cancel to create zero total disturbance; see *constructive interference*.

Dewpoint temperature − the temperature at which condensation begins.

Dew − condensation of water vapor into droplets of liquid on surfaces.

Diffraction − the bending of light around the edge of an opaque object.

Diffuse reflection − light rays reflected in many random directions, as opposed to the parallel rays reflected from a perfectly smooth surface such as a mirror.

Diopter − unit of measure of the refractive power of a lens.

Direct current − an electrical current that always flows in one direction only (i.e., from the positive terminal to the negative terminal).

Direct proportion − when two variables increase or decrease in the same ratio (at the same rate).

Dispersion − the splitting of white light into its component colors of the spectrum.

Displacement − a vector quantity for the change in the position of an object as it moves in a particular direction; also the shortest distance between the initial position and the final position of a moving body.

Distance − a scalar quantity for the length of the path traveled by a body irrespective of its direction.

Distillation − the vaporization of a liquid by heating and then the condensation of the vapor by cooling.

Doppler effect − an apparent change in the frequency of sound or light due to the relative motion between the source of the sound or light and the observer.

E

Echo − a reflected sound distinguished from the original sound, usually arriving 0.1 s or more after the original sound.

Einstein mass-energy relation − $E = mc^2$; E is the energy released, m is the mass defect, and c is the speed of light.

Elastic potential energy − the potential energy of a body by its configuration (i.e., shape).

Elastic strain − an adjustment to stress in which materials recover their original shape after stress is released.

Electric circuit – consists of a voltage source that maintains an electrical potential, a continuous conducting path for a current to follow, and a device where the electrical potential does work; a switch in the circuit is used to complete or interrupt the conducting path.

Electric current – the flow of electric charge; the electric force field produced by an electrical charge.

Electric field line – an imaginary curve tangent to which at any given point gives the direction of the electric field at that point.

Electric field lines – a map of an electric field representing the direction of the force that a test charge would experience; the direction of an electric field shown by lines of force.

Electric generator – a mechanical device that uses wire loops rotating in a magnetic field to produce electromagnetic induction to generate electricity.

Electric potential energy – potential energy due to the position of a charge near other charges.

Electrical conductors – materials that have electrons free to move throughout the material (e.g., metals); allows electric current to flow through the material.

Electrical energy – a form of energy from electromagnetic interactions.

Electric force – a fundamental force that results from the interaction of electrical charges; it is the most powerful force in the universe.

Electrical insulators – electrical nonconductors, or materials that obstruct the flow of electric current.

Electrical nonconductors – materials with electrons that do not move easily within the material (e.g., rubber); also called *electrical insulators*.

Electrical resistance – the property of opposing or reducing electric current.

Electrode – a conductor, which dips into an electrolyte and allows the electrons to flow to and from the electrolyte.

Electrolysis – the production of a chemical change using electricity. Electrolysis can be used to split up water into hydrogen and oxygen.

Electrolyte – water solution of ionic substances that will conduct an electric current.

Electromagnet – a magnet formed by a solenoid that can be turned on and off by turning the current on and off.

Electromagnetic force – one of four fundamental forces; the force of attraction or repulsion between two charged particles.

Electromagnetic induction – the process in which current is induced in a coil whenever there is a change in the magnetic flux linked with the coil.

Electromagnetic waves – the waves due to oscillating electrical and magnetic fields and do not need any material medium for their propagation; can travel through a material medium (e.g., light waves and radio waves); travel in a vacuum with a speed 3.0×10^8 m/s.

Electron – a subatomic particle that has the smallest negative charge possible; usually found in an orbital of an atom but is gained or lost when atoms become ions.

Electron configuration – the arrangement of electrons in orbits and sub-orbits about the nucleus of an atom.

Electron current – the opposite of conventional current; considers electric current to consist of a drift of negative charges that flows from the negative terminal to the positive terminal of a battery.

Electron pair – a pair of electrons with different spin quantum numbers that may occupy an orbital.

Electron volt – the energy gained by an electron moving across a potential difference of 1 V equals 1.60×10^{-19} Joules.

Electronegativity – the comparative ability of atoms of an element to attract bonding electrons.

Electroplating – where metal is covered with a layer of another metal using electricity.

Electrostatic charge – an accumulated electric charge on an object from a surplus of electrons or a deficiency of electrons.

Element – a pure chemical substance that cannot be broken down into anything simpler by chemical or physical means; there are over 100 known elements, the fundamental materials of which matter is made.

Endothermic process – the process in which heat is absorbed.

Energy – the capacity of a body to do work; a scalar quantity; the SI unit is the Joule; there are five forms: mechanical, chemical, radiant, electrical, and nuclear.

Equilibrium – the condition of a system when neither its state of motion nor its internal energy state tends to change with time; a balanced object is in equilibrium.

Escape velocity – the minimum velocity with which an object must be thrown upward to overcome the gravitational pull and escape into space; the escape velocity depends on the mass and radius of the planet/star, but not on the mass of the body being thrown upward.

Evaporation – a process of more molecules leaving a liquid for the gaseous state than returning from the gas to the liquid; can occur at any given temperature from the surface of a liquid; takes place from the surface of the liquid; causes cooling; faster if the surface of the liquid is large, the temperature is higher, and the surrounding atmosphere does not contain a large amount of vapor of the liquid.

Exothermic process – the process in which heat is evolved.

F

Fahrenheit scale of temperature – the ice-point (lower fixed point) is taken as 32 °F, and the steam-point (upper fixed point) is taken as 212 °F; the interval between these two points is divided into 180 equal divisions; the unit division on the Fahrenheit scale is 1 °F; the relationship relates the temperatures on the Celsius scale and the Fahrenheit scale, C/100 = (°F − 32) / 180; the temperature of a healthy person is 37 °C or 98.6 °F.

Farad – the SI unit of capacitance; the capacitance of a capacitor that, if charged to 1 C, has a potential difference of 1 V.

Faraday – the electric charge required to liberate a gram equivalent of a substance; 1 Faraday = 9,6485 coulomb/mole.

Fermat's principle – an electromagnetic wave takes a path involving the least time when propagating between two points.

First Law of Motion – every object remains at rest or in a state of uniform straight-line motion unless acted on by an unbalanced force.

Fluid – matter that can flow or be poured; the individual molecules of fluid can move, rolling by another.

Focus – the point to which rays that are initially parallel to the axis of a lens or mirror converge or from which they appear to diverge.

Force (F) – a push or pull which tends to change the state of rest or uniform motion, the direction of motion or the shape and size of a body; a vector quantity; the SI unit is a Newton, denoted by N; one N is the force which when acting on a body of mass 1 kg produces an acceleration of 1 m/s².

Force of gravitation – the force with which two objects attract by their masses; acts even if the two objects are not connected; an action-at-a-distance force.

Fossil fuels – formed from the remains of plants and animals that lived millions of years ago.

Fracture strain – an adjustment to stress in which materials crack or break because of the stress.

Fraunhofer lines – the dark lines in the spectrum of the sun or a star.

Freefall – the motion of a body falling to Earth with no other force except the force of gravity acting on it; free-falling bodies are weightless.

Freezing – the changing of a liquid to a solid state.

Freezing point – the temperature at which a phase change of liquid to solid occurs; the same temperature as the melting point for a given substance.

Frequency – the number of oscillations completed in 1 second by an oscillating body.

Frequency (of oscillations) – the number of oscillations made by an oscillating body per second.

Frequency (of waves) – the number of waves produced per second.

Friction – the force that resists the motion of one surface relative to another with which it is in contact; caused by the humps and crests of surfaces, even those on a microscopic scale; the area of contact is small, and the consequent high pressure leads to local pressure welding of the surface; in motion, the welds are broken and remade continually.

Fuel – any substance that burns in oxygen to produce heat.

Fuel rod – long zirconium alloy tubes are containing fissionable material for use in a nuclear reactor.

Fundamental charge – the smallest common charge known; the magnitude of the charge of an electron and a proton, which is 1.60×10^{-19} coulombs.

Fundamental frequency – the lowest frequency (longest wavelength) at which a system vibrates freely and can set up standing waves in an air column or on a string.

Fundamental properties – a property that cannot be defined in more straightforward terms other than to describe how it is measured; the fundamental properties are length, mass, time, and charge.

Fuse – a safety device in an electric circuit. If the current (i.e., the flow of charge) gets too high, the wire in the fuse melts, which breaks the circuit switching off the current.

G

g – a symbol representing the acceleration of an object in free fall due to the force of gravity; its magnitude is 9.80 m/s^2.

Gamma (γ) ray – a high energy photon of short wavelength electromagnetic radiation emitted by decaying nuclei (γ-decay).

Gases – a phase of matter composed of molecules that are relatively far apart moving freely in constant, random motion and have weak cohesive forces acting between them, resulting in the characteristic indefinite shape and indefinite volume of a gas.

Graham's Law of Diffusion – the diffusion rate of a gas is inversely proportional to the square root of its density.

Gram-atomic weight – the mass in grams of one mole of an element that is numerically equal to its atomic weight.

Gram-formula weight – the mass in grams of one mole of a compound that is numerically equal to its formula weight.

Gram-molecular weight – the gram-formula weight of a molecular compound.

Gravitational constant (G) – term in the equation for Newton's Law of Gravitation; numerically, equal to the force of gravitation, which acts between two bodies with a mass of 1 kg each separated by 1 m; the value of G is 6.67×10^{-11} Nm^2/kg^2.

Gravitational potential at a point – the amount of work done against the gravitational forces to move a particle of unit mass from infinity to that point.

Gravitational potential energy (PE) – the energy possessed by a body by its height from the ground; equals *mgh*.

Gravity – the gravitational attraction at the surface of a planet or other celestial body.

Greenhouse effect – the process of increasing the temperature of the lower parts of the atmosphere through redirecting energy back toward the surface; the absorption and re-emission of infrared radiation by carbon dioxide (CO_2), water vapor and a few other gases in the atmosphere.

Ground state – the energy state of an atom with its electrons at the lowest energy state possible for that atom.

H

Half-life – time required for one-half of the unstable nuclei in a radioactive substance to decay into a new element.

Halogens – the elements in group seven in the periodic table.

Hard water – water that finds it difficult to form lather with soap.

Heat – a form of energy that makes a body hot or cold; measured by the temperature-effect, it produces in any material body; the SI unit is the Joule (J).

Heisenberg uncertainty principle – states that there is a fundamental limit to the precision with which certain pairs of physical properties of a particle (i.e., **complementary** variables) can be known simultaneously (e.g., one cannot measure the exact momentum and position of a subatomic particle at the same time – the more certain is one, the less certain is the other).

Hertz (Hz) − unit of frequency; equivalent to one cycle per second.

Hooke's Law − within the elastic limit, stress is directly proportional to strain.

Horsepower − unit of power; 1 hp = 746 Watts.

Humidity − the ratio of water vapor in a sample of air to the volume of the sample.

Huygens' principle − each point on a light wavefront can be regarded as a source of secondary waves, the envelope of these secondary waves determining the position of the wavefront later.

Hypothesis − a tentative explanation of a phenomenon that is compatible with the data and provides a framework for understanding and describing that phenomenon.

I

Ice-point − the melting point of ice under 1 atm pressure; equal to 0 °C or 32 °F.

Ideal gas equation − $PV = nRT$.

Immiscible liquids – liquids that do not mix to form a solution, e.g. oil and water.

Impulse − equal to the product of the force acting on a body and the time for which it acts; if the force is variable, the impulse is the integral of Fdt from t_0 to t_1; the impulse of a force acting for a given time interval is equal to change in momentum produced over that interval; $J = m(v − u)$, assuming that the mass m remains constant while the velocity changes from v to u; the SI units are kg m/s.

Impulsive force − acts on a body for a short time but produces a large change in the momentum of the body.

Incandescent − matter emitting visible light because of high temperature (e.g., a light bulb, a flame from any burning source, the Sun).

Incident ray − line representing the direction of motion of incoming light approaching a boundary.

Index of refraction − the ratio of the speed of light in a vacuum to the speed of light in a material.

Indicator – a substance, which shows using a color change if a substance is acidic or basic.

Inertia − the property of matter that causes it to resist any change in its state of rest or of uniform motion; there are three kinds of inertia: the inertia of rest, the inertia of motion and the inertia of direction; the mass of a body is a measure of its inertia.

Infrasonic − sound waves at a frequency below the range of human hearing (less than 20 Hz).

Insulators − materials that are poor conductors of heat or electricity (e.g., wood or glass); materials with air pockets slow down the movement of heat because the air molecules are far apart.

Intensity − a measure of the energy carried by a wave.

Interference − the redistribution of energy due to the superposition of waves with a phase difference from coherent sources, resulting in alternate light and dark bands.

Intermolecular forces − interaction between molecules.

Internal energy – the sum of the kinetic energy and potential energy of the molecules of an object.

Inverse proportion – the relationship in which the value of one variable increases while the value of a second variable decreases at the same rate (in the same ratio).

Ion exchange – a method of removing hardness from water, and it replaces the positive ions that cause the hardness with H^+ ions.

Ion – a charged atom or group of atoms, e.g. Na+.

Ionic bond – a force of attraction between oppositely charged ions in a compound, resulting from a transfer of electrons.

Ionization – a process of forming ions from molecules.

Ionized – an atom or a particle with a net charge because it has gained or lost electrons.

Isobaric process – pressure remains constant.

Isochoric process – volume remains constant.

Isostasy – a balance or equilibrium between adjacent blocks of Earth's crust.

Isothermal process – in which temperature remains constant.

Isotope – atoms of the same element with the same atomic number (i.e., number of protons) but with a different mass number (i.e., number of neutrons).

J

Joule (J) – the unit used to measure work and energy; can also measure heat; 1 J = 1N·m.

Joule's Law of Heating – states that the heat produced when current (I) flows through a resistor (R) for a given time (t) is given by $Q = I^2Rt$.

K

Kelvin scale of temperature (K) – the lower fixed point is taken as 273.15 K (0 °C), and the steam-point (the upper fixed point) is taken as 373.15 K (100 °C); the interval between these two points is divided into 100 equal parts; each division is equal to 1 K.

Kelvin's statement of Second Law of Thermodynamics – it is impossible that, at the end of a cycle of changes, heat has been extracted from a reservoir, and an equal amount of work has been produced without producing some other effect.

Kepler's Laws of Planetary Motion – the three laws describing the motion of the planets.

Kepler's First Law – in planetary motion, each planet moves in an elliptical orbit, with the Sun located at one focus.

Kepler's Second Law – a radius vector between the Sun and a planet moves over equal areas of the ellipse during equal time intervals.

Kepler's Third Law – the square of the period of an orbit is directly proportional to the cube of the radius of the major axis of the orbit.

Kilocalorie (Kcal) – the amount of energy required to raise the temperature of 1 kg of water by 1 °C; 1 Kcal = 1,000 calories.

Kilogram – the fundamental unit of mass in the metric system of measurement.

Kinetic energy (KE) – possessed by a body due to its motion; $KE = \frac{1}{2}mv^2$, where m is mass and v is velocity.

L

Laser – a device that produces a coherent stream of light through stimulated emission of radiation.

Latent heat – energy released or absorbed by a body during a constant-temperature phase change.

Latent heat of vaporization – the heat absorbed when one gram of a substance changes from the liquid phase to the gaseous phase; also, the heat released when one gram of gas changes from the gaseous phase to the liquid phase.

Latent heat of fusion – the quantity of heat required to convert one unit mass of a substance from a solid state to a liquid state at its melting point without a change in its temperature; the SI unit is J kg^{-1}.

Latent heat of sublimation – the quantity of heat required to convert one unit of mass of a substance from a solid state to a gaseous state without a change in its temperature.

Law of Conservation of Energy – states that energy can neither be created nor destroyed but can be transformed.

Law of Conservation of Mass – mass (including single atoms) can neither be created nor destroyed in a chemical reaction.

Law of Conservation of Matter – matter can neither be created nor destroyed in a chemical reaction.

Law of Conservation of Momentum – the total momentum of a group of interacting objects remains constant in the absence of external forces.

Law of the lever – when a lever is balanced, the sum of the clockwise moments equals the sum of the anti-clockwise moments.

Lenz's Law – the induced current always flows in such a direction that it opposes the cause producing it.

Lever – a rigid body, which is free to turn about a fixed point called the fulcrum.

Light – a form of energy.

Light-year – the distance that light travels in a vacuum in one year (365.25 days); approximately 9.46×10^{15} m.

Line spectrum – an emission (of light, sound, or other radiation) spectrum consisting of separate isolated lines (discrete frequencies or energies); can be used to identify the elements in a matter of unknown composition.

Lines of force – lines drawn to make an electric field strength map, with each line originating on a positive charge and ending on a negative charge; each line represents a path on which a charge would experience a constant force; having the lines closer indicates a more energetic electric field.

Liquids – a phase of matter composed of molecules that have interactions stronger than those found in gas but not strong enough to keep the molecules near the equilibrium positions of a solid, resulting in the characteristic definite volume but the indefinite shape of a liquid.

Liter – a metric system unit of volume; usually used for liquids.

Longitudinal strain – the ratio of change in the length of a body to its initial length.

Longitudinal waves – the particles of the medium oscillate along the direction of propagation of a wave (e.g., sound waves).

Loudness – a subjective interpretation of a sound that is related to the energy of the vibrating source, related to the condition of the transmitting medium and the distance involved.

Lubricant – a substance capable of reducing friction (i.e., force that opposes the direction of motion).

Luminosity – the total amount of energy radiated into space each second from the surface of a star.

Luminous – objects that produce visible light (e.g., the Sun, stars, light bulbs, burning materials).

Lunar eclipse – when the Earth passes between the sun and the moon.

Lyman series – a group of lines in the ultraviolet region in the spectrum of hydrogen.

M

Magnetic domain – tiny physical regions in permanent magnets, approximately 0.01 to 1 mm, have magnetically aligned atoms, giving the domain an overall polarity.

Magnetic field – the region around a magnet where other magnetic objects experience its magnetic force; a model used to describe how magnetic forces on moving charges act at a distance.

Magnetic poles – the ends, or sides, of a magnet about which the force of magnetic attraction seems to be concentrated.

Magnetic quantum number – from the quantum mechanics model of the atom, one of four descriptions of the energy state of an electron wave; describes the energy of an electron orbital as the orbital is oriented in space by an external magnetic field, a kind of energy sub-sublevel.

Magnetic reversal – the changing of polarity of the Earth's magnetic field as the north magnetic pole and the south magnetic pole exchange positions.

Magnetic wave – the spread of magnetization from a small portion of a substance from an abrupt change in the magnetic field.

Magnification – the ratio of the size of the image to the size of the object.

Magnitude – the size of a measurement of a vector; scalar quantities that consist of a number and unit only.

Malleable – can be hammered into sheets (e.g., metals).

Malus Law – the intensity of the light transmitted from the analyzer varies directly as the square of the cosine of the angle between the plane of transmission of the analyzer and the polarizer.

Maser − microwave amplification by stimulated emission of radiation.

Mass (m) − the quantity of matter in a body; the SI unit is the kg; remains the same everywhere; a measure of inertia, which means resistance to a change of motion.

Mass defect − the difference between the sum of the masses of the individual nucleons of a nucleus and mass of nucleus.

Mass number − the sum of the number of protons and neutrons in a nucleus; used to identify isotopes (e.g., Uranium-238).

Matter − anything that occupies space and has mass.

Mean life − the average time during which a system (e.g., atom, nucleus) exists in a specified form.

Mechanical energy − the sum of the potential energy (PE) and the kinetic energy (KE) of a body; energy associated with the position of a body.

Mechanical waves − require a material medium for propagation (e.g., sound waves and water waves); also *elastic waves*.

Megahertz (MHz) − unit of frequency; equal to 10^6 Hertz.

Melting – the changing of a solid to a liquid state.

Melting point − the temperature at which a phase change of solid to liquid takes place.

Metal − matter having the physical properties of conductivity, malleability, ductility, and luster.

Meter − the fundamental metric unit of length.

MeV − a unit of energy; equal to 1.6×10^{-13} joules.

Millibar − a measure of atmospheric pressure equivalent to 1,000 dynes per cm^2.

Miscible fluids − fluids that can mix in any proportion.

Mixture − matter made of unlike parts with variable composition and separated into their parts by physical means.

Model − a mental or physical representation of something that cannot be observed directly; used as an aid to understanding.

Modulus of elasticity − the ratio of stress to the strain produced in a body.

Modulus of rigidity − the ratio of tangential stress to the shear strain produced in a body.

Mole − the amount of a substance that contains Avogadro's number of atoms, ions, molecules, or any other chemical unit; 6.02×10^{23} atoms, ions, or other chemical units.

Molecule – two or more atoms chemically combined.

Moment − a measure of the turning effect of a force. Moment of a force = force × perpendicular distance from the fulcrum.

Momentum − a measure of the quantity of motion; the product of the mass and the velocity of a body; SI units are kg·m/s.

Monochromatic light − consisting of a single wavelength.

N

Natural frequency − the frequency of oscillation of an elastic object in the absence of external forces; depends on the size, composition, and shape of the object.

Negative electric charge − one of two types of electric charge; repels other negative charges and attracts positive charges.

Negative ion − atom or particle that has a surplus or imbalance of electrons and a negative charge.

Net force − the resulting force after vector forces are summed; if a net force is zero, the vector forces have canceled, and there is no unbalanced force.

Neutralization − the reaction between an acid and a base to give salt and water.

Newton (N) − a unit of force defined as kg·m/s^2; 1 Newton is needed to accelerate a 1 kg mass by 1 m/s^2.

Newton's First Law of Motion − a body continues in a state of rest or uniform motion in a straight line unless an external (unbalanced) force acts upon it.

Newton's Law of Gravitation − the gravitational force of attraction acting between two particles is directly proportional to the product of their masses and inversely proportional to the square of the distance between them; the force of attraction acts along the line joining the two particles; real bodies having spherical symmetry act as point masses with their mass assumed to be concentrated at their center of mass.

Newton's Second Law of Motion − the rate of change of momentum is equal to the force applied; the force acting on a body is directly proportional to the product of its mass and acceleration produced by force in the body.

Newton's Third Law of Motion − for every action, there is an equal and opposite reaction; the action and the reaction act on two different bodies simultaneously.

Noise − sounds made up of groups of waves of random frequency and intensity.

Non-uniform acceleration − when the velocity of a body increases by unequal amounts in equal intervals of time.

Non-uniform speed − when a body travels unequal distances in equal intervals of time.

Non-uniform velocity − when a body covers unequal distances in equal intervals of time in a direction, or when it covers equal distances in equal intervals but changes its direction.

Normal (*N*) − a line perpendicular to the surface of a boundary.

Nuclear energy − the form of energy from reactions involving the nucleus.

Nuclear fission − the splitting of a heavy nucleus into more stable, lighter nuclei with an accompanying release of energy.

Nuclear force − one of four fundamental forces; a strong force of attraction that operates over short distances between subatomic particles; overcomes the electric repulsion of protons in a nucleus and binds the nucleus.

Nuclear fusion − a nuclear reaction of low mass nuclei fusing to form a more stable and more massive nucleus with an accompanying release of energy.

Nuclear reactor − a steel vessel in which a controlled chain reaction of fissionable materials releases energy.

Nucleons − a collective name for protons and neutrons in the nucleus of an atom.

Nucleus − the central, positively charged, dense portion of an atom; contains protons and neutrons.

O

Octet rule – during bonding, atoms tend to reach an electron arrangement with eight electrons in the outermost shell.

Ohm – unit of resistance; 1 ohm = 1volt/ampere.

Ohm's Law – the current flowing through a conductor is directly proportional to the potential difference across the ends of the conductor. At a constant temperature, the voltage across a conductor is proportional to the current flowing through it. Voltage = current × resistance (V = IR).

Open system – a system across whose boundaries matter and energy can pass.

Optical fiber – a long, thin thread of fused silica; used to transmit light; based on total internal reflection.

Orbital – the region of space around the nucleus of an atom where an electron is likely to be found.

Origin – the point on a graph where the x and the y variables have a value of zero at the same time.

Oscillatory motion – the to and fro motion (periodic) of a body about its mean position; also, *vibratory motion*.

Oxidation – the addition of oxygen or the losing of electrons.

P

Pascal – a unit of pressure equal to the pressure resulting from a force of 1 N acting uniformly over an area of 1 m^2.

Pascal's Law – states that the pressure exerted on a liquid is transmitted equally in all directions.

Paschen series – a group of lines in the infrared region in the spectrum of hydrogen.

Pauli exclusion principle – no two electrons in an atom can have the same four quantum numbers; a maximum of two electrons can occupy a given orbital.

Peltier effect – the evolution or absorption of heat at the junction of two dissimilar metals carrying current.

Period (of a wave) – the time for a wave to travel through a distance equal to its wavelength; denoted by T; period of a wave = 1/frequency of the wave.

Period (of an oscillation) – the time to complete one oscillation; does not depend upon the mass of the bob and amplitude of oscillation; directly proportional to the square root of the length and inversely proportional to the square root of the acceleration due to gravity.

Periodic wave – a wave in which the particles of the medium oscillate continuously about their mean positions regularly at fixed intervals of time.

Periodic motion – a motion that repeats at regular time intervals.

Permeability – the ability to transmit fluids through openings, tiny passageways, or gaps.

pH scale – a scale from 0 to 14. If the pH of a solution is 7 it is neutral; if the pH of a solution is less than 7 it is acidic; if the pH of a solution is greater than 7 it is basic.

Permanent hardness – hardness in water that cannot be removed by boiling. It is caused by calcium sulfate.

Phase – when particles in a wave are in the same state of vibration (i.e., in the same position and the direction of motion).

Phase change – the action of a substance changing from one state of matter to another; always absorbs or releases internal potential energy that is not associated with a temperature change.

Photons – quanta of energy in the light wave; the particle associated with light.

Photoelectric effect – the emission of electrons in some materials when the light of a suitable frequency falls on them.

Physical change – a change of the state of a substance but not in the identity of the substance.

Pitch (sound) – depends on the frequency of the wave.

Planck's constant – proportionality constant in the ratio of the energy of vibrating molecules to their frequency of vibration; a value of 6.63×10^{-34} J·s.

Plasma – a phase of matter; a hot highly ionized gas consisting of electrons and atoms that have been stripped of their electrons because of high kinetic energies.

Plasticity – the property of a solid whereby it undergoes a permanent change in shape or size when subjected to a stress.

Plastic strain – an adjustment to stress in which materials become molded or bent out of shape under stress and do not return to their original shape after the stress is released.

Polarized light – light whose constituent transverse waves are vibrating in the same plane.

Polaroid – a film that transmits only polarized light.

Polaroid or polarizer – a device that produces polarized light.

Positive electric charge – one of the two types of electric charge; repels positive and attracts negative charges.

Positive ion – atom or particle with a net positive charge due to an electron or electrons being torn away.

Positron – an elementary particle having the same mass as an electron but an equal and positive charge.

Potential difference (or *voltage*) – the force, which moves the electrons around the circuit. Unit is Volt (V).

Potential energy (PE) – possessed by a body by its position or configuration; see the *Gravitational potential energy* and *Elastic potential energy*.

Power – scalar quantity for the rate of doing work; the SI unit is Watt; 1 W = 1 J/s. The rate at which energy is converted from one form to another. Unit is Watts (W). Power = voltage × current (P = VI).

Pressure – a measure of force per unit area (e.g., kilograms per square meter (kg/m^2). Unit is Pascal (Pa).

Primary coil – part of a transformer; a coil of wire connected to a source of alternating current.

Primary colors – three colors (red, yellow, and blue) combined in various proportions to produce any other color. When red, green, and blue are combined, it results in white. For example, red + green + blue = white.

Principal quantum number – from the quantum mechanics model of the atom, one of four descriptions of the energy state of an electron wave; describes the main energy level of an electron regarding its most probable distance from the nucleus.

Principle of calorimetry – states that if two bodies of different temperature are in thermal contact, and no heat is allowed to go out or enter the system; heat lost by the body with higher temperature is equal to the heat gained by the body of lower temperature (i.e., the heat lost = the heat gained).

Products – chemicals produced in a chemical reaction.

Progressive wave – a wave that transfers energy from one part of a medium to another.

Projectile – an object thrown into space horizontally or at an acute angle and under the action of gravity; the path followed by a projectile is its trajectory; the horizontal distance traveled by a projectile is its range; the time is taken from the moment it is thrown until the moment it hits the ground is its time of flight.

Proof – a measure of ethanol concentration of an alcoholic beverage; double the concentration by volume (e.g., 50% by volume is 100 proof).

Properties – qualities or attributes that, taken together, are usually unique to an object (e.g., color, texture, size).

Proportionality constant – a value applied to a proportionality statement that transforms the statement into an equation.

Pulse – a wave of short duration confined to a small portion of the medium at any given time; also a *wave pulse*.

Q

Quanta – fixed amounts; usually referring to fixed amounts of energy absorbed or emitted by matter.

Quantum limit – the shortest wavelength; present in a continuous x-ray spectrum.

Quantum mechanics – model of the atom based on the wave nature of subatomic particles and the mechanics of electron waves; also *wave mechanics*.

Quantum numbers – numbers that describe the energy states of an electron; in the Bohr model of the atom, the orbit quantum numbers could be any whole number (e.g., 1, 2, 3, etc.); in the quantum mechanics model of the atom, four quantum numbers are used to describe the energy state of an electron wave (n, m, l, and s).

Quark – one of the hypothetical fundamental particles; has a charge with magnitudes of one-third or two-thirds of the charge on an electron.

R

Rad – a measure of radiation received by a material (radiation-absorbed dose).

Radiant energy – the form of energy that can travel through space (e.g., visible light and other parts of the electromagnetic spectrum).

Radiation – the emission and propagation of waves transmitting energy through space or some medium. Heat transfer through invisible rays, which travel outwards from the hot object without a medium.

Radioactive decay – the natural, spontaneous disintegration or decomposition of a nucleus.

Radioactive decay constant – a specific constant for a isotope that is the ratio of the rate of nuclear disintegration per unit of time to the total number of radioactive nuclei.

Radioactive decay series – series of decay reactions that begins with one radioactive nucleus that decays to a second nucleus that decays to a third nucleus and so on, until a stable nucleus is reached.

Radioactive Decay Law – the rate of disintegration of a radioactive substance is directly proportional to the number of undecayed nuclei.

Radioactivity – spontaneous emission of particles or energy from an atomic nucleus as it disintegrates.

Rarefaction – a part of a longitudinal wave where the density of the particles of the medium is less than the typical density.

Real image – an image generated by a lens or mirror that can be projected onto a screen.

Reactants – chemicals that react together in a chemical reaction.

Rectifier – converts alternating current to direct current.

Rectilinear motion – the motion of a body in a straight line.

Reduction – the removal of oxygen or the gaining of electrons.

Reflected ray – a line representing the direction of motion of light reflected from a boundary.

Reflection – the bouncing back of light from a surface.

Refraction – the bending of a light wave, a sound wave, or another wave from its straight-line path as it travels from one medium to another.

Refractive index – the ratio of the speed of light in a vacuum to that in the medium.

Relative density – (i.e., *specific gravity*) is the ratio of the density (mass of a unit volume) of a substance to the density of given reference material. *Specific gravity* usually means relative density concerning water. The term *relative density* is more common in modern scientific usage.

Relative humidity – the percentage of the amount of water vapor present in a specific volume of the air to the amount of water vapor needed to saturate it.

Resistance (R) – the opposition of a conductor to current (i.e., the flow of charge). A good conductor has a low resistance, and a bad conductor has a high resistance.

Resolving power – the ability of an optical instrument to produce separable images of different points of an object.

Resonance – when the frequency of an external force matches the natural frequency of the body.

Restoring force – the force which tends to bring an oscillating body back to its mean position whenever it is displaced from the mean position.

Resultant force – a single force, which acts on a body to produce the same effect on it as done by other forces collectively; see *balanced forces*.

Reverberation – apparent increase in the volume of sound caused by reflections from the boundary surfaces, usually arriving within 0.1 seconds after the original sound.

Rigid body – an idealized extended body whose size and shape are fixed and unaltered when forces are applied.

S

Salt – when a metal replaces the hydrogen of an acid.

Saturated air – air in which an equilibrium exists between evaporation and condensation; the relative humidity is 100 percent.

Saturated solution – the apparent limit to dissolving a given solid in a specified amount of water at a given temperature; a state of equilibrium exists between dissolving solute and solute coming out of solution.

Scalar quantity – a physical quantity described entirely by its magnitude.

Scientific law – a relationship between quantities; usually described by an equation in the physical sciences; describes a broader range of phenomena and is more important than a scientific principle.

Scientific principle – a relationship between quantities within a specific range of observations and behavior.

Second – the standard unit of time in the metric and English systems of measurement.

Second Law of Motion – the acceleration of an object is directly proportional to the net force acting on that object and inversely proportional to the mass of the object.

Secondary coil – part of a transformer; a coil of wire in which the voltage of the original alternating current in the primary coil can be stepped up or down by electromagnetic induction.

Secondary colors – formed when two primary colors (i.e., red, green, blue) are mixed. The three secondary colors are yellow, magenta, and cyan. For example, red + green = yellow; red + blue = magenta; and blue + green = cyan.

Second's pendulum – a simple pendulum whose period on the surface of the Earth is 2 seconds.

Semiconductors – elements whose electrical conductivity is intermediate between that of a conductor and an insulator.

Shear strain – the ratio of the relative displacements of one plane to its distance from the fixed plane.

Shear stress – the restoring force developed per unit area when deforming force acts tangentially to the surface of a body, producing a change in the shape of the body without any volume change.

Siemens – the derived SI unit of electrical conductance; equal to the conductance of an element with a resistance of 1 ohm; also written as ohm^{-1}.

Simple harmonic motion – the vibratory motion occurs when the restoring force is proportional to the displacement from the mean position and is directed opposite to the displacement.

Simple pendulum – a heavy point mass (a small metallic ball) suspended by a light inextensible string from frictionless rigid support; a simple machine based on the effect of gravity.

Snell's Law – states that the ratio of *sin i* to *sin r* is a constant and is equal to the refractive index of the second medium concerning the first.

Solar eclipse – happens when the moon passes between the sun and the Earth.

Solenoid – a cylindrical coil of wire that becomes electromagnetic when current passes through it.

Solids – a phase of matter with molecules that remain close to fixed equilibrium positions due to strong interactions between the molecules, resulting in the characteristic definite shape and definite volume of a solid.

Solution – a mixture of a solute (usually a solid) and a solvent (usually a liquid).

Sonic boom – sound waves that pile up into a shock wave when a source is traveling at or faster than the speed of sound.

Sound – a form of energy.

Specific gravity – see *relative density*.

Specific heat – the amount of heat energy required to increase the temperature of 1 g of a substance by 1 °C; each substance has its specific heat value.

Speed – a scalar quantity for the distance traveled by a body per unit of time; if a body covers the distance in time, its speed is given by distance/time; SI units are m/s.

Spin quantum number – from the quantum mechanics model of the atom, one of four descriptions of the energy state of an electron wave; describes the spin orientation of an electron relative to an external magnetic field.

Stable equilibrium – a body is in stable equilibrium if, when slightly moved, its center of gravity rises.

Standing waves – the condition where two waves of equal frequency traveling in opposite directions meet and form stationary regions of maximum displacement due to constructive interference and stationary regions of zero displacement due to destructive interference.

State of motion – when a body changes its position concerning a fixed point in its surroundings; the states of rest and motion are relative to the frame of reference.

State of rest – when a body does not change its position concerning a fixed point in its surrounding; the states of rest and motion are relative to the frame of reference.

Steam-point – the temperature of steam over pure boiling water under 1 atm pressure; taken as the upper fixed point (100 °C or 212 °F) for temperature scales.

Stefan-Boltzmann Law – the amount of energy radiated per second per unit area of a perfectly black body is directly proportional to the fourth power of the absolute temperature of the surface of the body.

Sublimation – the changing of a solid directly to a gas. (Iodine is an example of a substance that sublimes).

Superconductors – some materials in which, under certain conditions, the electrical resistance approaches zero.

Super-cooled – water in the liquid phase when the temperature is below the freezing point.

Supersaturated – containing more than the average saturation amount of a solute at a given temperature.

Surface tension – the property of a liquid due to which its surface behaves like a stretched membrane.

Suspension – a mixture of a liquid and a finely divided insoluble solid.

T

Temperature – a measure of the hotness or coldness of a body; according to the molecular model, measures the average kinetic energy of the molecules; heat flows from a body at a higher temperature to a body at a lower temperature.

Temporary hardness – hardness in water that can be removed by boiling. It is caused by calcium hydrogen carbonate.

Tensional stress – the opposite of compressional stress; when one part of a plate moves away from another part that does not move.

Tesla (T) – SI unit of magnetic flux density; the magnetic flux density of a magnetic flux of 1 Wb through an area of 1 m^2.

Thermal capacity – the quantity of heat required to raise the temperature of the body by one degree (1 K or 1 °C).

Thermal equilibrium – when two bodies in contact are at the same temperature, and there is no heat flow between them; also, the average temperature of the bodies in thermal equilibrium.

Thermal expansion – the increase in the size of an object when heated.

Thermometer – a device used for the measurement of temperature; the mercury thermometer is commonly used.

Third Law of Motion – when two objects interact, the force exerted on one object is equal in size and opposite in direction to the force exerted on the other object; forces always occur in matched pairs that are equal and opposite.

Titration – the process of adding one solution from a burette to a measured amount of another solution to find out exactly how much of each is required to react.

Total internal reflection – a condition where all light is reflected from a boundary between materials; occurs when light travels from a denser to a rarer medium, and the angle of incidence is greater than the critical angle.

Transformation of energy – converting one form of energy into another (e.g., when a body falls, its potential energy is converted to kinetic energy).

Transverse wave – a wave in which the particles of the medium oscillate in a direction perpendicular to the direction of propagation of the wave (e.g., water waves, light waves, radio waves).

Trough – the point of maximum negative displacement on a transverse wave.

U

Ultrasonic – sound waves too high in frequency (above 20,000 Hz) to be heard by the human ear.

Unbalanced forces – when some forces act on a body and the resultant force is not zero.

Uniform acceleration – when the velocity of a body increases by equal amounts in equal intervals of time.

Uniform circular motion – the motion of an object in a circular path with uniform speed; accelerated motion.

Uniform speed – when a body travels equal distances in equal intervals of time.

Uniform velocity – when a body travels along a straight line in a direction with equal distance in equal time intervals.

Universal Law of Gravitation – every object is attracted to every other object with force directly proportional to the product of their masses and inversely proportional to the square of the distance between the centers of the two masses.

Unpolarized light – light consisting of transverse waves vibrating in all possible random directions.

Unstable equilibrium – a body is in unstable equilibrium if, when slightly moved, its center of gravity falls.

V

Van der Waals force – general term for weak attractive intermolecular forces.

Valency – the number of electrons an atom wants to gain, lose, or share to have a full outer shell.

Vapor – the gaseous state of a substance that is generally in a liquid state.

Vector quantity – a quantity that needs magnitude and direction to describe it.

Velocity (v) – distance traveled by a body in a direction per unit time; the displacement of the body per unit time; a vector quantity; the SI units are m/s.

Vibration – a back and forth motion that repeats itself.

Virtual image – an image formed when the reflected or refracted light rays appear to meet; this image cannot be projected on a screen.

Volt (V) – a unit of potential difference equivalent to joules/coulomb.

Voltage drop – the difference in electric potential across a resistor or other part of a circuit that consumes power.

Volume – the amount of space an object occupies.

W

Watt (W) – SI unit for power; equivalent to joule/s.

Wave – a disturbance or oscillation that moves through a medium.

Wavelength (λ) – the distance between the two nearest points on a wave in the same phase; the distance between two adjacent crests or two adjacent troughs.

Wave (mechanical) – a periodic disturbance produced in a material medium due to the vibratory motion of the particles of the medium.

Wave mechanics – alternate name for quantum mechanics derived from the wavelike properties of subatomic particles.

Wave motion – the movement of a disturbance from one part of a medium to another involving the transfer of energy but not the transfer of matter.

Wave period – the time required for two successive crests (or successive regions) of the wave to pass a given point.

Wave velocity – the distance traveled by a wave in one second; depends on the medium through which it passes.

Weight − the force with which a body is attracted towards the center of the Earth; the SI unit is N; the gravitational units are kg·wt and g·wt; the weight of a body is given by *mg*. Weight = mass × acceleration due to gravity.

Weightlessness − the state when the apparent weight of a body becomes zero; objects while falling freely under the action of gravity are seemingly weightless.

Wien's Displacement Law – states that for a black body, the product of the wavelength corresponding to its maximum radiance and its absolute temperature is constant.

Work − work is done when a force acting on a body displaces it; work = force × displacement (W = Fd) in the direction of the force; work is a scalar quantity; the SI unit is Joule.

Y

Young's modulus of elasticity − the ratio of normal stress to the longitudinal strain produced in a body.

Z

Zeeman effect − the splitting of the spectral lines in a spectrum when the source is exposed to a magnetic field.

Zeroth Law of Thermodynamics – states that if body A is in thermal equilibrium with body B, and B is also in thermal equilibrium with C, then A is necessarily in thermal equilibrium with C.

Customer Satisfaction Guarantee

Your feedback is important because we strive to provide the highest quality prep materials. Email us comments or suggestions.

info@sterling–prep.com

We reply to emails – check your spam folder

Highest quality guarantee

Be the first to report a content error for a $10 reward
or a grammatical mistake to receive a $5 reward.

College study aids by Sterling Test Prep

Cell and Molecular Biology Review

Organismal Biology Review

Cell and Molecular Biology Practice Questions

Organismal Biology Practice Questions

Physics Review (Paert 1 and 2)

Physics Practice Questions (Vol. 1 and 2)

Organic Chemistry Practice Questions

United States History 101

American Government and Politics 101

Environmental Science 101

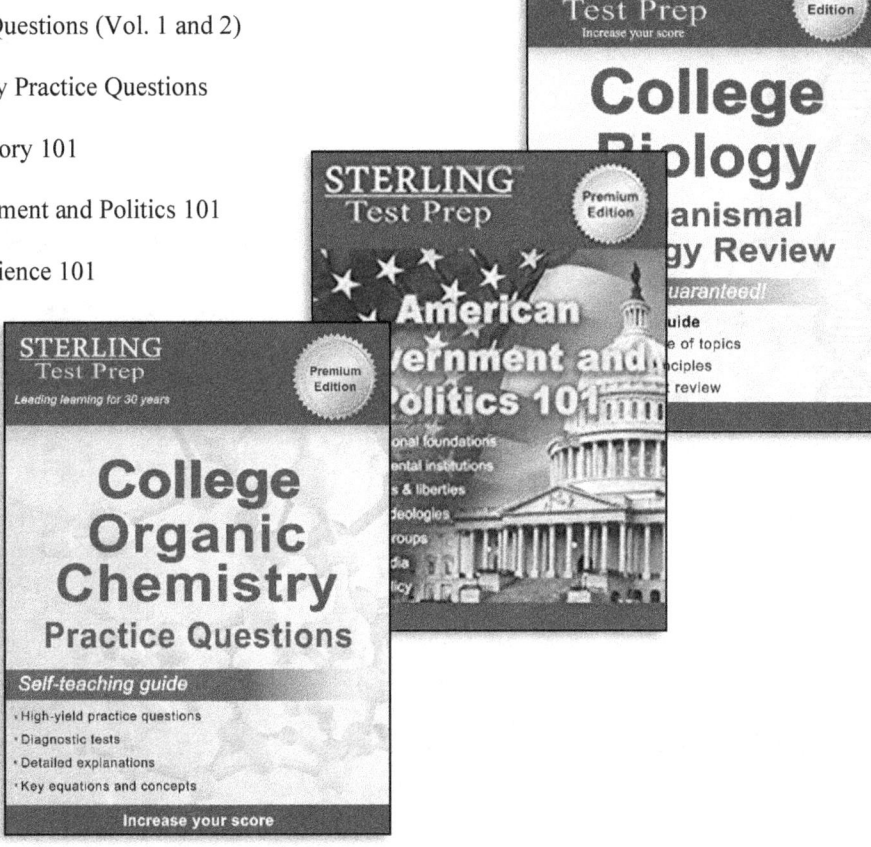

Visit our Amazon store

College Level Examination Program (CLEP)

Biology Review

Biology Practice Questions

Chemistry Review

Chemistry Practice Questions

Introductory Business Law Review

College Algebra Practice Questions

College Mathematics Practice Questions

History of the United States I Review

History of the United States II Review

Western Civilization I Review

Western Civilization II Review

Social Sciences and History Review

American Government Review

Introductory Psychology Review

Visit our Amazon store

www.ingramcontent.com/pod-product-compliance
Lightning Source LLC
Chambersburg PA
CBHW081342070526
44578CB00005B/693